The Universe at large presents a unique survey of key questions outstanding in contemporary astronomy and cosmology.

In this timely volume, 11 of the world's greatest living astronomers and cosmologists present their views of what problems must be addressed by future research. Allan Sandage presents a 23-point plan to reach a full understanding of the large-scale structure of the Universe; Geoffrey Burbidge looks at the future of the quasi-Steady State alternative to the Big Bang; active galactic nuclei (AGNs) are discussed by E. Margaret Burbidge, Donald Osterbrock, and Malcolm Longair; Igor Novikov, Donald Lynden-Bell, Martin Rees, and Rashid Sunyaev look at the physics of black holes; and Bernard Pagel and Hubert Reeves concentrate on what we don't yet understand about elements in the cosmos.

This book provides a unique review of our current understanding in astronomy and cosmology, and a host of ideas for profitable future research – for graduate students and researchers.

The Universe at large

The Universe at large

Key issues in astronomy and cosmology

edited by

Guido Münch, Antonio Mampaso, and Francisco Sánchez

 CAMBRIDGE
UNIVERSITY PRESS

PUBLISHED BY THE PRESS SYNDICATE OF THE UNIVERSITY OF CAMBRIDGE
The Pitt Building, Trumpington Street, Cambridge CB2 1RP, United Kingdom

CAMBRIDGE UNIVERSITY PRESS
The Edinburgh Building, Cambridge CB2 2RU, United Kingdom
40 West 20th Street, New York, NY 10011-4211, USA
10 Stamford Road, Oakleigh, Melbourne 3166, Australia

First published 1997

Printed in the United Kingdom at the University Press, Cambridge

Typeset in 9/14pt Lucida [KW]

A catalogue record for this book is available from the British Library

ISBN 0 521 55367 9 hardback
ISBN 0 521 58944 4 paperback

Contents

Contents

Preface

In November 1992, during the return trip to Mexico City after a meeting held to commemorate the 50th Anniversary of Tonantzintla Observatory's foundation, Guido Münch and Antonio Mampaso reflected on the fact that some of the astrophysical problems considered during the meeting, albeit under discussion decade after decade, nevertheless still remained unsolved. Some of those issues are widely recognized as crucial to present-day astronomy, and numerous astronomers have been working towards their solution. Others, however, appear to be very concrete and even harmless riddles – often limited to a specific area of astrophysics – and are occasionally tackled by investigators, in most cases only to 'rediscover' the underlying phenomenon. Curiously enough, though, they seem to defy all attempts to explain them.

Most astronomers strive to unravel the apparent mysteries of the Universe. But what could be more exciting than to understand the puzzles that have remained for so long unexplained? It must be recognized that PhD supervisors can hardly be expected to offer their students one of those unsolved key problems which, unheard of in textbooks, run the risk of falling into oblivion from one generation to the next. In the belief that every experienced investigator hides in his or her desk drawers one – or many – of these mysteries, it seemed to us that it would be both amusing and interesting to ask some of the wisest astronomers whether they would consider the idea of getting together to discuss and, finally, put on paper their unsolved problems for publication in a book. At the same time, we believed they would also provide us with a clear view of the great key problems facing astrophysics at the end of the present century.

Upon our return to the Instituto de Astrofísica de Canarias, following the excited endorsement by its Director, Francisco Sánchez, of our idea to organize such a meeting, we began to take preliminary steps towards the organization of the project. To start with, it became clear that the book would be especially valuable if it were to include the actual debate among the 'wise men', were they to present orally their contributions in a conference convened for the purpose. The immediate problem of finding financial support for such an unusual reunion was promptly settled upon securing the Foundation Banco de Bilbao-Vizcaya's backing. In order to tackle the more difficult problem of finding ideal and willing contributors to the book, we started by communicating our idea by letter to three of the greatest *maestros*, asking for their sincere opinion; we were ready to admit that the whole idea was impractical or even purposeless. But almost contrary to our expectations we received enthusiastic answers, encouraging us to proceed, promising participation and, which was most valuable at the time, suggesting the names of other colleagues who could effectively contribute to the success of the project. Of the astronomers to whom were sent additional invitations only two declined. At the same time, the responses received suggested, on the whole, a programme definitely oriented towards observational cosmology, thus providing some degree of homogeneity. Once the list of contributors to the book was definitive and the conference programme had been outlined, it became clear that a few young promising investigators would greatly profit by attending the conference. The Foundation Banco de Bilbao-Vizcaya's assistance was also crucial in this respect, awarding generous grants to ensure participation of a dozen up-and-coming Spanish investigators.

As plans to hold the conference unfolded, we received many requests for invitations and offers to contribute papers, which we almost invariably declined, obeying our self-imposed guidelines limiting the number of participants to under fifty and no communications other than invited papers. From the outset we intended to avoid at all costs the pitfalls of most contemporary meetings and 'workshops': far too many delegates and even more contributors! However, realizing that only a small number of persons would actually be attending, it was imperative to publish the developments of the conference in a book, rather than in the more informal and abbreviated fashion used for meeting proceedings. Only in this manner would we leave for the scientific community at large a lasting record of this unique encounter. Notwithstanding some rather pessimistic comments that we heard along the way, such as: 'Your speakers will never write their manuscripts; they are simply too busy!', our original plan was eventually fulfilled.

The editors find great pleasure in thanking Rafael Rebolo for his assistance during the planning phases of the conference, and Mark Kidger for contributing significantly to the transcription of the video tapes, as well as to the final texts of the typescripts. The enthusiastic support of the project secretary, Ms Monica Murphy, is also gratefully acknowledged. Without the dedicated collaboration of these individuals, staging the conference and preparing the book typescript would have been very hard indeed.

The Editors

1 Astronomical problems for the next three decades

Allan Sandage

1.1 Introduction

Rare as it is, a prophetic gathering of minds such as this is not without precedent. A congress, later to become famous, was held in Paris in August of 1900. Its purpose was to set out the programme 'for the new century' in mathematics. A principal speaker at that Second International Congress of Mathematics was the legendary Göttingen mathematician David Hilbert, then but 38 years old. Hilbert, in the midst of a profoundly active period where he was working on new problems in the calculus of variations, on the foundations of geometry, and on his search for an axiomatic method to prove the necessity of the truth of all of mathematics, was invited to give the major address.

In her excellent biography of Hilbert, Constance Reid (1986) writes

> The new century seemed to stretch out before him as invitingly as a blank sheet of paper and a freshly sharpened pencil.
>
> He would like to make a speech that was appropriate to the significance of the occasion. Perhaps he could discuss the direction of mathematics in the coming century in terms of certain important problems upon which mathematicians could concentrate their efforts. What was (Herman) Minkowski's opinion?

Minkowski was enthusiastic and supported the idea but

> did not fail to point out that there were objections to this subject. Hilbert would probably not want to give away his own ideas for solving certain problems.

There was no reply from Hilbert to the urgings by Minkowski, and as late as June (the Congress was to be in August) he still had not produced a lecture. Then in the middle of July Minkowski received the proof sheets of Hilbert's lecture that was to become such an important document for the progress of mathematics.

It is in the spirit of Hilbert's address that Münch, Mampaso, and Sánchez have organized and edited this book. Their hope is that 11 could partially do for astronomy what Hilbert did alone for mathematics nearly a century ago.

1.1.1 Hilbert's 23 problems

In the Paris lecture Hilbert set out 23 individual problems,

> the solution of which, he was confident, would contribute greatly to the advance of mathematics in the coming century. (Reid 1986.)

He began his lecture with three prescient paragraphs, parts of which are as follows:

> Who of us would not be glad to lift the veil behind which the future lies hidden; to cast a glance at the next advances of our science in its development during future centuries...
>
> History teaches the continuity of the development of science. We know that every age has its own problems, which the following age either solves or casts aside as profitless and replaces by new ones...
>
> The close of a great epoch not only invites us to look back into the past but also directs our thought to the unknown future.

Hilbert's 23 problems have had significant influence on the development of pure mathematics over the last hundred years. They were taken up as challenges by others. Any mathematician solving a Hilbert problem was, and still is, on the way to relative fame and often academic fortune.

To get a flavour of the game we list the 23 set-outs, taken from the summary of Hilbert's lecture printed in *L'enseignement mathematique*, Vol. 2, 1900, pp. 349–55, but copied here from Constance Reid's biography. Those that are starred are the problems that Hilbert discussed in introductory detail at the Congress. The others were left without comment.

(i) *Cantor's problem of the cardinal numbers in the continuum.

(ii) *The compatibility of the arithmetic axioms.

(iii) The equality of volumes of two tetrahedra of equal bases and equal altitudes.

(iv) The problem of the straight line as the shortest distance between two points.

(v) Lie's concept of a continuous group of transformations without the assumption of the differentiability of the functions defining the group.

(vi) *The mathematical treatment of the axioms of physics.

(vii) *The irrationality and the transcendence of certain numbers.

(viii) *Problems of prime numbers (including the Riemann hypothesis).

(ix) The proof of the most general law of reciprocity in any number field.

(x) The determination of the solvability of a Diophantine equation.

(xi) The problem of quadratic forms with any algebraic numerical coefficients.

(xii) The extension of Kronecker's theorem of Abelian fields to any algebraic realm of rationality.

(xiii) *The proof of the impossibility of the solution of the general equation of the 7th degree by means of functions of only two arguments.

(xiv) The proof of the finiteness of certain complete systems of functions.

(xv) A rigorous foundation of Schubert's enumerative calculus.

(xvi) *The problem of the topology of algebraic curves and surfaces.

(xvii) The expression of definite forms by squares.

(xviii) The building up of space from congruent polyhedra.

(xix) *The determination of whether the solutions of 'regular' problems in the calculus of variations are necessarily analytic.

(xx) The general problem of boundary values.

(xxi) *The proof of the existence of linear differential equations having a prescribed monodromic group.

(xxii) *Uniformization of analytic relations by means of automorphic functions.

(xxiii) The further development of the methods of the calculus of variations.

A few of these such as (iii), (iv), and (xvii) seem elementary enough, but their apparent simplicity must be illusory. Guido, a mathematician as well as an astrophysicist, has undoubtedly been at home with these problems for much of his career. But for ordinary astronomers they probably make as much sense as the problems which the present contributors have submitted may seem to make to mathematicians, chemists, or even physicists.

1.1.2 Predicting the future

Most predictions of the future in books such as *The Next 100 Years* (Brown, Bonner, and Weir 1963) or the daily astrological columns are out of date in a time that is usually 10% of the time to their goal.[†] Astrological columns depend on predictions so general that many can be said to come true even the same day. Such predictions have been called Motherhood statements because, although they are obvious and therefore need not be made, they are made anyway.

It is customary when putting together a book such as this to include a list of desirable action items. Too often many of these also turn out to be Motherhood statements such as, 'Need more data', 'What we really need is a correct theory next time', 'We could solve the problem if we understood it', 'The Time Allocation Committee does not understand the problem, and the future will come only when they are replaced', or 'The Hubble constant will be known when the present proponents of both sides have passed to their various rewards'.

Such predictions are not very useful, and Münch, Mampaso, and Sánchez have decreed that no Motherhood statements are to be a part of the present record. Yet, has it ever been possible to predict the future in more concrete terms? Yes, to some limited degree and over some small time interval. How is it done?

It is like a Taylor expansion about the present starting epoch, with future time as the interval. The first term of the series is the present state of play. The future is then divined by adding the first derivative times the time interval to this term, to which is then added the next derivative (the curvature) times the interval squared, and then the derivative of the curvature times the interval cubed, etc. Pretty soon the error made in the extrapolation via the series becomes so large that the prediction is useless for large future times unless all the derivatives are superbly well known.

---

† Ten years after the appearance of *The Next 100 Years*, a conference was held at Caltech to amend the 100-year predictions in this book.

What Münch, Mampaso, and Sánchez required was that the contributors supply *all the necessary* derivatives so that the Taylor expansion can be tolerably accurate for the next 30 years. Is that, in fact, reasonable?

Questions such as Hilbert's supply only the first derivative. The higher terms are largely unknown because the curvature is a changing thing with time, depending on how the initial state has changed with the discoveries made by adding the second term, and the third derivative is then changed again by the discoveries concerning the second derivative made, again, in the same time interval, and so on. Nevertheless, this is already some progress, because progress depends on change. As Heraclitus said:

> no one can step in the same river twice.

Rapid change largely belies prediction for large intervals into the future. The present rate of change is now so large in astronomy that three decades is perhaps a decade or so beyond any frontier that can be seen with even faint precision. Said differently, we see through the glass only darkly, hardly beyond more than the first derivative.

The plan of this report is to give first derivatives of a few selected astronomical problems. And, of course, knowledge of the first term is given by the past history of a subject. You do not know where you are going until you know where you have been.

1.2 Triumphs of the past leading to the present state

Consider the discoveries and new interpretations in astronomy made since say 1940 as (1) a guide to how well the advances could have been predicted in the 1930s, and (2) as the first term in the expansion relating to predictions of the state of play in the year 2025.

Four areas that became the 'new astronomy' after 1940 are now the minimum backbone of the subject taught to beginning university students. This gospel is often known even before they enter university because the concepts are now part of the general culture. These, and the history of how each of the revolutions came about during the lifetimes of the contributors, and even some of the younger readers, are too well known (or should be) to require comment. They are listed here to remind us of parts of astronomical literacy.

(A) Stellar populations as age indicators (1940–).
(B) Stellar evolution leading to age dating of galaxies (1950–).
(C) Chemical evolution of galaxies (1957–).
   Elements made in stars.
   Recycling through the ISM.
   Discovery of stellar metallicity differences.
   Age–metallicity relations.
   Metallicity–size (mass) relations.
(D) Astrophysical processes leading to:
   QSOs.
   AGNs.
   Neutron stars (pulsars).
   Radio sources (cosmic rays).
   X-ray sources.
   γ-ray sources.
   Black holes.
   The Sunyaev–Zel'dovich effect (up-Compton of the $3°$ MWB).
   Gravitational lenses.

Could essays on any of these topics have been written in 1930? No; hardly one of these subjects had seen a glimmer of its later birth. Even the first derivative of a Taylor expansion would have failed for them.

Will it be different for the next 30 years? In some ways, yes, if 'all the major fundamental facts and astrophysical processes are now known, and if all we need do is to fill in the details with examples that will be discovered'. Although these premises are unlikely (all the processes and combinations thereof may not be known), nevertheless, we can set out unsolved problems *within our presently perceived framework* of current problems. These all have their own first term of the series.

1.3 The first derivative of a present-day Taylor expansion: 23 astronomical problems

Imitation is often the mother of progress. One can do no better than attempt an imitation of Hilbert in the intent of his 1900 Paris lecture. Therefore, in the spirit set forth by Münch, Mampaso, and Sánchez, here follow 23 problems whose solutions seem possible in the next 30 years.

The problems are all astronomical (not astrophysical). They are divided into three broad areas as:

  (A) The Hubble sequence of galaxy classification.
  (B) The Galaxy.
  (C) The Universe; practical cosmology.

  We list the problems without initial comment. Those with asterisks are discussed in sketchy detail in the final sections.

  1.3.1 A listing of the problems

  (A) Understanding the Hubble sequence.

  (i) Is the sequence due to evolution or initial conditions?
  (ii) *Parameters that do and do not vary along it.
  (iii) *Cause of the sequence width (van den Bergh luminosity classes).
  (iv) *Is spiral structure totally kinematic? (The role of rotation.)
  (v) *Is the initial star-formation rate (SFR) the principal driver?
  (vi) Cosmogony of the density–morphology relation.
  (vii) Role (if any) of 'mergers' (definition thereof).
  (viii) Origin and age of the dust (AGB stars)?

  (B) Stellar evolution and the Galaxy.

  (ix) Age, kinematics, and chemical distribution functions of all the Galactic components.
  (x) The cosmogony of the distributions.
  (xi) The sequence of events (appearance of the early Galaxy).
  (xii) *The age–metallicity relation at different Galactic positions. (The wave of death from the centre outwards. The Galaxy is a living thing.)
  (xiii) The mass function from stars to rocks. Rocks between the stars?
  (xiv) Star counts to map the halo and the thick disk (vast modern database management capabilities).

  (C) The Universe: practical cosmology.

  (xv) *Is the expansion real?
         (a) The Tolman SB $(1 + z)^4$ test.
         (b) Time dilation with SNe: $t(z) = t(0)/(1 + z)$.
         (c) The MWB temperature: $T(z) = T(0)(1 + z)$.
  (xvi) Evolution in the look-back time (primeval galaxies).

(xvii) *The distance scale.
  (a) Calibration of all indicators.
  (b) The bias problems of all indicators.
  (c) Why some methods are wrong.
(xviii) q_0.
(xix) Explain the excess $N(m)$ counts.
(xx) The nature of the dark matter: Ω.
(xxi) Are there significant velocity deviations from the pure cosmological expansion?
(xxii) The IGM (gas, dust, and rocks between the galaxies?).
(xxiii) The formation time for large-scale structure (are clusters and groups old or young?).

1.4 Sketchy details of the questions that have asterisks

We now give galloping commentaries on a few highlights of what is known concerning answers to the starred questions in the list of 23 in the above section.

1.4.1 Questions concerning the Hubble sequence

Q.(ii) *Parameters that do and do not vary along the sequence*
The galaxy classification sequence was devised by Hubble based on (a) the presence or absence of disks dividing the E and the spiral types, (b) the size of the bulge relative to the disk, (c) the openness of the spiral pattern as measured by the tangent angle of a logarithmic spiral, and (d) the degree of resolution into stars. The unsolved problem is to identify the reason for the continuum along the sequence, based on the presently unknown physics of galaxy formation.

It is widely believed that a continuum exists in the classification sequence (Sandage and Bedke 1995) because the adjacent classes seem closely related. By extension, therefore, all classes are related through the continuum. If so, there must be a master parameter (or a small set of such parameters) that spreads the galaxies along the sequence. The problem is to identify those parameters that determine, through the formation process, the present-day morphological properties of the sequence.

No one believes that galaxies look today as they did just after they were formed. Primeval galaxies probably did not fit the present-day classification continuum. How then did the galaxies change into the present forms? Did the primeval galaxies 'evolve' along basically the present Hubble sequence, or did they fall onto the sequence, each from a more primitive state, 'landing' at their present place in the classification according to some parameter such as the amount of hydrogen left over after the initial collapse (Sandage, Freeman, and Stokes 1970, hereafter SFS)? The collapse occurred either with some regularity (Eggen, Lynden-Bell, and Sandage 1962, hereafter ELS), or more chaotically (Searle and Zinn 1978) within a collapsing envelope of regularity, or, at the other extreme, in complete chaos.

Present-day galaxies show variations of particular physical parameters that are systematic along the modern classification sequence. The obvious way to begin to search for the 'master parameter' that governs the formation process is to enumerate the variation of trial parameters along the sequence. Reviews have been given recently by Roberts and Haynes (1994) and by Buta *et al.* (1994) with, however, somewhat different results that nevertheless are central to the problem.

The physical parameters known to vary systematically along the classification sequence are (a) colour, (b) the contents of HI, HII, CO, and dust, (c) the pitch angle of the spiral arms, (d) rotational velocity, (e) present-day star-formation rate, (f) surface brightness, and (g) mass-to-light ratio. The parameters that do not vary by large factors are (1) the mean B band absolute magnitude of E, S0, Sa, Sb, and early Sc galaxies, and (2) the mass inside the radius to a surface brightness of 25 mag $arcsec^{-2}$.

These statements apply only for the mean values. Within each morphological type there is a distribution of the values of each parameter, and these distributions have large overlap from class to class. Hence, the dispersion of each distribution defines the *vertical* spread about the ridge line of the classification continuum (i.e. the centre line of Hubble's tuning fork diagram). The vertical spread among the spirals is quantified by the various van den Bergh (1960a, b) 'luminosity classes' (L). We know that the mean absolute magnitudes of the various L classes vary systematically, becoming fainter with increasing L values (described in Question (iii) below). Hence, besides a 'master parameter' that determines the gross Hubble type (T), there are other parameters that spread the galaxies of a given type into the continuum of L values (van den Bergh 1982; Sandage 1996b).

One can speculate, as contained in Question (v), that the parameter that determines the Hubble type is the size of the initial density fluctuation, $\delta\rho/\rho$, that alternatively determines the star-formation rate with time, whereas the parameter that determines the L class within a given type is the mass; it determines the rotational velocity set up by any given tidal torque that generates the rotation.

The fact that so many physical parameters vary systematically along the Hubble sequence is strong evidence that the classification sequence does have a fundamental significance. A striking case is the variation of the neutral-hydrogen and the dust content in galaxies in the earliest part of the sequence. Figure 1.1 shows the percentage of detections in both radio 21-cm emission and in the IR dust detection for E, S0, and Sa galaxies (Hogg, Roberts, and Sandage 1993). The S0 galaxies are divided into three groups depending on the subtlety (S) or prominence (P) of the S0 criteria. The Sa galaxies are divided into six subtypes depending on where in the very wide Sa morphological box, from very early (VE) to late (L), a given Sa galaxy lies. The progressive increase in detectability as the Hubble type becomes 'later', even among the subtypes of S0 and Sa, proves that morphology alone (which is the sole basis of the classification) does in fact track the hidden 'master parameter' that we seek.

That the *detection rate* is significant, i.e. that its significance is not overcome by selection effects, is shown in Fig. 1.2, also from Hogg *et al.* (1993), where the observed fluxes of the HI (21-cm) radiation and the IR flux have been changed into total masses. Plotted is the log of the ratio of the HI + H$_2$ masses to the optical luminosity in solar units. (The H$_2$ masses are inferred in the usual way from the CO flux.) Again, the continuity along the classification sequence is evident.

The current literature contains enough comprehensive data on rotational velocities (see Rubin *et al.* 1985 with previous references; Bottinelli *et al.* 1990), masses (Rubin *et al.* 1985), absolute magnitudes (Sandage and Tammann 1987; Sandage 1996a, b), neutral-hydrogen and dust content (Roberts and Haynes 1994), and star-formation rates (see Gallagher, Hunter, and Tutukov 1984; Sandage 1986; Kennicutt 1989) that a comprehensive analysis of the data as functions of T and L

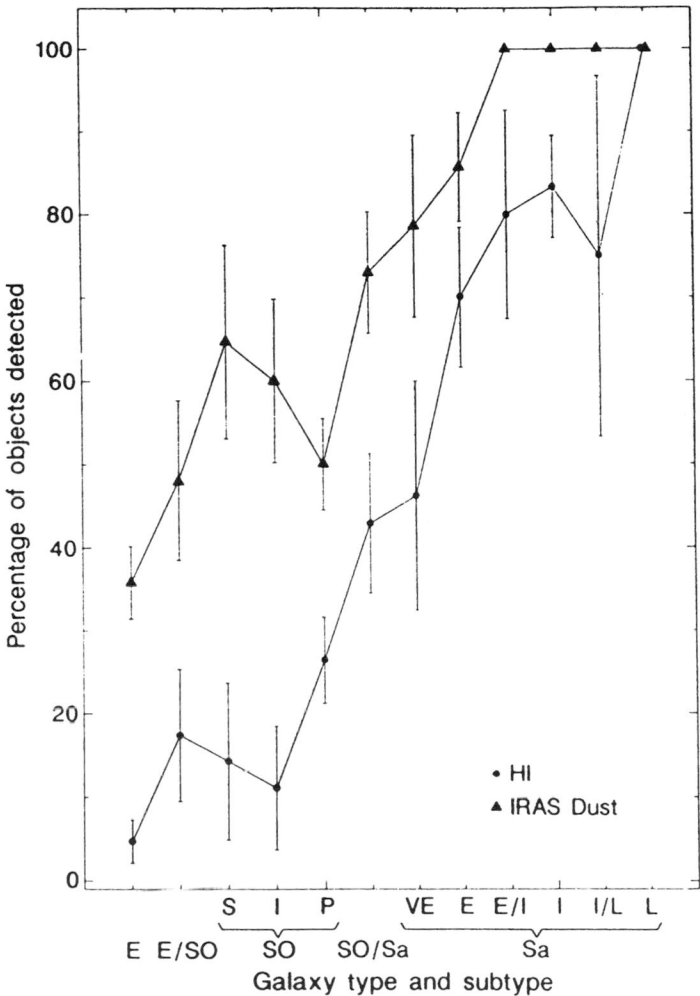

Fig 1.1. Detection rate of 21-cm radiation and of IR flux from dust as a function of Hubble subtype. Diagram from Hogg *et al.* (1993).

along and perpendicular to the Hubble sequence is ripe for a new attack to discover the master parameter and its allies. This is a crucial project that could be finished in several years, begun with data already available and continued by obvious survey observing programmes, suitably organized.

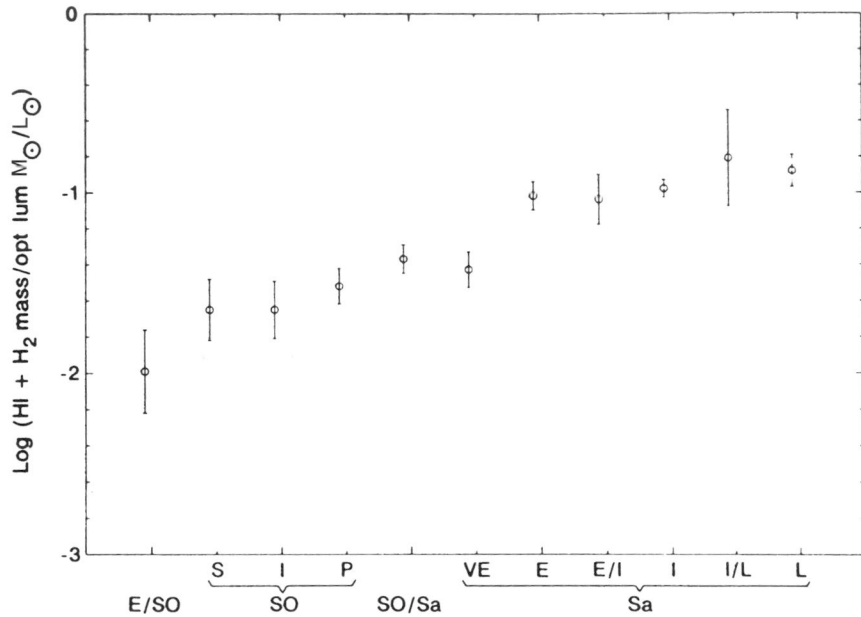

Fig 1.2. The log of the (HI + H$_2$) mass-to-light ratio as a function of Hubble subtype. Diagram from Hogg *et al.* (1993).

Q.(iii) *The cause of the sequence width (the van den Bergh luminosity classes)*

Van den Bergh's (1960a, b) major addition to the Hubble classification is a finer division within each Hubble type according to the regularity of the spiral pattern. In his discovery papers he proved that the luminosity classes, which order the spiral pattern within each type by their regularity, also order the galaxies by luminosity, but only in the mean. Figures 1.3 and 1.4 show the distribution of absolute magnitude of Sb and Sc spirals in the *Revised Shapley–Ames Catalogue* (Sandage and Tammann 1987, the RSA). Although the individual dispersions of absolute magnitude in each *L* class are large (Sandage 1996a), giving large overlap between the classes, nevertheless the mean absolute magnitudes track the *L* values. The question to be investigated is what is the main parameter that determines the *L* class. It seems likely that it is rotational velocity (and therefore a function of mass). The time is ripe for a comprehensive study of this question of the role of mass to determine *L*, but not necessarily *T*. (The overlap in the

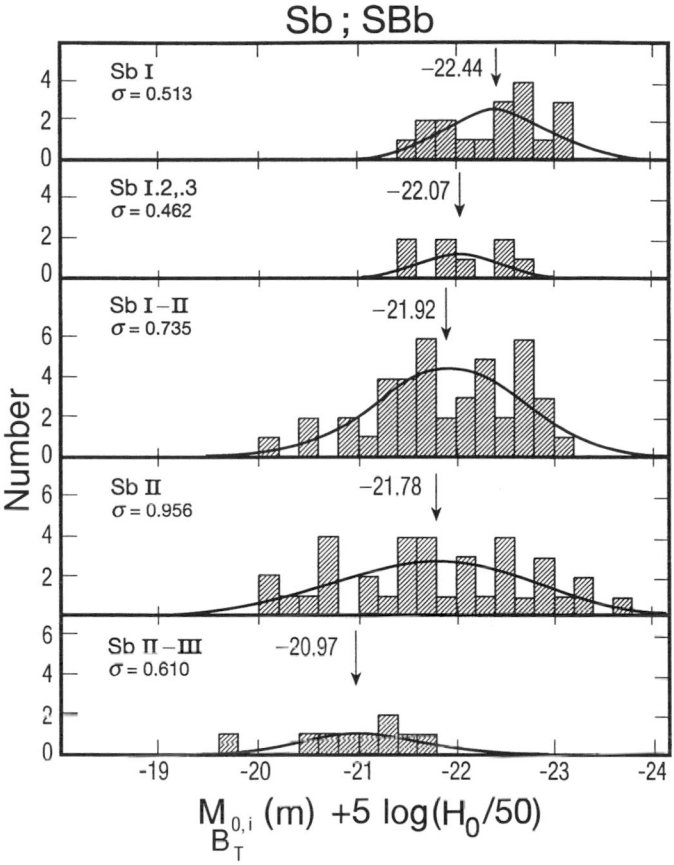

Fig 1.3. The distribution of absolute magnitudes for Sb galaxies in the RSA as a function of van den Bergh luminosity classes. Diagram from Sandage (1996a).

mass distributions for different *T* values is almost complete according to Rubin *et al.* 1985.)

Q.(iv) *Is spiral structure totally kinematic? (The role of rotation)*
In one of the central papers on this subject, yet virtually uncited, Kennicutt (1981) obtained tight correlations between the pitch angle of the spiral pattern and the rotational velocity. Figure 1.5 is Kennicutt's correlation. The right-hand panel separates the data by Hubble type, showing that, on average, Sc galaxies have the smallest rotational velocities whereas Sa galaxies have the highest, but again with very large overlap. No study of possible correlations was made by Kennicutt of the variation of mass with

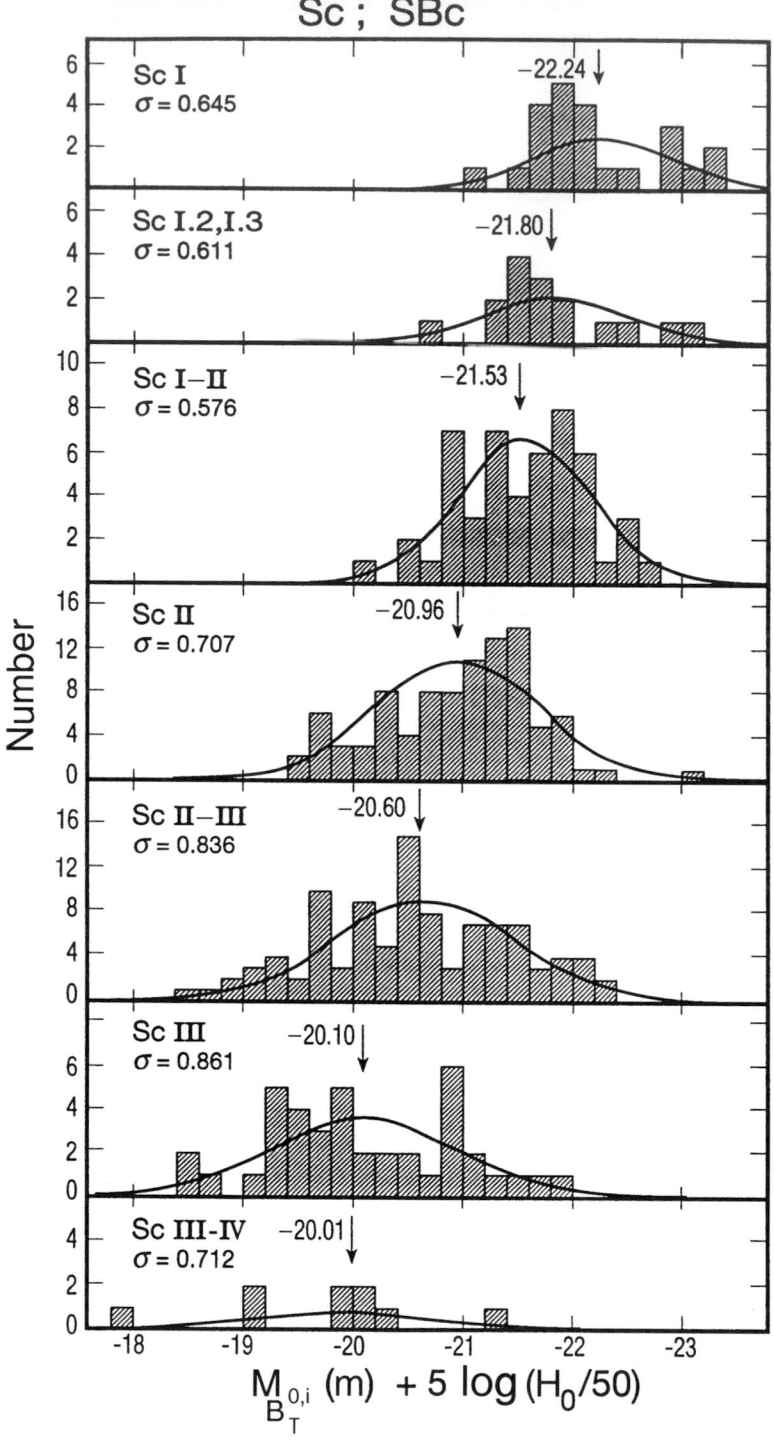

Fig 1.4. The distribution of absolute magnitudes for Sc galaxies in the RSA as a function of van den Bergh luminosity classes. Diagram from Sandage (1996a).

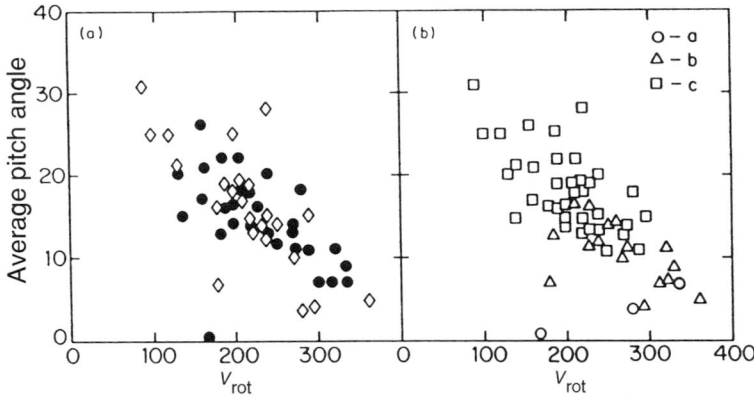

Fig 1.5. The correlation of rotational velocity with pitch angle of the spiral pattern as determined by Kennicutt (1981). The left panel shows the data separated by the method with which the rotation velocity was measured (filled circles by optical spectrosopy; unfilled diamonds by 21-cm line widths). The right panel separates the data by Hubble type. Diagram from Kennicutt (1981).

the L class, or of rotational velocity with L. That would be an obvious place to start now.

But again, there is considerable overlap in the distributions. This is also seen in the large overlap between the Sa, Sb, and Sc galaxies as summarized by Rubin *et al.* (1985). The same overlap is shown in Fig. 1.6, also taken from Kennicutt (1981), showing that the pitch angles, measured by fitting logarithmic spirals to the images, are correlated with estimated Hubble types, verifying one of the three criteria used by Hubble in his classifications. Figures 1.5 and 1.6 show that rotation either 'determines' the pitch angle or is related to it by an unknown hidden variable that is itself related to both. Again, the time is ripe to correlate the entire available database on rotation velocities, pitch angles, van den Bergh luminosity classes, masses, luminosities, and bulge-to-disk ratios.

Q.(v) *Is the initial star-formation rate the principal driver?*
The picture of the formation of our Galaxy as set out by ELS is taken as the starting point, but generalized to the complete Hubble sequence including both the diskless (E) galaxies and the spirals. The parameter of interest is the star-formation rate as a function of epoch, compared with the rate of

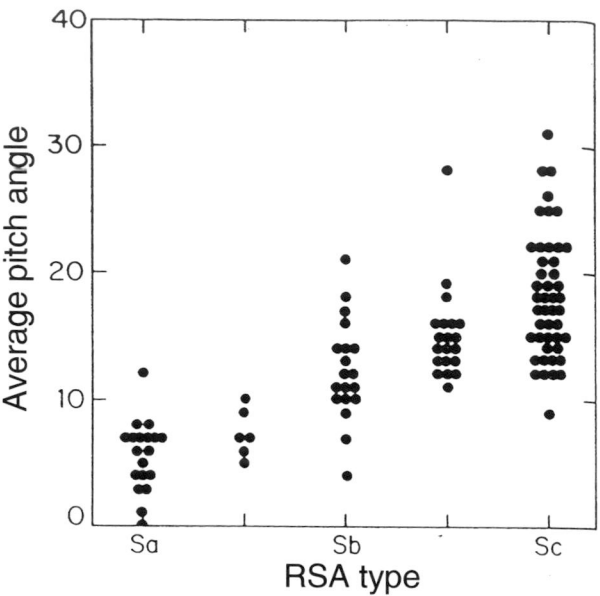

Fig 1.6. Correlation between morphological type from the RSA1 and the pitch angle of the spiral pattern as determined by Kennicutt (1981).

the free-fall collapse (i.e. collapse without pressure support) from the larger volume of a protogalactic cloud.

If stars form out of the gas in the collapsing protogalactic cloud and if they use up 'all' of the gas in a time that is shorter than the free-fall time, no disk will form because no dissipation of energy by gas–gas collisions can occur (there is little or 'no' gas left after the collapse is complete). This is because star–star collisions have a time scale longer than the age of the Universe, and hence the system made of stars is essentially collisionless. Therefore, the presence or absence of a disk depends on whether the time to convert gas into stars is longer or shorter than the free-fall collapse time, t_{ff}. This time, of course, goes as the reciprocal of the square root of the mass density, i.e. it is simply a form of Kepler's law, or even closer, of the $P\sqrt{\rho} = \mathrm{const}$ law for pulsating variables. (Consider the dimensions, using Newton's gravity force law, for a proof.)

The larger the density of a particular mass blob in the initial Harrison–Zel'dovich fluctuation spectrum (or something similar to it), the shorter will be its time of free-fall collapse to form a particular galaxy. Therefore,

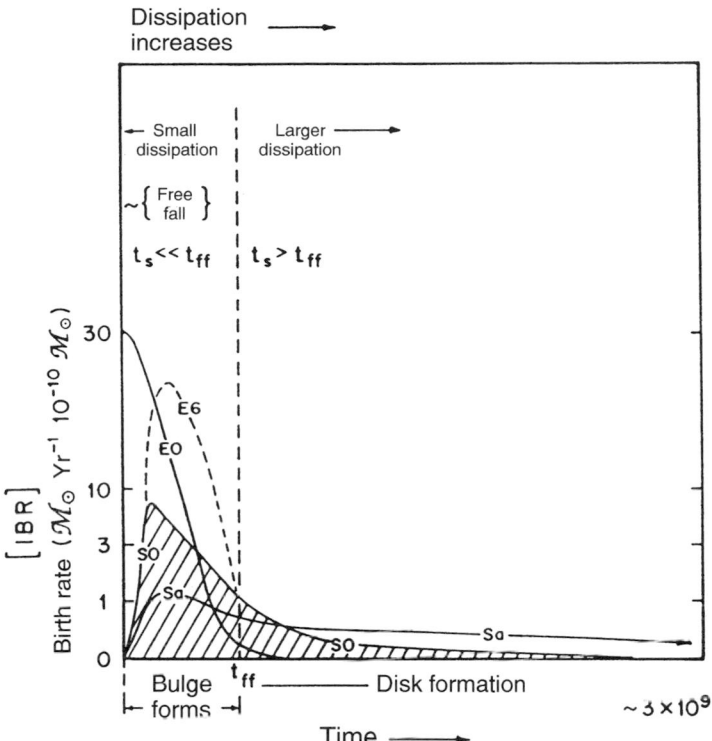

Fig 1.7. Speculation of how the rate of star formation varies as a function of time for E, S0, and Sa galaxies. The epoch of formation is at the origin on the left. An epoch of $\sim 3 \times 10^9$ years is at the border on the right. Diagram is from Sandage (1986).

the ratio of the time, t_s, for most of the gas to turn into stars to the time t_{ff} determines whether disks or diskless systems form. This was a central implied premise of ELS.

Figure 1.7 is a schematic of bulge and/or disk formation (or both) in galaxy types of E, S0, and Sa. Clearly, the bulge-to-disk ratio depends on the t_s/t_{ff} ratio. There is only small dissipation (i.e. there is no appreciable heating of the gas by gas–gas collisions; collisions do not occur in free fall) in that part of the history to the left of the vertical dashed line in the diagram. When dissipation occurs by gas–gas collisions, the structure is then partially held up against the collapse by pressure support, and the collapse is slower than the free-fall time, depending on the rate at which the gas can cool by radiation via atomic processes. The gas that has not formed into stars during the last phases of the collapse forms a disk, in which younger stars are subsequently formed. The strength of the disk relative to the bulge (increasing along the sequence from Sa through Sm) again depends on the t_s/t_{ff} ratio, which again depends on the $\delta\rho/\rho$ density contrast. This is shown schematically in Fig. 1.8, where again the star-formation rate is plotted as a function of time. The disks of Sm galaxies dominate over the bulge (there is no bulge), whereas the opposite is true for Sa galaxies.

Because collapse time depends on the density contrast of the initial fluctuation relative to the background density, we can suppose that $\delta\rho/\rho$ is indeed the 'master parameter' that spreads the galaxies along the Hubble sequence. This leads naturally to the conjecture that the morphological type of the final galaxy (after the collapse is complete) depends on the amount of gas left over in the disk (SFS 1970) following the free-fall era. No gas, or at most only a little gas (however, recent observations show that even E galaxies sometimes have very weak disks) is left in E galaxies. Increasing amounts of gas, and hence increasing present-day star-formation rates, are left in the later morphological types.

Figures 1.7 and 1.8, taken from Sandage (1986), have been made by generalizing the history of the star-formation rate set out in a prescient paper by Gallagher, Hunter, and Tutukov (1984). They could ingeniously estimate the SFR for various Hubble types at three epochs in past time. The diagrams also summarize the main premise of ELS, stated before, that the galaxies form by collapse from a larger volume, and that the present Hubble type is determined by how much gas is left in the disk after the collapse is complete (SFS).

A series of pictures of the collapsing halo with condensations that may be protoglobular clusters (each of mass $10^8 M_\odot$; the star-formation efficiency is perhaps no higher than 1% according to the well-known results of Larson (1987, 1988, 1990)) is shown in Fig. 1.9 taken from Zurek, Quinn, and Salmon (1988, ZQS). Note the lumpy nature of the distribution of 'stars' and 'clusters' in the left middle frame at a redshift of $z = 1.03$. These lumps are the Searle–Zinn fragments in an ELS envelope. They are not the 'merger' of independent fragments. Rather, they are the 'micro-development' within an ELS generally collapsing envelope caused by a

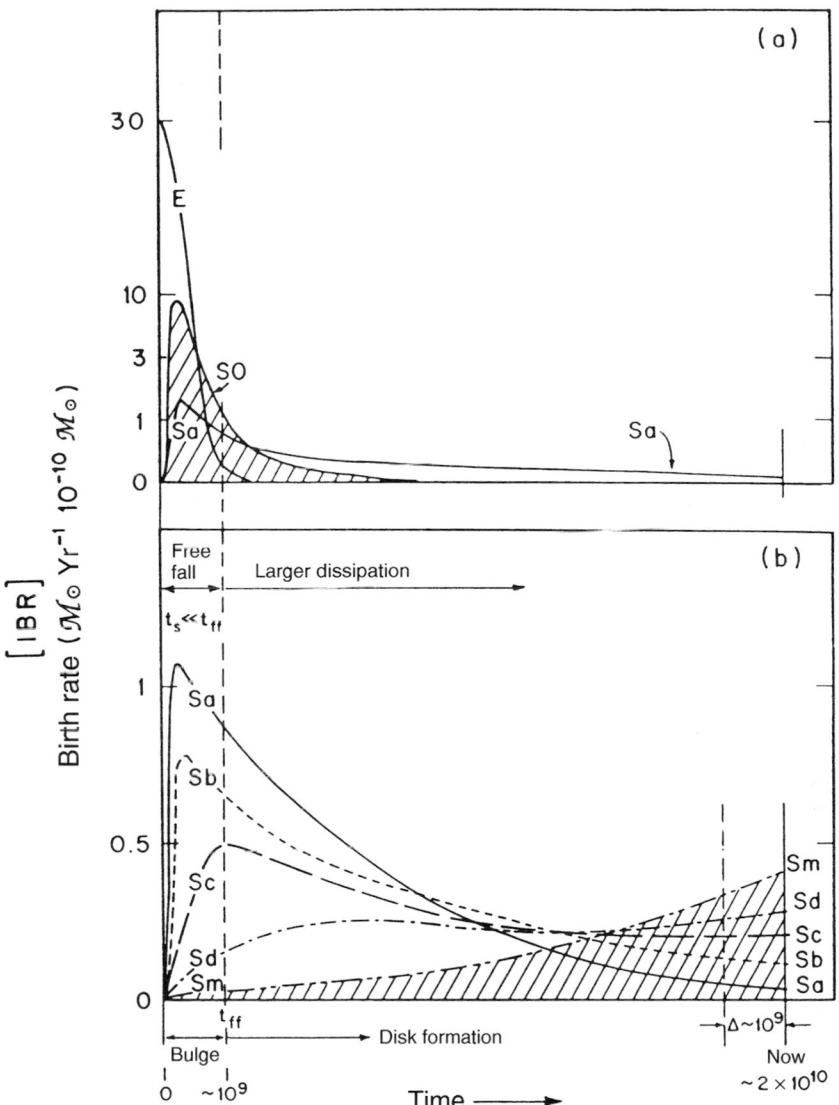

Fig 1.8. *Top:* Similar to Fig. 1.7 but with less detail and with an extension to the present epoch at the edge of the right-hand border. *Bottom:* Speculation of the variation of the rate of star formation with time for spirals along the entire disk sequence. The build up of the disk of Sm galaxies is hatched.

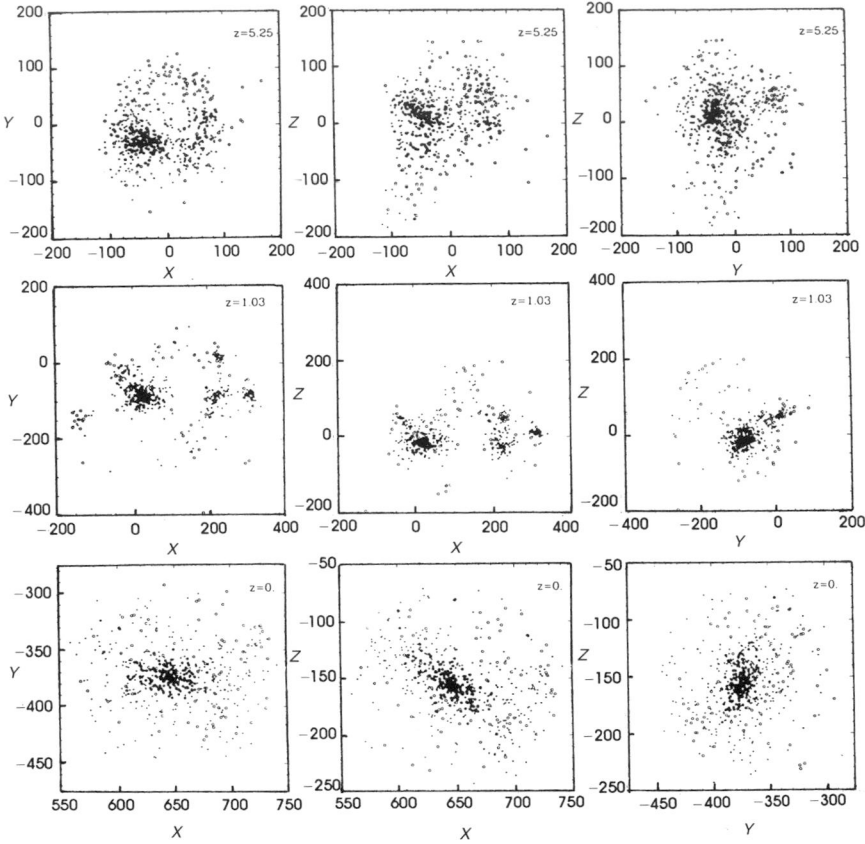

Fig 1.9. *N*-body simulation by Zurek *et al.* (1988) of a collapsing protogalaxy at three different epochs after the initial collapse, measured by the redshift (look-back time). The projections on the three *X*, *Y*, *Z* axes are shown.

hierarchy of density contrasts within the initial larger density perturbation that itself became the entire Galaxy. This consequence of the ZQS simulation in Fig. 1.9 is the reason for the statement (Sandage 1990) that the Searle–Zinn model of the halo, if it is only that the halo was formed by 'fragments falling into equilibrium', is ELS with noise and not a separate process.

The point has been misunderstood in the literature. What is generally implied there is that mergers have taken place between *separate initial galaxies* to form the present Galaxy. This is the picture of Schweizer

(1986), of which Searle and Zinn is a subset. However, the ZQS simulation in Fig. 1.9 belies such a Schweizer–Searle–Zinn picture as the merger of separate galaxies, or even separate fragments to form the Galaxy. The fragments were already present in the initial density fluctuation of the protogalaxy, being the fine structure of the density profile of the proto-galaxy itself.

A critique of what ELS said, what they should have said but did not, what they did say but should not have, what the critics have said that ELS said but which they did not, and what the critics have concluded incorrectly concerning the time scale and similar properties of the collapse is set out elsewhere (Sandage 1990). For example, age differences between globular clusters are permitted in ELS if a hierarchy of density contrast lumps $\delta\rho/\rho$ exist within the collapsing ELS envelope, a point the critics continue to miss. There are five characteristic properties of the Hubble sequence that are explained by Fig. 1.8:

(i) The bulge-to-disk ratio is a function of Hubble type.
(ii) The disk surface brightness varies systematically with Hubble type. The disk surface brightness is related to the integral under the dashed curve for each galaxy.
(iii) Integrated colour at the present epoch varies progressively with bulge-to-disk ratio and Hubble type.
(iv) The mean age of the disk at the present epoch is a function of Hubble type.
(v) The present-day star-formation rate per unit mass for Sc, Sd, and Sm galaxies is much higher than for S0, Sa, and Sb galaxies.

1.4.2 Questions concerning the Galaxy

Q.(xii) *The age–metallicity relation at different Galactic positions. (The wave of death from the centre outwards. The Galaxy is a living thing.)*

Of the many surprises seen in research in the past 45 years, none was more unexpected or even stark in its implications for Galactic evolution than the shape of the age–metallicity relation (AMR) as it emerged from the early data.

If all the chemical elements heavier than helium are made in the stars and are then recycled through the interstellar medium by stellar death (supernovae) or by stellar winds, then the average metal abundance in any particular part of the Galaxy must be a function of time. This requirement for a chemical build up with time was first proposed by Hoyle (1953) in the introduction to his paper on stellar nucleosynthesis. It was such a radical idea at the time that we all thought it to be a speculation that would fade away. Rather, it became the fundamental premise of the standard model of Galactic chemical evolution. It has been proved many times over by the data, once the basic method of age dating the stars (Schönberg and Chandrasekhar 1942; Sandage and Schwarzschild 1952) via the turn-off from the main sequence had been developed and widely applied to clusters and to field stars.[†]

The surprise was the extreme steepness of the AMR in the earliest history of the Galaxy, at least interior to the solar circle. The steepness shows that the star-formation rate and subsequent supernova activity, at least in the solar neighbourhood, and presumably everywhere inside the solar circle (see Fig. 1.16 later) was extremely high within only a few billion years after the end of the collapse phase.

The first clues centred about the age dating of the two oldest open clusters known in the late 1950s and early 1960s, together with a determination of their heavy element abundance. The oldest known well-documented disk cluster until 1962 was M67. An early colour–magnitude diagram (Johnson and Sandage 1955) showed a main-sequence termination point near $M_V = +3.5$, giving an age of about 5 Gyr. A more complete three-colour study, where the effects of reddening and metallicity on the colours could be separated (Eggen and Sandage 1964), showed beyond doubt that the metal abundance of the individual stars in M67 was nearly solar even at this great age. However, the age was only about one-third of that ascribed, even at that time, to the halo globular clusters, so the result, although interesting, was not paradigm changing.

However, the pattern of old age and high metallicity continued with the disk open cluster NGC 188, initially age dated at about 10 Gyr (Sandage

---

[†] That, of course, is a story in itself of how to age date a particular field star that is still near the main sequence. There is a long history of ingenious photometric methods and their calibration, all set in place since 1960 involving, at one time or another, perhaps 20% of all stellar astronomers. It is a brilliant history that should be told someplace in a review.

1962), and again with a near solar metallicity determined from three-colour photometry and spectra (Eggen and Sandage 1969). Although the ages of these two clusters have been in moderate controversy now for 20 years, with the age estimates changing by nearly a factor of 1.5 (i.e. as low as 3.5 Gyr and 7 Gyr for M67 and NGC 188 respectively), the relatively high age and high metallicity for both are well established.

An early age–metallicity diagram (AMD) (Sandage 1968), reproduced in Fig. 1.10, shows how rapid the increase of [Fe/H] must have been in the earliest period of Galactic evolution. This schematic shows the rapid collapse of the halo on the much compressed time scale of ELS at 2×10^8 years. We now would spread that to perhaps as much as ~1 Gyr for the bulk of the globular cluster system (van den Berg, Bolte, and Stetson

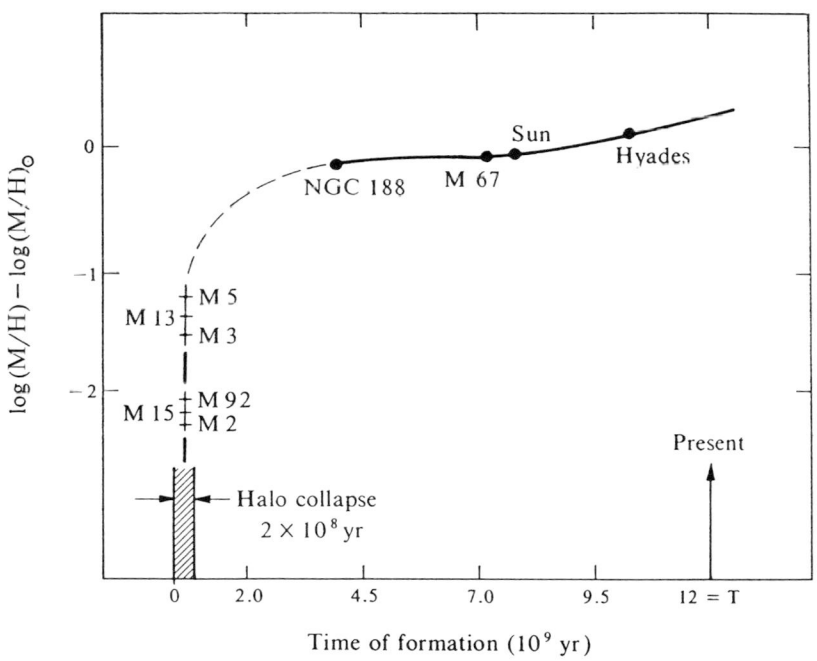

Fig 1.10. An early version of the age–metallicity relation (Sandage 1968) based on an assumed age of the Galaxy of 12 Gyr. The rapid increase of the abundance of heavy elements at early epochs argues for an almost sudden synthesis of nuclides heavier than hydrogen, either during the collapse phase of the Galaxy or very soon thereafter.

1990), but still the collapse had to be 'rapid' in the sense of ELS such that the motion towards what became the Galactic Centre had to be large compared with the motion orbiting about it. Otherwise, the large eccentricities of the old 'high-velocity stars' (actually the stars of low angular momentum about the Galactic Centre) cannot be explained (ELS).

A more modern version of an AMR, with somewhat reduced ages for M67 and NGC 188 and with a somewhat greater age for the Galaxy and for the halo collapse interval, is given in Fig. 1.11 (Sandage 1982). A mean [Fe/H]–age correlation for field stars is shown as the crosses, taken from Twarog (1980). Although the steepness of the rise of [Fe/H] is less severe than in Fig. 1.10, the initial enrichment is still extraordinarily rapid. It even

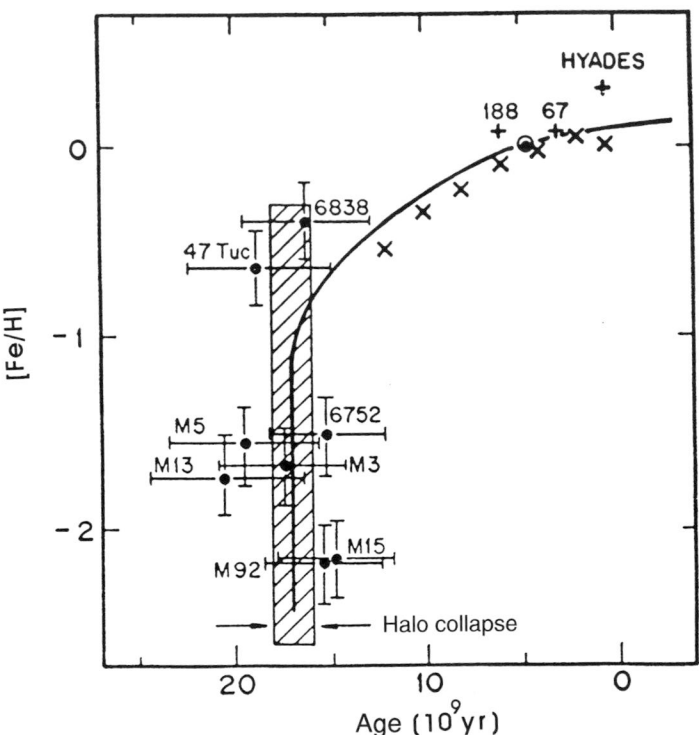

Fig 1.11. A later version (Sandage 1982) of the AMR showing also the field star data of Twarog (1980). The age of the Galaxy was assumed here to be 17 Gyr. However, modern data suggest either 12 or 13 Gyr for the Galactic age (see the text).

becomes more like Fig. 1.10 when the ages of the globular clusters are reduced to 13 or even 12 Gyr (Sandage 1993; Mazzitelli, d'Antona, and Caloi 1995), keeping M67 and NGC 188 as plotted.

A most important development was made when the AMRs of the Large Magellanic Cloud (LMC) and the Small Magellanic Cloud (SMC) were shown to be fundamentally different (Smith and Stryker 1986; Mateo 1987; Smith, Searle, and Manduca 1987) than the shapes in either Fig. 1.10 or Fig. 1.11. Figure 1.12 shows the Smith and Stryker summary for SMC and LMC, but with the shape of the relation for the Galaxy superimposed from Fig. 1.11. The interpretation is that the enrichment to 'incompletion' of the ISM is nearly finished in the solar neighbourhood, but is still only on the way to completion in both the LMC and the SMC. Clearly, the chemical evolution is far advanced in our region of the Galaxy compared with that in the Clouds.

At this stage a most remarkable discovery was made by Geisler (1987) in regions of the Galaxy where the AMR *resembles those in the LMC and SMC*. Figure 1.13 (top) shows colour–magnitude diagrams for some of Geisler's 'low'-metallicity, intermediate-age open clusters. Note that *they are all outside* the solar circle, in the lower-density regions of the anticentre sectors.

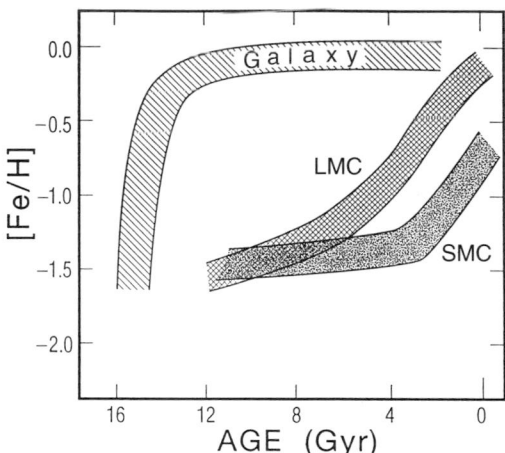

Fig 1.12. The degree of completion of the chemical evolution in the LMC and SMC compared with the general form of the evolution of the metal enrichment history in the Galaxy taken from Fig. 1.11. Data for SMC and LMC are from Smith and Stryker (1986).

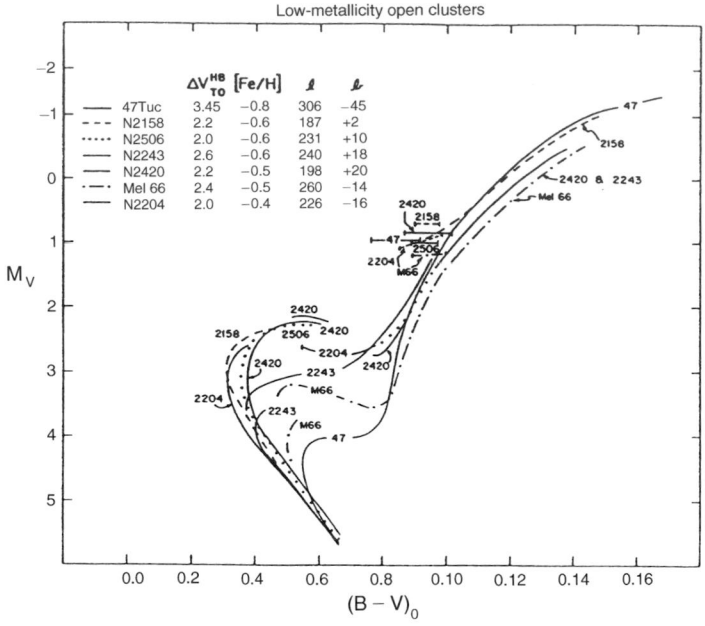

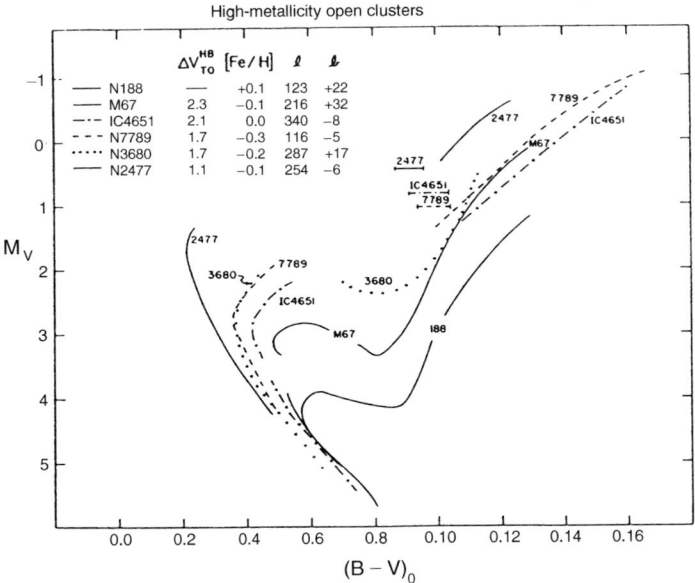

Fig 1.13. Colour–magnitude diagrams for intermediate-to-old open clusters in the Galaxy divided into a solar-like metallicity group (bottom) and a lower metallicity group, many from Geisler (1987) (top). Ages of these open clusters range from 2 to 9 Gyr. (From Sandage 1988aa.)

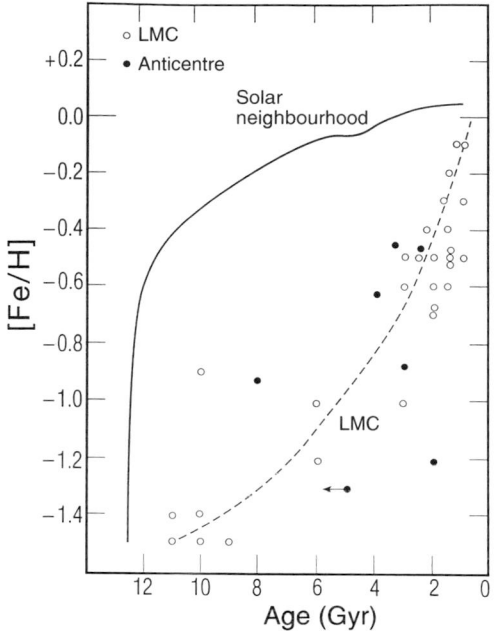

Fig 1.14. The AMR for the LMC (unfilled circles), the solar neighbourhood from Fig. 1.11, and the Geisler anticentre clusters (filled circles). Adapted from Geisler (1987).

Figure 1.14 is a summary of the presently available data. The schematic form for the AMR in the solar neighbourhood is the solid line at the left, again from Figs. 1.10 and 1.11. Geisler's anticentre clusters are filled circles. Clusters in the LMC are unfilled circles. The dashed line is the mean LMC relation put through the unfilled circles. The diagram is similar to Fig. 1.12 but now includes the clusters in the outer regions of the Galaxy.

The conclusion is that the outer Galactic regions are in a 'younger' chemical evolution state than regions at and inside the solar circle. Hence, the Galaxy evolves chemically at different rates depending on Galactocentric distance.

Early evidence suggested that many of the bulge stars are supermetal rich (Whitford 1978; Whitford and Rich 1983; Rich 1988, 1990). That picture received support in many of the 1990 ESO workshop papers, cited in the references as Jarvis and Terndrup (1990) (see, for example, the papers

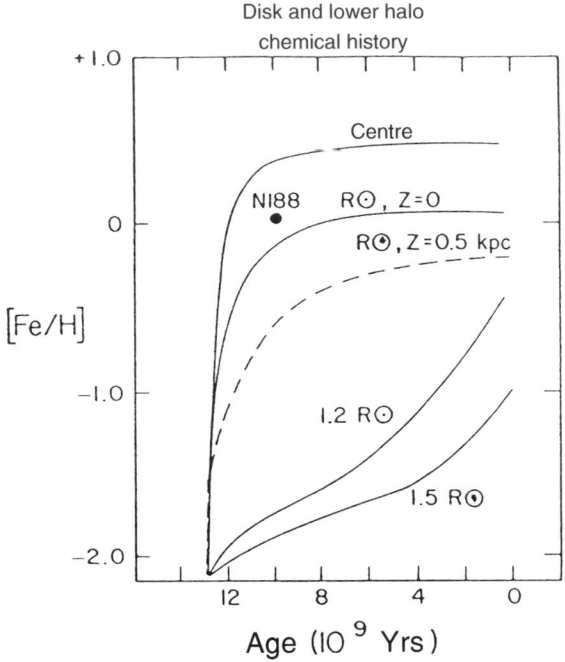

Fig 1.15. Schematic of the different shapes of the AMR for different regions of the Galaxy, assuming that stars in the bulge are supermetal rich. The maximum enrichment for the bulge stars is assumed, as determined by Rich (1988). Refinements may be required (McWilliam and Rich 1994).

in that report by Barbuy and Grenon 1990; Geisler and Friel 1990; Renzini and Greggio 1990; and Whitford, Terndrup, and Frogel 1990).[†]

Figure 1.15 is a schematic of the inferred form of the AMR at different Galactocentric distances based on these data. Note the similarity of the curves for positions 1.2 and 1.5 times the radius of the solar circle to the curves for LMC and SMC in Fig. 1.12.

† There is, however, some contrary evidence, together with reasons why the broad-band and low-spectral-resolution evidence may be deceptive, as set out in McWilliam and Rich (1994). Until these data and analysis relative to the earlier conclusions just cited can be adjudicated, the level of 'supermetallicity' of stars near the Galactic Centre may still be an open question. Clearly, supermetal rich stars exist in the bulge, but their percentage of the total, and therefore the mean, metallicity of the bulge is not yet definitive. Evidently, this is a key project for the future, having direct bearing on the cosmogony of the formation of the Galactic Bulge.

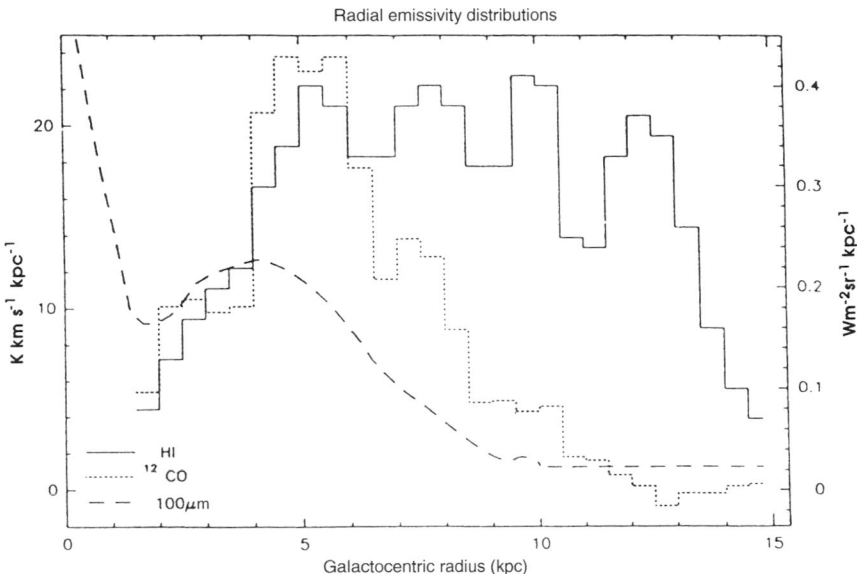

Fig 1.16. Radial dependences of emissivity from HI, CO, and dust in the Galaxy. Diagram from Burton and Deul (1987).

Figures 1.14 and 1.15 can be understood if the star-formation rate in the inner Galaxy was very high in earlier epochs, causing the metallicity yield from the ISM to reach 'completion'. Conversely, the percentage of 'enrichment completion' is progressively lower in the outer Galactic regions. This would be the case if most of the gas out of which stars form has been used up in the centre but is still abundant in the outer regions. Figure 1.16 from Burton and Deul (1987) shows precisely this situation.

The history shown by combining Figs. 1.14–16 can be described as the centre of the Galaxy having died to star formation, the process having used up most of the star-forming material. A wave of death is still sweeping outwards with time. The Galaxy seems to be a living thing that is still coming of age.

Finally, the consequence of Fig. 1.15 is that (1) there must be a gradient in [Fe/H] with Galactocentric distance and (2) the slope of that gradient must be a function of time. That the first prediction is true was discovered in external galaxies by Searle (1971), and then found in the Galaxy by many

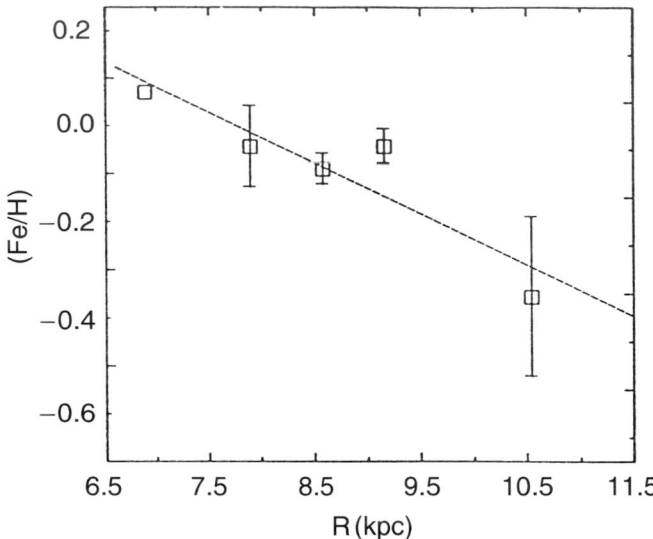

Fig 1.17. Summary by Cameron (1985) of the chemical gradient in the Galaxy as he determined it using 38 open clusters at different Galactocentric distances.

astronomers using many types of object. Examples are Hawley (1977) and Shaver *et al.* (1983) using HII regions, Torres-Peimbert and Peimbert (1977) using planetary nebulae, Mayor (1976) using dF and dG stars, and Janes (1979), Panagia and Tosi (1981), and Cameron (1985) using star clusters of different ages and Galactic positions. Cameron's data for intermediate-age clusters that occur just beyond the solar circle are shown in summary in Fig. 1.17. This shows the beginning of the trend farther beyond the solar circle and as a function of age found by Geisler, shown in Fig. 1.14. Figure 1.18 shows the [Fe/H] gradient derived by Harris (1981) for Cepheids adopting the distance to the centre of the Galaxy at $R_0 = 8.5$ kpc.

1.4.3 Questions concerning practical cosmology and the Universe

Q.(xv) *Is the expansion real?*

Although there is no credible alternative to the Friedmann–Lemaître–Robertson explanation of the redshift as a change of the scale factor of the metric with time (i.e. an expansion of the manifold), there has been only relatively meager direct evidence (in the spoken version of the lecture, but

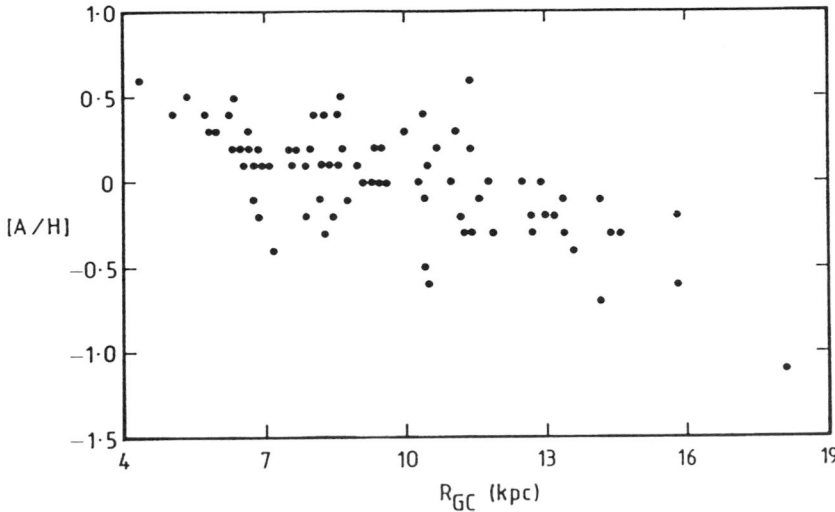

Fig 1.18. The gradient in [A/H] in the Galactic disk as derived from Cepheid variables by Harris (1981).

see below) for it. The only known direct proofs for the reality of the expansion are the three listed as the subtitles here.

(a) The Tolman *SB* test. The Tolman surface brightness test has been known in principle since Tolman's (1930) discovery paper, expanded in his textbook (Tolman 1934, eq. 189.5). A heuristic explanation of the effect is as follows. Take two identical luminous objects whose average surface brightness is the same. Put one in the expanding Universe and keep the other stationary relative to the observer. The object that is moving with a redshift z relative to the observer will have an observed $\langle SB \rangle$ that is fainter by $(1 + z)^4$ than the one that is stationary. This is composed of two factors. (A) The apparent area of the moving object is increased by $(1 + z)^2$ owing to the effect of aberration on each of the observed linear dimensions. (B) The received energy per unit area is reduced by two factors of $(1 + z)$ owing to the redshift itself. One factor comes from the decrease in the energy of each photon by the frequency shift (this is the energy effect because $E = h\nu$). The other comes from a decrease in the rate of arrival of the photons because the path length of the beam is continuously increasing (the number effect). Combining the aberration and the two flux effects gives the $(1 + z)^4$ dependence. This standard theory is explained in many places elsewhere, either in the way suggested above (e.g. Sandage 1995b, Lectures 4 and 6) or by more fundamental methods.

Although simple in principle, the test has proved difficult to realize because galaxies as the test objects do not have uniform *SB*. Rather the *SB* varies by a factor of 1000 in their projected images. The technical problems of dealing with this variation centre on defining an appropriate *metric* (rather than isophotal) area over which to measure the average *SB* for E galaxies at different redshifts. This problem has finally been solved in principal by using so called Petrosian (1976) metric diameters with which to define an appropriate area, and by finding the manner in which the $\langle SB \rangle$ so measured varies with the absolute magnitude of the E galaxies (Sandage and Perelmuter 1990a, b, 1991, hereafter SP-I, II, or III).

The solution to this last problem is essential because the *SB* is not a constant for all E galaxies. It depends on the intrinsic size and/or total luminosity of the ellipticals in such a strong way that all data must be corrected to a standard set of conditions (either a standard radius or a standard absolute magnitude).

Figure 1.19 (Sandage and Perelmuter 1991, SP-III) shows the variation of $\langle SB \rangle$ for galaxies in the Virgo, Fornax, and Coma clusters as a function of their absolute magnitudes (on the long distance scale, equivalent to $H_0 = 50$, but using individual cluster distance moduli). Five different Petrosian radii have been used, together with the effective radius in the upper left panel.

A Petrosian radius is defined as the radius where the ratio of the *SB* averaged over that radius to the *SB* at that radius is a particular value, such as 1.3, 1.5, 1.7, 2.0, and 2.5 mag, as in Fig. 1.19. The very valuable invariant properties of Petrosian radii are discussed extensively elsewhere (SP-I–III; Kron 1995; Sandage 1995b).

Figure 1.19 sets out a practical way to perform the Tolman *SB* test. Although the $\langle SB \rangle$ varies with absolute magnitude, the form of the function in Fig. 1.19 is well defined. A cluster of galaxies at high redshift, in the absence of evolution in the look-back time, will have the same form as in this diagram, but the ordinate would be fainter by $(1 + z)^4$ (in magnitudes) if the expansion is real. If, then, we were to measure a number of cluster members deep into the cluster's luminosity function so as to reproduce the shape of the variation in Fig. 1.19, the test would be complete.

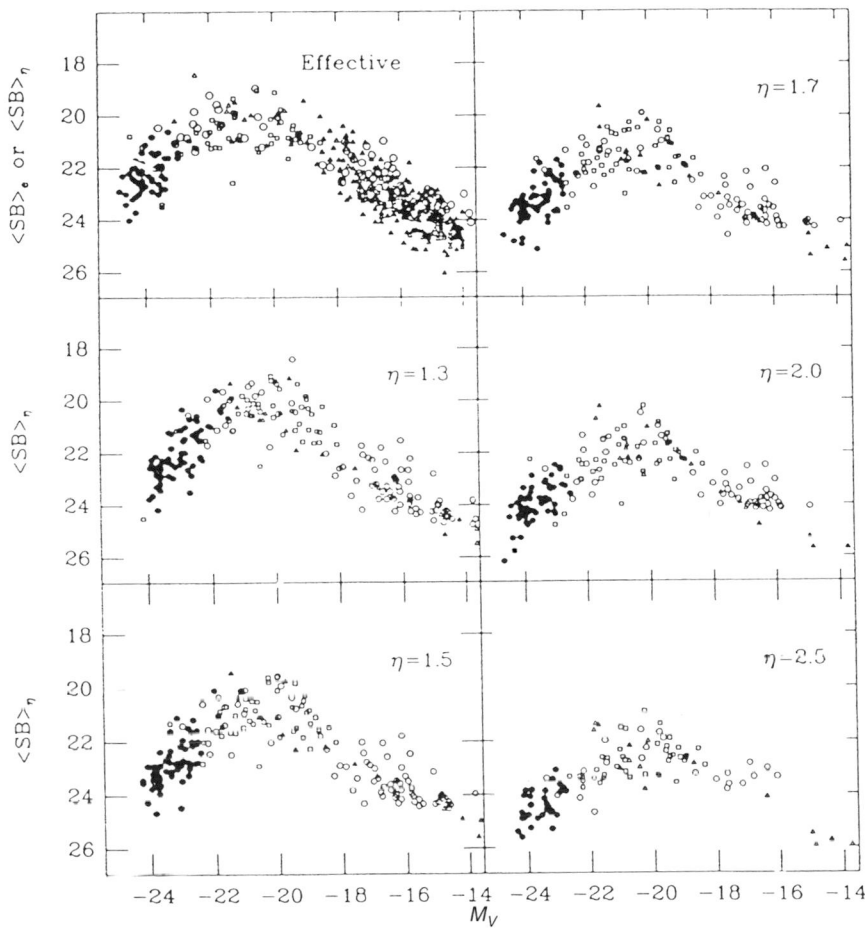

Fig 1.19. The variation of the surface brightness of E galaxies of different absolute magnitudes in the Virgo, Fornax, and Coma clusters. The solid dots are the first few ranked cluster galaxies. The *SB* data are averaged over five different Petrosian radii (η = 1.3, 1.5, 1.7, 2.0, and 2.5 mag) and over the 'effective' radius. To apply the Tolman surface brightness test requires reduction of the data at given redshifts to a fixed (standard) absolute magnitude, or, better, a mapping of the variation shown here over a range in absolute magnitude using a number of cluster galaxies in given clusters at high redshift. Diagram from Sandage and Perelmuter (1991).

The test is also feasible, but with less weight, if we were to measure the $\langle SB \rangle$ of the first few ranked cluster galaxies at several Petrosian radii (or minimally at only one) and, from a calculation of the absolute magnitude (or the metric radius), then correct the data to a standard (fixed) value of M or R.

This latter procedure was done by SP-III (1991) using data from the literature reported by Djorgovski and Spinrad (1981, DS) with the result shown in Fig. 1.20. The objects studied by DS were first-ranked cluster galaxies at various redshifts as large as $z \sim 1$. They used a Petrosian radius of 2 mag over which to average the SB. Figure 1.20, upper left panel, shows the DS data, but not reduced to the standard condition of a fixed linear radius (as from Fig. 6 of SP-III, 1991). The expected Tolman $(1 + z)^4$ signal is the dashed line. The solid line is the least squares solution. The other three panels show the data reduced to a fixed linear radius according to the calibration of the $(\langle SB \rangle, R)$ relation in Fig. 6 of SP-III, calculated for different q_0 values that define the spatial curvature.

If the DS data were definitive, Fig. 1.20 would be a very strong test of the Tolman prediction. The proof that the expansion is real would be decisive.

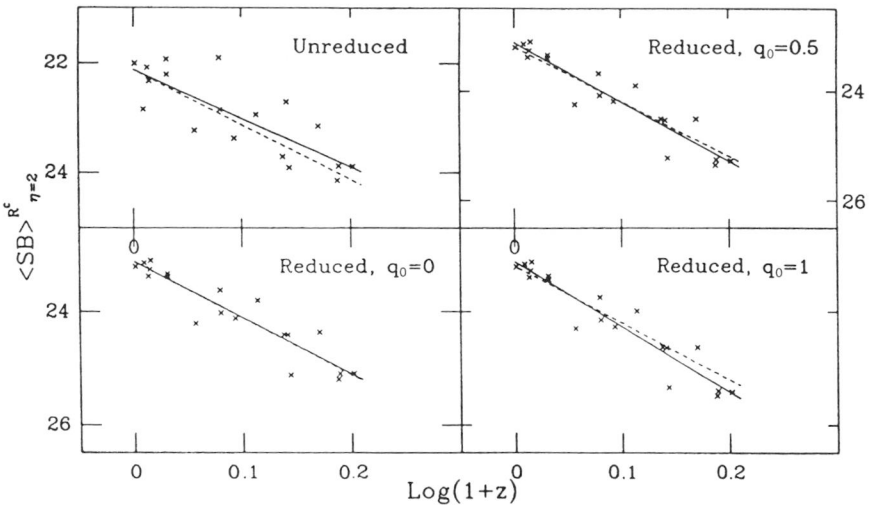

Fig 1.20. Test for the Tolman (1930, 1934) surface brightness $(1 + z)^4$ effect in first-ranked cluster E galaxies at high redshifts using the data of Djorgovski and Spinrad (1981). Diagram from Sandage and Perelmuter (1991).

However, the authors did not make the test with their own data, stopping short by showing only the Hubble (m, z) diagram and the angular Petrosian 2.0 mag diameter vs z diagram separately. They state that their data are preliminary, primarily showing the possibilities of using Petrosian measures of metric angular size for the angular diameter test of the standard model.

The entire machinery is now in place to perform the Tolman test in a definitive manner. What is required is a programme to observe perhaps 20 galaxy clusters over a redshift span of say $z = 0.03$ to $z = 0.7$ (of which there are now many clusters known). If the luminosity grasp into each of these clusters could be, say, 4 mag, then a crucial part of the parameter space of Fig. 1.19 would be sampled, and from that, the position along the ordinate would be determined for each of the galaxies in each of the clusters. These SB values should be $(1 + z)^4$ fainter than the ridge lines in Fig. 1.19 if the expansion is real. Of course, evolution in the look-back time must be considered. A discussion elsewhere (SP-III, 1991) suggests ways to test for it and to correct the data thereby.

That the test is feasible according to the available flux levels has been calculated by SP-III. The instruments at the Observatorio del Roque de los Muchachos of the IAC are ideal in their characteristics to perform the test. It is hoped that astronomers in the cosmology section of the IAC will undertake the definitive Tolman test with their unique and unprecedented facilities on La Palma. The proof of the expansion in this way is surely a key project for the next decade or so.

(b) Time dilation. The time dilation test, that all clocks observed by us at large redshifts will appear to keep time at a rate $(1 + z)$ slower, has been known since it was proposed by Wilson (1939) using supernovae of type Ia. Such objects have a remarkably fixed shape to their light curve, with discontinuities at fixed times that can be used as time markers.

Figure 1.21 shows the standard template curve, built up by adding the data for 22 individual light curves of 'Branch normal' (Branch, Fisher, and Nugent 1993) type Ia supernova. No stretching of the time scale along the abscissa has been done to combine these data. Figure 1.21 is the template used as a standard shape in the Basel Atlas of type Ia supernovae (Leibundgut *et al.* 1991). The top curve shows B-band photometric data.

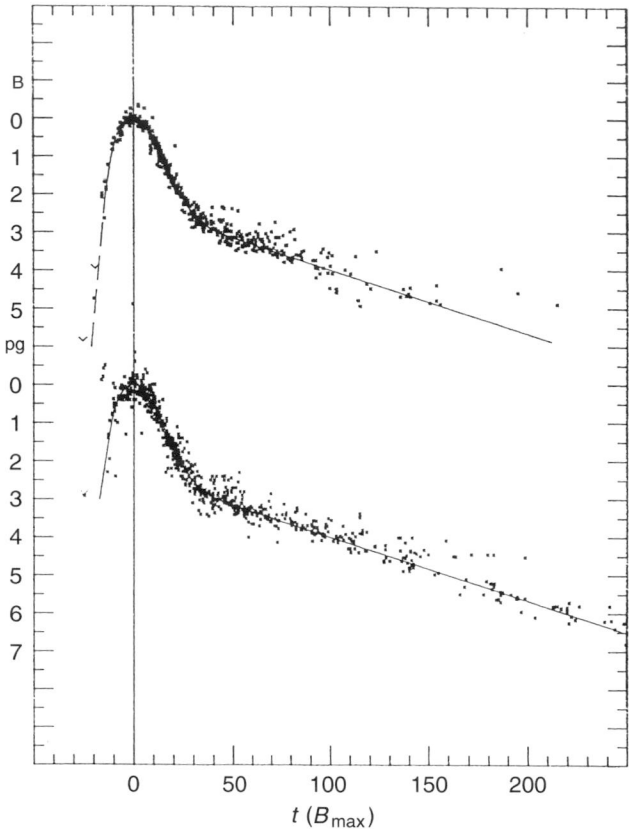

Fig 1.21. Template light curve made up of 22 individual light curves for 'Branch normal' supernovae of type Ia. Diagram from the summary by Cadonau, Sandage, and Tammann (1985) as used in the Basel SNe Ia Atlas (Leibundgut *et al.* 1991).

The bottom curve shows earlier, somewhat less accurate, photographic m_{pg} archival data.

Wilson reasoned that if type Ia supernovae could be observed at large redshifts, their light curves would be stretched in the time axis by the factor of $(1 + z)$. Until recently, no supernovae had been found at adequately large redshifts to test the prediction.

This position was stated in the spoken version of this chapter. However, the situation has changed dramatically in the meantime. A programme of finding SNe Ia at large redshifts, led by Perlmutter at Berkeley, has begun to produce quite wonderful results. Although this group's first published paper (Perlmutter *et al.* 1995) on a single SN Ia at $z = 0.458$ does not address the time dilation problem directly, eight other high redshift SNe discovered later by them have, for the first time, apparently shown the time dilation (Perlmutter, private communication). Details remain to be seen, but the preliminary data seem convincing.

(c) The MWB temperature. It has always been known that, as the Universe expands, the temperature of the microwave background must decay as $1 + z$ (Tolman 1934, eq. 171.6, combined with 171.2; see also Bahcall and Wolf 1968). A practical test came to be realized with an understanding concerning a calculation by McKellar (1941) using the observations by Adams (1941, 1942) of the interstellar CN absorption lines. One of these lines is 2 eV above the ground state, requiring a radiation temperature of $\sim 3\,\mathrm{K}$ in the Galactic interstellar medium to excite the upper state. Neither McKellar nor Adams made more of their observations, but it was later conjectured (see Thadeus 1972 for a review) that this excitation energy is supplied by the 3-K microwave background of the relic radiation. Hence, if suitable absorption lines can be found in objects of high redshift, the prediction that the temperature required to excite these lines is higher than in the Galaxy by the factor of $(1 + z)$ can be tested, provided that the radiation is in fact the Alpher–Herman (1948, 1950; Alpher, Folin, and Herman 1953) radiation left over from the Big Bang.

The test has been conducted over many years, but each attempt gave only upper limits to the temperature, $T(z)$, such that if $T(z) = T(0)(1 + z)^n$, only an upper limit could be put on n. The prediction of the standard Friedmann–Lemaître–Robertson–Walker model is that $n = 1$. The last best value for n from an upper-limit test was $n < 1.73$ (Meyer *et al.* 1986).

In the spoken version of this chapter, I stated that the test is crucial but that it was yet to produce a positive verification that $n = 1$. However, readers will see from the discussion section that the test may now have been done successfully. Hubert Reeves points out, both in the discussion of this chapter and as a line item in the discussion at the end of his own contribution, that the report by Songaila *et al.* (1994) of an observation with the Keck 10-m telescope confirms the prediction at a redshift of $z = 1.776$, deriving thereby $n = 0.98 \pm 0.11$. This most important result would be a direct proof of the reality of the expansion, and the validity of the hot beginning of creation from the standard Big Bang model, or something substantially like it.

It can be expected that confirmation and a strengthening of the test will be a key project in the next ten years.

Q.(xvii) *The distance scale*

Finally, we sketch in the briefest manner the current status of the extragalactic distance scale, with the consequent value of the Hubble constant.

It is not the purpose of this chapter to set out in the necessary detail, as in a lawyer's brief, the case for either the short distance scale with $H_0 \approx 85$ or the long scale with $H_0 \approx 50$. This has been done in many places, and, at the moment, all the briefs are contentious.

(xvii)a. Calibration of all indicators. Influential but prejudicial (because of their selectivity and omissions) reviews of the short scale by Jacoby *et al.* (1992) and van den Bergh (1992, 1993) persuaded many astronomers in the United States that the evidence for the short scale was overwhelming. The three independent methods that were claimed there to give $H_0 \approx 85$ were (a) the Tully–Fisher 21-cm line-width correlation with absolute magnitude, (b) the Ford–Jacoby–Ciardullo planetary nebula method, and (c) the Tonry–Schneider surface brightness fluctuation method. The agreement seemed impressive to most astronomers. In view of this apparentness, no credence was given to nine methods that are also in internal agreement amongst themselves in support of the long distance scale with $H_0 \approx 50$ (Sandage and Tammann 1995b; Sandage 1995b for summaries).

Reviews over the years showing the validity of the long distance scale have been given by Tammann and Sandage (1981, 1995b), Tammann (1986, 1987, 1988, 1992), Sandage and Tammann (1995b). Individual papers supporting the long scale via the supernova methods include Branch (1995), Branch and Khokhlov (1995), Branch and Tammann (1992), Saha *et al.* (1994, 1995, 1996), Tammann and Sandage (1995a, b). Papers supporting the long distance scale via the Tully–Fisher method corrected for observational selection bias include Bottinelli *et al.* (1986a, b, 1988), Kraan-Korteweg, Cameron, and Tammann (1988), Sandage (1988a, b, 1994a, b, 1995a), Fouque *et al.* (1990), Federspiel, Sandage, and Tammann (1994), Sandage, Tammann, and Federspiel (1995), Theureau *et al.* (1997), and Federspiel, Tammann, and Sandage (1997). Criticisms of the PN and the SBF methods are in Tammann (1992, 1993). The first criticism of selection effects in the PN method was made by Bottinelli *et al.* (1991).

However, despite whatever strong evidence exists at any given time that favours either the long or the short scale, the debate over the distance scale will end only when the methods that lead to one or the other of the scales are decisively shown to be defective.

The purposes of this chapter on this question are (1) to discuss the current status of the method to determine H_0 through supernovae of type Ia that require $H_0 \approx 55$, (2) to contrast this value obtained therefrom with the values claimed to be near 85 in the reviews of Jacoby *et al.* (1992) and others, and (3) to show the pernicious effect of observational selection bias, leading incorrectly to a short distance scale ($H_0 \approx 85$), if uncorrected. More detailed critiques have been set out elsewhere, including Teerikorpi (1975a, b), Bottinelli *et al.* (1986a, b), Kraan-Korteweg, Cameron, and Tammann (1988), Sandage (1988a, b, 1993, 1995b, 1996a, b), Tammann (1992), Federspiel, Sandage, and Tammann (1994), Sandage and Tammann (1995a, b), Tammann and Sandage (1995a, b), Sandage, Tammann, and Federspiel (1995), Federspiel, Tammann, and Sandage (1997), and Theureau *et al.* (1997), some of which have been cited above.

(xvii)a The method through supernovae of type Ia. Before criticizing the methods that lead to the short distance scale ($H_0 \approx 85$), we set out the most persuasive case to date that requires the long scale with $H_0 \approx 50$. The method centres around supernovae of type Ia (SNe Ia) as standard candles.

The history of ideas that SNe Ia are excellent standard candles started with Kowal (1968) with his demonstration of a tight Hubble diagram (apparent magnitude at maximum vs redshift). A modern proof that SNe Ia are among the best standard candles known is the tightness of the Hubble redshift–apparent magnitude diagram shown in Fig. 1.22 (Tammann and Sandage 1995a), using data where observations were begun not more than five days after the epoch of maximum light. The dispersion is small at $\sigma(M) = 0.34$ mag in B and 0.33 mag in V. These are the smallest dispersions of any known photometric distance indicator. For example, the Tully-Fisher LW–M correlation has a dispersion that averages

$\sigma(M) \approx 0.7$ mag (Bottinelli *et al.* 1988; Kraan-Korteweg *et al.* 1988; Sandage 1988b, 1994b; Fouque *et al.* 1990).

With a dispersion as low as $\sigma(M) \approx 0.3$ mag, we need only obtain a calibration of the absolute magnitude of a few SNe Ia to calibrate Fig. 1.22 for the value of the Hubble constant at the 10% level. To date, five calibrations exist via Cepheid distances to four galaxies that are the parents of five prototypical type Ia SNe. Details for the first three are in Saha *et al.* (1994, 1995). Calibrations of the two additional SNe Ia in NGC 4496 and NGC 4536 have been completed (as of May 1995). They support the conclusions from the first three that the mean absolute magnitude at the maximum of SNe Ia is close to $\langle M_B \rangle = -19.5$ (Saha *et al.* 1996), requiring, via Fig. 1.22, that $H_0 \approx 55\,\mathrm{km\,s^{-1}\,Mpc^{-1}}$. The data as they exist at this writing (1997) for the first five calibrations in the programme are set out in Table 1.1.

One sees here the inconsistency with the short scale by noting that de Vaucouleurs (1979, his Table 9) requires $\langle M_B \rangle = -18.5$ for $H_0 \approx 100$, using the final result from his armada of methods. Pierce (1994), using Tully–Fisher, requires $\langle M_B \rangle = -18.7$ for his value of $H_0 = 87$. Hence, the SNe Ia data contradict both these versions of the short distance scale.

Critics of the SN result have claimed that SNe Ia are not good standard candles, despite the decisive proof of Fig. 1.22, and the additional multiple proofs by Branch and Tammann (1992), and the addition to the twinning data by Rood (1994). The answer to critics such as Phillips (1993) and Hamuy *et al.* (1995) has been made by showing (Tammann and Sandage 1995a) that their putative variation of SNe Ia light-curve shape has only a

Table 1.1. *Absolute magnitude at maximum for five type Ia supernovae calibrated using Cepheid distance to the parent galaxies.*

Galaxy	SN	M_B (max)	M_V (max)
IC 4182	1937C	-19.65 ± 0.18	-19.64 ± 0.13
NGC 5253	1895B	-19.80 ± 0.30	————
NGC 5253	1972E	-19.55 ± 0.23	-19.50 ± 0.21
NGC 4536	1981B	-19.24 ± 0.20	-19.22 ± 0.20
NGC 4496	1960F	-19.50 ± 0.23	-19.59 ± 0.23
mean		-19.55 ± 0.11	-19.49 ± 0.11
rms		$0.^{\mathrm{m}}21$	$0.^{\mathrm{m}}19$

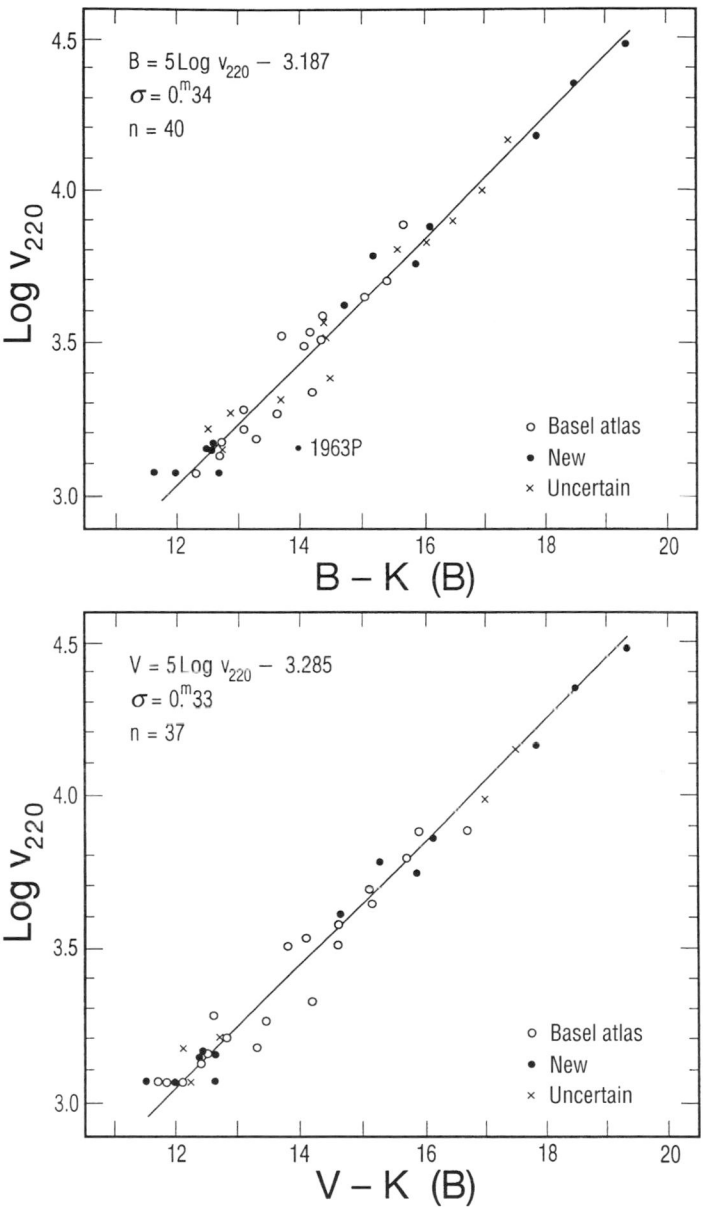

Fig 1.22. Hubble diagram in *B* and *V* (corrected for the *K* term for the effects of redshift) for 43 SNe Ia using literature data for the SNe light curves that contain observations that begin no later than five days after the adopted epoch of maximum light. Diagram from Tammann and Sandage (1995a).

small, if any, effect on the value of H_0 obtained in their reductions using such a claimed variation. A decisive proof leading to the same conclusion by a different method, involving the variation of SNe Ia envelope expansion velocities, has been given by Fisher *et al.* (1995).

Calibration of other distance indicators, such as (a) the TF method, (b) the turnover luminosity of the globular cluster luminosity function, (c) the distance to the Virgo cluster used as a standard rod, (d) the mean absolute magnitude of the van den Bergh luminosity classes, and (e) the PN method when cognizance is taken of the sloping bright end for the PN luminosity function, all require the long distance scale when the effects of selection bias are considered. Details are in the cited references, the most detailed of which are Tammann (1992), Sandage and Tammann (1995b), and Sandage (1995b, Lectures 8–10).

Criticism of the Freedman *et al.* (1994) announcement is given in the discussion section. Two of their precepts are incorrect, reducing their value of H_0 by at least 20%, but more likely by a factor of 1.5. Their comment on H_0 was inappropriate because the reports in the public media were taken to suggest a solution to H_0 that causes 'a crisis in cosmology' with regard to the time scale. No such crisis exists. Rather, the importance of the first experiment of the Freedman *et al.* 'key *HST* project' was a demonstration that *HST* can find Cepheid distances to galaxies with $(m - M)$ as large as ≈ 32 at photometric levels of $V \sim 28$, which in fact is the extent of what their *HST* experiment, and others like it, did.

(xvii)b, c Bias problems and why some methods have produced wrong results. We finally come to the other side of the coin to show in greater detail how bias problems, uncorrected for, have led incorrectly to the short distance scale.

There is a large literature on the problems caused by observational selection bias and how such bias adversely affects the determination of the distance scale. The principal papers from Europe are from the Finnish, French, and Swiss groups. Particularly important early papers are those by Teerikorpi (1975a, b, 1984, 1987, 1990), Bottinelli *et al.* (1986a, b, 1988), Fouque *et al.* (1990), and Kraan-Korteweg *et al.* (1988). Later papers include those by Federspiel *et al.* (1994), Sandage, Tammann and Federspiel (1995), Federspiel, Tammann, and Sandage (1997) and Theureau *et al.* (1997). The salient points showing the pernicious effects of uncorrected bias are in Sandage (1988a, b, 1995b, Lecture 10 for a summary with literature references to 1994). Additional results are in Sandage (1996a, b).

It is our position that all results that have led to the short distance scale to date using the TF method suffer from either a neglect or an underestimate of the effects of observational selection bias. Correcting for the bias, both for field galaxies (Sandage 1988b; Federspiel *et al.* 1994) and for cluster galaxies (Kraan-Korteweg *et al.* 1988; Fouque *et al.* 1990; Sandage, Tammann, and Federspiel 1995), requires the long distance scale with $H_0 \approx 50$.

The main point of argument centres about the intrinsic dispersion of the TF method, but similar considerations concerning the intrinsic dispersion apply equally to each of the other photometric methods mentioned previously. The bias arises because, unless the intrinsic dispersion of the indicator is zero, the nearby objects in a set are common, whereas the distant objects must be uncommon (brighter than average) to enter the catalogues that are cut at a particular apparent magnitude. The difference between the proper mean $\langle M_0 \rangle$ and the mean at a given redshift $\langle M(z) \rangle$ increases progressively with redshift. Hence, using photometric distances uncorrected for this selection effect will produce a distance scale in which the Hubble constant seems, incorrectly, to increase outwards (Teerikorpi 1975a, b; Sandage 1988a, 1994a), and where its mean value is too high. The size of the effect depends on the size of the intrinsic dispersion of the distance indicator. It is here that the debate rests. A vital activity in the future to solve this central problem must be to devise observational programmes to determine decisively the intrinsic dispersions of all distance indicator methods such as Tully–Fisher, planetary nebulae, supernovae, surface brightness fluctuations, and combinations of Hubble types with van den Bergh luminosity classes such as the de Vaucouleurs (1979) Λ_c index. In addition, the physics of the expanding envelope method for type II SN used by Kirshner (1994) and his group must be understood so as to prevent the current mistake (Schmidt *et al.* 1992, 1994) of using the wrong dilution (Wagoner) factor (see Branch 1995) from being made again.

The debate on the value of H_0 will be closed only when the severity of the effect of observational selection bias is agreed between all parties. There is presently a large denial by the short-distance-scale astronomers of the severity of the effect, and conversely, much emphasis is put on it by the proponents of the long scale. Clearly, the determination of the intrinsic dispersion of each of the distance indicators is one of the key problems in observational cosmology for the coming decades. It may indeed take the entire 30 allotted years.

Acknowledgements

In what better place could the most remarkable conference on which this book is based be held than a paradise that is host to two highly prophetic mountain observatories: it is in the Canary Islands that the future will be invented, organized through the offices of the remarkable new Instituto de Astrofísica de Canarias.

References

Adams, W.S.: 1941, ApJ, **93**, 11

Adams, W.S.: 1942, ApJ, **97**, 105

Alpher, R.A., Follin, J.W., and Herman, R.C.: 1953, Phys Rev, **93**, 1347

Alpher, R.A., and Herman, R.C.: 1948, Phys Rev **74**, 1737

Alpher, R.A., and Herman, R.C.: 1950, Rev Mod Phys, **22**, 153

Bahcall, J.N., and Wolf, R.A.: 1968, ApJ, **152**, 701

Barbuy, B., and Grenon, M.: 1990, in Jarvis and Terndrup (1990), p. 83

Bottinelli, L., Gouguenheim, L., Fouque, P., and Paturel, G.: 1990, A&AS, **82**, 391

Bottinelli, L., Gouguenheim, L., Paturel, G.A., and Teerikorpi, P.: 1986a, A&A, **156**, 157

Bottinelli, L., Gouguenheim, L., Paturel, G.A., and Teerikorpi, P.: 1986b, A&A, **166**, 393

Bottinelli, L., Gouguenheim, L., Paturel, G.A., and Teerikorpi, P.: 1988, ApJ, **328**, 4

Bottinelli, L., Gouguenheim, L., Paturel, G.A., and Teerikorpi, P.: 1991, A&A, **252**, 550

Branch, D.: 1995, Lecture at Seventh Marcel Grossmann Meeting, Stanford, July 1994, eds. M. Kaiser and R. Jantzen (Singapore: World Scientific), in press

Branch, D., and Khokhlov, A.M.: 1995, Phys Rep, **256**, 53

Branch, D., and Tammann, G.A.: 1992, Ann Rev A&A, **30**, 359

Branch, D., Fisher, A., and Nugent, P.: 1993, AJ, **106**, 2383

Brown, H. Bonner, J., and Weir, J.: 1963, *The Next 100 Years* (New York: Viking)

Burton, W.B., and Deul, E.R.: 1987, in *The Galaxy*, eds. G. Gilmore and B. Carswell, NATO ASI Series C, Math and Physical Sciences, Vol. 207 (Dordrecht: Reidel), p. 141

Buta, R., Mitra, S., de Vaucouleurs, G., and Corwin, H.G.: 1994, AJ, **107**, 118

Cadonau, R., Sandage, A., and Tammann, G.A.: 1985, in *Supernovae as Distance Indicators*, Lecture Notes in Physics, No. 224, ed. N. Bartel (New York: Springer), p. 151

Cameron, L.M.: 1985, A&A, **147**, 47

de Vaucouleurs, G.: 1979, ApJ, **227**, 729

Djorgovski, S.B., and Spinrad, H.: 1981, ApJ, **251**, 417

Eggen, O.J., and Sandage, A.: 1964, ApJ, **140**, 130

Eggen, O.J., and Sandage, A.: 1969, ApJ, **158**, 669

Eggen, O.J., Lynden-Bell., D., and Sandage, A.: 1962, ApJ, **136**, 748 (ELS)

Federspiel, M., Sandage, A., and Tammann, G.A.: 1994, ApJ, **430**, 29

Federspiel, M., Tammann, G.A., and Sandage, A.: 1997, ApJ, in press (FTS97)

Fisher, A., Branch, D., Hoflich, P., and Khokhlov, A.: 1995, ApJL, **447**, 273

Fouque, P., Bottinelli, L., Gouguenheim, L., and Paturel, G.: 1990, ApJ, **439**, 1

Freedman, W. *et al.*: 1994, Nature, **371**, 757

Gallagher, J.S., Hunter, D.A., and Tutukov, A.V.: 1984, ApJ, **284**, 544

Geisler, D.: 1987, AJ, **94**, 84

Geisler, D., and Friel. E.D.: 1990, in Jarvis and Terndrup (1990), p. 77

Hamuy, M., Phillips, M.M., Maza, J., Suntzeff, N.B., Schommer, R.A., and Aviles, R.: 1995, AJ, **109**, 1

Harris, H.C.: 1981, AJ, **86**, 707

Hawley, S.A.: 1977, BAAS, **9**, 374

Hogg, D.E., Roberts, M.S., and Sandage, A.: 1993, AJ, **106**, 907

Hoyle, F.: 1953, ApJS, **1**, 121

Jacoby, G. *et al.*: 1992, PASP, **104**, 599

Janes, K.A.: 1979, ApJS, **39**, 135

Jarvis, B.J., and Terndrup, D.M. (eds.): 1990, *Bulges in Galaxies* (Garching: ESO)

Johnson, H.L., and Sandage, A.: 1955, ApJ, **121**, 616

Kennicutt, R.C.: 1981, AJ, **86**, 1847

Kennicutt, R.C.: 1989, ApJ, **344**, 685

Kirshner, R.P.: 1994, Harvard Mag, Vol. 97 (Nov/Dec), p. 20

Kowal, C.T.: 1968, AJ, **73**, 1021

Kraan-Korteweg, R.C., Cameron, L.M., and Tammann, G.A.: 1988, ApJ, **331**, 620

Kron, R.G.: 1995, in *The Deep Universe: Evolution in the Galaxy Population*, eds. B. Binggeli and R. Buser (Berlin: Springer), Lecture 6

Larson, R.B.: 1987, in *Nearly Normal Galaxies*, ed. S.M. Faber (New York: Springer), p. 26

Larson, R.B.: 1988, in *Globular Cluster Systems in Galaxies*, IAU Symp. 126, eds. J.E. Grindlay and A.G.D. Philip (Dordrecht: Kluwer), p. 311

Larson, R.B.: 1990, in *Physical Processes in Fragmentation and Star Formation*, eds. R. Capuzzo-Dolcetta, C. Chiosi, and A. Di Fazio (Kluwer Acad. Pub.), Astrophys. and Space Sci. Lib., **162**, p. 389

Leibundgut, B., Tammann, G.A., Casonau, R., and Cerrito, D.: 1991, A&AS, **89**, 537 (Basel SNe Ia Atlas)

Mateo, M.: 1987, in *Globular Cluster Systems in Galaxies*, IAU Symp. 126, eds. J.E. Grindlay and A.G.D. Philip (Kluwer: Dordrecht), p. 557

Mayor, M.: 1976, A&A, **48**, 301

Mazzitelli, I., d'Antona, F., and Caloi, V.: 1995, A&A **302**, 38

McKellar, A.: 1941, Pub DAO, Victoria BC, **7**, 251

McWilliam, A., and Rich, R.M.: 1994, ApJS, **91**, 749

Meyer, D.M., Black, J.H., Chafee, F.H., Foltz, C., and York, D.G.: 1986, ApJ, **308**, L37

Panagia, N., and Tosi, M.: 1981, A&A, **96**, 306

Perlmutter, S. *et al.*: 1995, ApJL, **440**, L41

Petrosian, V.: 1976, ApJ, **209**, L1

Phillips, M.M.: 1993, ApJ, **413**, L105

Pierce, M.: 1994, ApJ, **430**, 53

Reid, C.: 1986, *Hilbert, a Biography* (New York: Springer)

Renzini, A., and Greggio, L.: 1990, in Jarvis and Terndrup (1990), p. 47

Rich, R.M.: 1988, AJ, **95**, 828

Rich, R.M.: 1990, in Jarvis and Terndrup (1990), p. 65

Roberts, M.S., and Haynes, M.P.: 1994, Ann Rev A&A, **32**, 115

Rood, H.: 1994, PASP, **106**, 170

Rubin, V.C., Burstein, D., Ford, W.K., and Thonnard, N.: 1985, ApJ, **289**, 81

Saha, A., Labhardt, L., Schwengeler, H., Macchetto, F.D., Panagia, N., and Tammann, G.A.: 1994, ApJ, **425**, 14

Saha, A., Sandage, A., Labhardt, L., Schwengeler, H., Tammann, G.A., Panagia, N., and Macchetto, F.D.: 1995, ApJ, **438**, 8

Saha, A. *et al.*: 1996, ApJS, **107**, 693

Sandage, A.: 1962, ApJ, **135**, 333

Sandage, A.: 1968, in *Galaxies and the Universe*, ed. L. Woltjer (New York: Columbia University Press), p. 75

Sandage, A.: 1982, ApJ, **252**, 574

Sandage, A.: 1986, A&A, **161**, 89

Sandage, A.: 1988aa, in *Calibration of Stellar Ages*, ed. A.G. Davis Philip, (Schenectady, NY: L. Davis Press), p. 43

Sandage, A.: 1988a, ApJ, **331**, 583

Sandage, A.: 1988b, ApJ, **331**, 605

Sandage, A.: 1990, J Roy Soc Canada, **84**, 70

Sandage, A.: 1993, AJ, **106**, 719

Sandage, A.: 1994a, ApJ, **430**, 1

Sandage, A.: 1994b, ApJ, **430**, 13

Sandage, A.: 1995a, in *Particles, Strings and Cosmology, PASCOS95*, eds. T. Bagger, G. Domohos, A. Falh, and S. Kovesi-Domahos (Singapore: World Scientific), p 317

Sandage, A.: 1995b, in *The Deep Universe, Practical Cosmology: Inventing the Past*, eds. B. Binggeli and R. Buser, (Berlin, New York: Springer), Lectures 4 and 6

Sandage, A.: 1996a, AJ, **111**, 1

Sandage, A.: 1996b, AJ, **111**, 18

Sandage, A., and Bedke, J.: 1995, *Carnegie Atlas of Galaxies* (Washington, DC: Carnegie Institution of Washington), Pub. No. 638

Sandage, A., and Perelmuter, J.-M.: 1990a, ApJ, **350**, 481

Sandage, A., and Perelmuter, J.-M.: 1990b, ApJ, **361**, 1

Sandage, A., and Perelmuter, J.-M.: 1991, ApJ, **370**, 455

Sandage, A., and Schwarzschild, M.: 1952, ApJ, **116**, 463

Sandage, A., and Tammann, G.A.: 1987, *A Revised Shapley–Ames Catalogue of Bright Galaxies* (Washington, DC: Carnegie Institution of Washington), Pub. No. 635

Sandage, A., and Tammann, G.A.: 1995a, ApJ, **446**, 1

Sandage, A., and Tammann, G.A.: 1995b, Two Lectures on the Hubble Constant at the *3rd School D. Chalonge* (Erice Sept. 1994), ed. R. Sánchez (Dordrecht; Kluwer), p 403

Sandage, A., Freeman, K.C., and Stokes, N.R.: 1970, ApJ, **160**, 831 (SFS70)

Sandage, A. Tammann, G.A., and Federspiel, M.: 1995, ApJ, **452**, 1

Schönberg, M., and Chandrasekhar, S.: 1942, ApJ, **96**, 161

Schmidt, B.P., Kirshner, R.P., and Eastman, R.G.: 1992, ApJ, **395**, 366

Schmidt, B.P. *et al.*: 1994, ApJ, **432**, 143

Schweizer, F.: 1986, Science, **231**, 193

Searle, L.: 1971, ApJ, **168**, 327

Searle, L., and Zinn, R.: 1978, ApJ, **225**, 357

Shaver, P.A., McGee, R.X., Newton, L.M., Danks, A.C., and Pottasch, S.R.: 1983, MNRAS, **204**, 53

Smith, H.A., and Stryker, L.L.: 1986, AJ, **92**, 328

Smith, H.A., Searle, L., and Manduca, A.: 1987, in *Globular Cluster Systems in Galaxies*, IAU Symp. 126, eds. J.E. Grindlay and A.G.D. Philip (Dordrecht: Kluwer), p. 563

Songaila, A. *et al.*: 1994, Nature, **371**, 43

Tammann, G.A.: 1986, in *Observational Cosmology*, IAU Symp. 124, eds A. Hewitt, G. Burbidge, and L.Z. Fang (Dordrecht: Reidel), p. 151

Tammann, G.A.: 1987, in *Relativistic Astrophysics*, ed. M.P. Ulmer, (Singapore: World Scientific), p. 8

Tammann, G.A.: 1988, in *The Extragalactic Distance Scale*, eds. S. van den Bergh and C.J. Pritchet, ASP Conference Ser. No. 4, p. 282

Tammann, G.A.: 1992, Physica Scripta, **T43**, 31

Tammann, G.A.: 1993, in *Planetary Nebulae*, IAU Symp. 155, eds. R. Weinberger and A. Acker (Dordrecht: Kluwer), p. 515

Tammann, G.A., and Sandage, A.: 1981, in *Highlights in Astronomy*, ed. R.M. West (Dordrecht: Reidel), Vol. 6, p. 301

Tammann, G.A., and Sandage, A.: 1995a, ApJ, **452**, 16

Tammann, G.A., and Sandage, A.: 1995b, in *The Local Velocity Field and the Hubble Constant*, IAU Symp. 168, eds. M. Kafatos and Y. Kondo (Dordrecht: Kluwer), p 163

Teerikorpi, P.: 1975a, A&A, **45**, 117

Teerikorpi, P.: 1975b, Obs Mag, **95**, 105

Teerikorpi, P.: 1984, A&A, **141**, 407

Teerikorpi, P.: 1987, A&A, **173**, 39

Teerikorpi, P.: 1990, A&A, **234**, 1

Thaddeus, P.: 1972, Ann Rev A&A, **10**, 305

Theureau, G., Hansi, M., Ekholm, T., Bottinelli, L., Gouguenheim, L., Paturel, G., and Teerikorpi, P.: 1997, A&A, in press

Tolman, R.C.: 1930, Proc Nat Acad Sci, **16**, 511

Tolman, R.C.: 1934, *Relativity, Thermodynamics and Cosmology* (Oxford: Clarendon), p. 467

Torres-Peimbert, S., and Peimbert, M.: 1977, Rev Mexicana A&A, **2**, 181

Twarog, B.A.: 1980, ApJ, **242**, 242

van den Berg, D., Bolte, M., and Stetson, P.B.: 1990, AJ, **100**, 445

van den Bergh, S.: 1960a, ApJ, **131**, 215

van den Bergh, S.: 1960b, ApJ, **131**, 558

van den Bergh, S.: 1982, PASP, **94**, 745

van den Bergh, S.: 1992, PASP, **104**, 861

van den Bergh, S.: 1993, Rev Mexicana A&A, **157**, 230

Whitford, A.E.: 1978, ApJ, **226**, 777

Whitford, A.E., and Rich, R.M.: 1983, ApJ, **274**, 723

Whitford, A.E., Terndrup, D.M., and Frogel, J.M.: 1990, in Jarvis and Terndrup (1990), p. 19

Wilson, O.C.: 1939, ApJ, **90**, 634

Zurek, W.H., Quinn, P.J., and Salmon, J.K.: 1988, ApJ, **330**, 519

1.5 **Discussion**

Question[†]

(Osterbrock): Talking about galaxy evolution and the sequence of galaxies, you showed the images from Kristian, which are very interesting indeed; I think that they were taken to a very low level of surface brightness with the *Space Telescope*. What do you think the appearance

---

† The questions here about the Kristian *HST* deep frames concern Q(xvi), not discussed in this written record, but which was developed in the spoken lecture.

of those galaxies would be compared with nearby bright galaxies in your atlases, if you were to go to a similar level of surface brightness?

Answer

(Sandage): With the surface brightness being essentially independent of distance, except for the $(1 + z)^4$ term, there is a whole range of surface brightnesses that you can get from the *HST* by digging into the data set, so you can certainly make the parameters the same by various stretches of a digital record. Even so, it is clear that none of these galaxies look like the nearby ones. What Kristian says is that these are completely different morphological forms to the normal nearby ones. Now that is the crucial aspect; how do you quantify it? As a morphologist you just look at it. So, as a morphologist, I believe that these are abnormal; that I have never seen anything locally like them, except, that is, for the 4% of nearby galaxies that Hubble originally called irregular. So, with the *HST*, instead of 4%, one is getting >50% of types that would be called irregular.

Question

(M. Burbidge): Those galaxies in the Kristian imaging are definitely very peculiar, but I was struck by the fact that there is a region at a distance of only $10\,000\,\mathrm{km\,s^{-1}}$ Hubble recession velocity (the main part of the Hercules cluster) which does have some weird pairings and peculiar galaxies, so it is not impossible that, in nearby space, there are things that look somewhat like those very distant ones.

Answer

(Sandage): In the Hercules cluster there are certainly spirals and there are interacting spirals, but I would say, morphologically, that they are more similar to the standard Hubble spirals than any of these Kristian galaxies. It is true that Hercules is a very peculiar cluster in the fact that there are spirals in the centre; it is a low-density cluster and Coma does not have such things, but that is the density–morphology relationship. I think that the spirals in Hercules are essentially Hubble normal. Now that is not what you said, but I guess we have a slight disagreement there.

[49]

Question

(Lynden-Bell): I just wanted to ask whether your rocks are made of hydrogen and helium? It would be very important if your rocks were made of silicon. One would expect that the basic constituent would be hydrogen plus around 30% helium. Are you thinking that there are many objects of planetary, or cometary, size containing a significant fraction of the mass, or do you mean 'rocks' when you say 'rocks'?

Answer

(Sandage): I mean rocks. I was a student at Caltech when the famous and extraordinarily controversial man Fritz Zwicky was there. You know what Zwicky said, that, according to Boltzmann and all of this wonderful stuff of equipartion of energy via $e^{h\nu/kT}$, you can predict what the mass function is all the way from clusters of galaxies to galaxies, to stars, to planets, to rocks, to atoms, and it's an exponential function going all the way to grains. Well, we all thought he was crazy, but we were students; what did we know? The fact that has always impressed me is that, every time an astronomer goes for the stellar mass function, it turns down just where hydrogen ceases to burn. Now, how does the fragmentation hypothesis know the details of the nuclear physics, so that the mass function knows something about the cross sections for thermonuclear reactions? So, it has always seemed to me that Zwicky was probably right, that, at least into the brown-dwarf region, one would, a priori, think that there should be no connection between where the stellar mass function decays and where stars begin to cease. So the brown dwarfs are crucial, and then where do you go from there to planets? That's a problem for the future. What do you think?

Question

(Lynden-Bell): I deduce that you mean 'balls of hydrogen' when you say 'rocks'?

Answer

(Sandage): No, I can put my hand through a ball of hydrogen, but I cannot put my hand through a rock, so there is a real difference in the state of the matter. I guess I was wondering if there are asteroid-like rocks between the stars left over from the star-formation process.

Question
(Lynden-Bell): Can you put your hand through a brown dwarf?

Answer
(Sandage): I do not know, I've never seen one. Rebolo was looking for brown dwarfs – why don't you (Rebolo) join in this conversation? I know it's not time yet, but . . .

Comment
(R. Rebolo): Yes, we are searching for brown dwarfs in young open clusters. It's the obvious place to search for them and very recently we have found, I do not know if it's a brown dwarf, but it is an M9 dwarf in the Pleiades cluster and we only know of four M9 dwarfs in the whole sky.

Question
(Sandage): M9? The spectroscopists never went that far!

Answer
(Rebolo): Now we are reaching that. There are special classifications. These classifications were developed in the United States by Kirkpatrick. He is a young man. In fact, I have even seen one M9.5 cluster star classified.

Question
(Sandage): What comes next?

Answer
(Rebolo): M10! So we are working on this object, but I cannot guarantee it is a brown dwarf. However, we have spectroscopic tests to carry out.

Question

(Sandage): Is not the definition of a brown dwarf 'a star that essentially does not radiate'?

Answer

(Rebolo): The definition is 'an object that cannot produce stable hydrogen burning'.

Question

(Sandage): So, it is still visible . . .

Answer

(Rebolo): Yes, it radiates even at visible wavelengths, but very little. The temperatures for brown dwarfs close to the substellar limit are estimated to be around 2000 or 1500 K, so the peak is in the K band.

Question

(Reeves): When you mention the three undone tests for the Big Bang, you mention this test about the temperature rising when we go back in distance. You did not mention the recent detection, by Songaila, I think, in a quasar at $z = 1.99$ and a temperature of about 8 K. Was there a reason why you do not believe it?

Answer

(Sandage): I guess I did not know about it! But I had known about the experiments by Meyer *et al.* that had been going on many years, and they did not get down far enough in termperature.

Comment

(Reeves): This new experiment reports a value of $T \sim 7.6$ K. At the z it was just about the correct prediction. This was reported in Nature around June (1994, **371**, 43).

Question

(Sandage): What was the signal to noise?

Answer

(Reeves): That I do not know.

Question

(Sandage): This is a Hawaii group?

Answer

(Reeves): Yes, Songaila is a Hawaii group.

Comment

(Sandage): It would be very nice to have a confirmation.

Question

(Novikov): I understand that you tried to restrict yourself to astronomical problems and not to go into the details of problems related to astrophysics. Still, I believe that there are a few general problems which are common to astronomy, astrophysics, and probably physics too, and probably nobody is going to discuss them here. So, allow me to list a few of them and ask for your comments on this list. The first is the problem of the origin of the Universe itself; we understand that it is not a question of philosophy, it is a matter of astronomy, astrophysics and physics. The second is, what can we say about the structure of the Universe on the largest scales? The third, what is the origin of the matter in the Universe? Once again, this concerns the physics of the very early Universe. The fourth, what was there before the beginning of the expansion of the Universe? And the last from this short list: what is the future of the Universe? I believe that, from some point of view, the future is more important for everybody and for our Universe than its past. And I believe personally that the future is related to today's events more closely than was thought. What would your comments be?

Answer

(Sandage): By 'future' do you mean the feedback that quantum mechanicians now claim?

Comment

(Novikov): Yes, but not only that; very shortly I believe that it will be possible to have a non-trivial topology of the structure of space-time, including the possibility of flying from the future into the past.

Question

(Sandage): Can that mean that you could go back into the past and murder your mother?

Answer

(Novikov): I cannot, but in the future somebody probably could go from the future into the past. The question of murder is, though, another matter!

Question

(Sandage): But then you would not be here!

Answer

(Novikov): No, because, you see, it has been proven that you cannot receive somebody from the future if you do not have a time machine now. So it should be constructed now if we want to receive somebody from the future. Just because we cannot see anybody from the future, or something from the future, it does not mean that the construction of a time machine in the future is impossible.

Question

(Sandage): You get killed going through the worm hole, don't you?

Answer

(Novikov): Probably; we can discuss this after my lecture.

Question

(Longair): Just two questions. One is: I have always been impressed by the way in which many of these correlations improve when you go to the near-infrared waveband. For example, the period–luminosity

[P–L] relations for RR Lyrae stars appears to get narrower in dispersion in the infrared waveband. Similarly, one finds very good correlations for luminosity against redshift for luminous galaxies in the near infrared. What is your view as to whether or not one should be attempting this type of distance measurement in the near infrared, rather than using the traditional optical one.

Answer

(Sandage): To get a correct answer all experiments must give the same answer. So we have produced an answer in the wavelengths that we ancients have been able to observe in for the last 30 years, and we are looking forward greatly to your measurements at these other wavelengths to see whether we are right or wrong. But, I do not think the dispersion in the wavelengths that we have been working on is so large that it vitiates the conclusions.

Question

(Longair): Could you compare your classical approach with the more physical methods which are now being developed, things like the Sunyaev–Zel'dovich effect and the gravitational lens effects? I have always had this philosophical problem, that the methods that use standard candles depend on an act of faith, that nothing subtle happens to the objects which are used in the relationship. Physical methods do have the advantage of getting out to large distances directly. I would value your comment on what the relative merits of these methods are.

Answer

(Sandage): As an observer, I will argue with you, saying that we used the P-L relation for 40 years without knowing why it worked. You have checks and balances if you are an observer, to see whether the correlations continue. Take the fact that the Hubble diagram itself is so narrow in first-ranked ellipticals. We do not know why the first-ranked ellipticals are such a narrow part of the luminosity function, but you can show that it is the case, just like in experimental nuclear physics. Before much was known about the nucleus, many of the results of, say, radioactivity, such as the decay rates, could be measured time and time and time again without contradiction. That is the same as for the P–L relation. I think that astronomers are permitted to use something that they can repeat with internal tests, without understanding. You use gravity all the time, but you do not understand gravity.

Question

(G. Burbidge): Allan, just for clarification, something that you discussed early on in your talk: the tests of the expansion hypothesis using the surface brightness effect. I have been going around for some years saying that you and – I have forgotten whom – with J. M. Perelmuter (ApJ, **370**, 455 (1991)), had demonstrated the correctness of this around 1990, but you implied that it still had to be done; in other words, do you mean that your result was not valid?

Answer

(Sandage): No, I did not mean that it was not valid. We thought that we had overcome the technical difficulty of how to define a surface brightness across an image, where the surface brightness varies by a factor of a thousand. So the technical aspect of the experiment is, I think, in place. The question is applying that technical apparatus to the data in the litera-ture. We did not go to the telescope and get new data; we used Spinrad and Djorgovsky's data (ApJ, **251**, 417 (1981)), which were photographic and not taken for that purpose. We had to take on faith what the table in the journal article gave. That indeed showed a very strong $(1 + z)^4$ relation. Did they feed that into their data table, or not? It is not clear from the way the paper was written. I believe the test can be made technically now, but it has not been carried out yet. It is now a very simple experiment, with CCDs, to get what is called the Petrosian diameter (ApJ, **209**, L1 (1976)), and with the enormous facilities of the telescopes now available, it could be done in four or five nights.

Question

(A. Aragon): One question comes to mind when presenting the Kristian data of the quality you showed us from the *HST*, which perhaps come from very high redshifts: if these galaxies are at a redshift above one and if the $(1 + z)^4$ scaling of surface brightness is true, then that is quite a huge factor in surface brightness. We are comparing the morphological results for these objects with results of heavily saturated plate material for nearby objects, for very bright galaxies, and when one looks into the details, are we sure that we are comparing like with like? Are we comparing the same kind of samples? That worries me a lot when people claim that things are different.

Answer

(Sandage): It is not true that the plate material is heavily saturated, because all you need to do is to take a shorter exposure on photographic plates. Now all you modern people do is to change the stretch on the database in your computer digital record. But we have a whole range of photographic exposures, for example, in the standard classification of the Hubble sequence; you saw that in the four frames of different stretches in the CCD Kristian database.

Comment

(Aragon): That is true, but if these galaxies are at very high redshifts, there is a very large factor in surface brightness, and I am not really sure that we are actually reproducing the experiment in the local Universe. One should look very carefully into ways of repeating the same experiment, perhaps with large area linear detectors, so that we actually know that, in the statistics, we are comparing like with like. I do believe that these things are completely different, but, when people tell me fractions and numbers, I am not sure that we know them to within a pretty huge uncertainty factor.

Answer

(Sandage): I am so pleased to hear you say that, because all of you people, rather than the 11 of us that are so ancient, will solve these problems, and if you do not have questions and question us, you will not do the experiment. So please, do the experiment!

Detailed inspection of Kristian's frames with Aragon's $(1 + z)^4$ factor in mind shows that there are few, if any, classical Hubble types in the small diameter images that are similar to the larger galaxies of standard morphology, but simply decreased by a factor of, say, 16 in surface brightness.

Most of the faint small images are *irregular* rather than regular, but at faint surface brightnesses many look like the Zurek, Quinn, and Salmon (ApJ, **330**, 519 (1988)) collapsar.

Comment

(Aragon): I wish that I could have a 5 × 5-degree CCD detector one of these days!

Answer

(Sandage): You do not need a 5 × 5-degree CCD detector to classify, for example, 5 arcmin nearby Shapley–Ames galaxies.

Comment

(Aragon): But I need a big detector if I want to get a decent sample which is comparable, not to classify one galaxy – I think that I can do that fairly easily – it's just to get the numbers, and the wide field is necessary to get the numbers.

Question

(Sandage): All you are interested in is the statistics?

Answer

(Aragon): I think so.

Answer

(Sandage): OK.

Question

(C. Gallart): I would like to refer to the images by Kristian. My intuition would tell me that the blue galaxy that you show is just like a group of galaxies, maybe something like the Local Group. Maybe I am confused about the scale, but do you have any reason to believe that it is a single interacting galaxy?

Answer

(Sandage): OK, the question is: in that last frame by Kristian, there was a galaxy that had a central core and a family of blobs around it. Could that be a group, or was that a single galaxy in the process of formation, such as simulated by Zurek *et al.* (1988) [Fig. 1.9]? We really do not know. What Kristian tells me is that, at his estimate of the distance, and I do not know how he got it, the scale there is that of one single galaxy, instead of the 1-Mpc scale of the Local Group (where M31 is about 1 Mpc from us). So, it is his belief that this is a part of a single aggregate instead of what will become even a compact Hickson group. But it is a very fair question, and it is open and I do not know the precise answer to your question.

Question

(J. Beckman): I would like your detailed comments, rather than general comments, on the kind of errors that must, in your opinion, enter into the most recent estimates of distance scale, the ones recently reported by Wendy Freedman and colleagues, using Cepheids in nearby galaxies and the similar results obtained by the group at the CFHT. I would like to hear detailed comments about the errors implied.

Answer

(Sandage): We have no argument about the Freedman distance M100. What we do have an argument about is the following. They used the wrong redshift for the core of the Virgo cluster. They used $1400\,\mathrm{km\,s^{-1}}$, whereas the experiment of tying the core of Virgo into the distant expansion field is unequivocal in giving 1179, instead of 1400 (Jergen and Tammann, A&A, **276**, 1 (1993)), where the cosmic velocity of Virgo is freed from all local velocity anomalies by their ingenious method of finding v_{cosmic}. That already reduces the Hubble constant of Freedman *et al.* by 19%. The second problem is that they have one single spiral, and are then divining the distance of the elliptical core of the cluster from that spiral alone. The elliptical core is the only one that can be tied to the external expansion field relative to the CMB. The spirals, by the density-morphology relation, are spread in an outer halo, and so the distance to a single spiral, whose detailed relation to the core is not known, is not a determination. If you take the difference in the redshift between 1400 and 1179, you would get $68\,\mathrm{km\,s^{-1}\,Mpc^{-1}}$. So, even if we ignore the second part (where the spiral is, relative to the core), we are not arguing about 100 versus 50, that was the old argument, as late as 1990; no one is arguing about a hundred. Now the argument is between, at most, 80 and 50 but, if you use the kinematics properly, we claim it's between 68 and 50. Then you have at least three tenths of a magnitude uncertainty for the position of the spiral relative to the core, making their number 68 plus or minus 17. (Note added in proof: Recent data [FTS97] have shown beyond doubt that the distance of the Virgo cluster core is 21.8 ± 2 Mpc, giving $H_0 = 54 \pm 5$.) The second experiment you talk about, by Pierce *et al.* (Nature, **371**, 385 (1995)) is technically beyond the edge.

Question

(H. Zinnecker): I would like to follow up the earlier discussion on the rocks and ask you again, whether you think that the rocks really would have such a steep mass function that they could account for the missing mass in the Universe. That sounds very unlikely, as there would be more material in heavy elements than in hydrogen and helium. I think that is what Donald Lynden-Bell meant. There could be rocks, lots of rocks, but it should be a shallow-mass function, so there is not a lot of mass in them. Would you like to comment on that?

Answer

(Sandage): I think it's a problem for the future. We would really like to know what the slope is, but I do not know how to go about finding it. I have no idea what experiments you would need to do.

Comment

(Zinnecker): For a start, we would have to find a case for a brown dwarf which is beyond doubt.

Answer

(Sandage): That is not a rock!

Comment

(Zinnecker): Yes, but let us start first with hydrogen and helium in very low-mass systems. I think that the only way to get there is to obtain dynamical information; you have to find very low-mass objects in triple or binary systems. There was a recent paper by Christof Leinert, in Germany, who found a triple system, two of which are in a close orbit; it is not far away, so you can resolve them by speckle interferometry and get a speckle orbit. They revolve with a two-year period. The combined luminosity of that system is very low already, so if you split them into two, it becomes even lower, but still they are not convinced that it is a brown dwarf; they could still be at the edge. But more important is your question as to whether or not there could be a feature imprinted by hydrogen burning onto the mass function. You could make a physical case for whether or not the mass function turns down as a consequence of the hydrogen burning setting in.

Answer

(Sandage): The mass function of asteroids has nothing to do with the nuclear cross section for hydrogen burning.

Comment

(Zinnecker): That is true, but one would maintain that these objects have been formed in the disks of systems, and only a very minute fraction of the heavy elements are incorporated in solar systems. You have to have gravity in the first place, in order to form plenty of brown dwarfs. There have been papers which have shown that there is a minimum Jeans mass which is a hundredth of the solar mass. So, I think that it is difficult to find ways to form rocks in sufficient quantities in the way that you would like.

Question

(A. González): Could you please comment on what you believe is the importance of large scale tidal fields in the process of galaxy formation. Why is it that tidal fields are important? If they are strong enough they can produce an eruption of perturbations and, probably, galaxy formation will occur from the fusion of several small clouds, rather than one single object. This is probably what we are observing from the Hubble pictures. What can you comment about this?

Answer

(Sandage): I do not know how to answer your question in detail, but the zero order approximation for galaxy formation is just the Harrison–Zel'dovich spectrum, where you have $\delta\rho/\rho$. This is the fluctuation spectrum, such that the collapse then begins in each of these regimes, without necessarily having any connection of the regimes. That is not true where you have a cluster, but you have many galaxies outside the nuclear clusters that are at the intersections of the sheets, so I would think that the principal physics is just a very high $\delta\rho/\rho$, collapsing to a stable configuration. Then, how do you start rotation? You have to have tidal torques to do that. So, it's much more complicated than just the initial formation picture, and this detail that you are putting in may in fact be very important. Surely, to get rotation of the spirals, you need to have interactions, and we know rotation of the spirals is a crucial aspect of the Hubble sequence? I wonder if Donald would have clearer ideas than these?

Question

(Lynden-Bell): I basically agree with what you have said, Allan. I think that interaction and tides are important, and I think that the initial collapse goes on at the free-fall time. So, I think that what you are saying is not in contradiction with the idea that galaxies formed basically in the free-fall time. It's just that the free-fall time of the outer parts may be somewhat longer, and the outer parts, in particular, may get more affected by the tides.

Comment

(A. González): Yes, I am thinking of a recent paper by Bertschinger, who says that the large-scale tidal fields could be more important than the field produced by the nearby galaxies around the density perturbations. The large-scale tidal field, at probably up to 20–30 Mpc, would be more important around the perturbations.

Question

(Rees): One final comment. You have said a great deal about galaxy morphology as something to be understood in the future, but there is the basic question of why galaxies exist. What I mean by this is that we do know physically why stars exist in the mass range between 0.1 and $50M_\odot$: that is a stable hydrogen burning regime; but we do not know in the same way why galaxies exist with a characteristic mass of 10^{11}–$10^{12}M_\odot$. We know that it is partly an astrophysical problem, partly a cosmological and initial conditions problem, but it is something to be explained, and we would like to know why galaxies end above some mass and clusters begin. Why is the Coma cluster not one great, amorphous galaxy? That is clearly a broader perspective of galaxy morphology, which is another problem for the future. I do not know if you would agree with that?

Answer

(Sandage): Yes, and that is a point where you need to understand the physics before you can use it. One place where your comments are really very germane is, why is there an upper mass to the brightest cluster galaxy? Except for the CDs, you do not get brighter than -23 with ordinary ellipticals. The luminosity function is almost vertical, and the question is, why is that?

2 New vistas in cosmology and cosmogony

Geoffrey Burbidge

2.1 Introduction

In this chapter, I would like to discuss the modern history of the major ideas and observations in extragalactic astronomy and attempt to extrapolate into the future.

It is appropriate to start the discussion with the situation as it appeared some 70 years ago in 1925. This is a time span a decade or so longer than the average age of the pundits assembled here. It also means that this book's contributors have all participated in the work that has gone on in the second half of this period.

2.2 1920–55

The 1920s was the era when the first major discoveries in extragalactic astronomy were made. They were, firstly, that the basic constituents of the Universe are galaxies of stars – the 'island Universe' concept, which was established largely on the basis of the work by Curtis, Shapley, Lundmark, and especially Hubble. Secondly, the redshift–distance relation was established, by several astronomers, culminating in the famous paper by Hubble (1929). From this it was (to us in retrospect) a small step to interpret the redshift as being due to expansion, and thus to relate the observed redshift–distance relation to the concept of an expanding

Universe, which corresponded to a solution of Einstein's equations obtained by Friedmann first and independently by Lemaître. In the West it was the work of Lemaître which became known, in part because it was brought to the attention of Eddington, who heavily promoted and publicized the concept of an expanding Universe. The prime movers in developing this concept were clearly Einstein, Friedmann, Lemaître, Hubble, Tolman, and Eddington. Attempts to argue that the redshift was not evidence for expansion were made by Zwicky and McMillan and others, but they were disregarded, though it took 60 years before Sandage and Perelmuter (1990) established that the redshifts of galaxies are truly expansion shifts.

By the early 1930s the expanding Universe was generally accepted, and what came with it fairly obviously was the idea that there must have been a beginning and that the Universe has a finite age.

The time scale associated with the age was $\sim H_0^{-1} \simeq 2 \times 10^9$ years, where H_0 is the measured rate of expansion, which Hubble and Humason had determined to be $550 \, \mathrm{km \, s^{-1} \, Mpc^{-1}}$. It was realized that this could be compared with the ages obtained from radioactive decay. From the Pb/U ratio the age was $\sim 1.3 \times 10^9$ years (Holmes and Lawson 1927), but from the $^{235}\mathrm{U}/^{238}\mathrm{U}$ ratio (Aston 1929; Rutherford 1929) an age of $\sim 3 \times 10^9$ years was obtained. At that time all of these ages were much less than the ages of the stars, which were estimated to be $\sim 10^{13}$ years (Eddington 1924; Jeans 1929).

The discrepancy of the two time scales, 10^9 years and 10^{13} years, remained until Bethe's work on stellar energy generation, in the late 1930s, when it was realized that only a small fraction of the mass of the core ($\sim 10^{-2}$) of a star ($\sim$10% of the total mass) must be burned before the star evolved. These arguments led to a reduction of all of the stellar ages to values of $\sim 3 \times 10^9$ years. As the value of the Hubble constant has been corrected downward over the last 50 years, from 550 to 180, to 75, and to $50 \, \mathrm{km \, s^{-1} \, Mpc^{-1}}$, age determinations for the stars from stellar evolutionary arguments and the age of the elements based on radioactivity have stabilized to values in the range 12–17×10^9 years for the oldest stars and 12–15×10^9 years for the age of the elements. There is currently a debate raging concerning the value of H_0. Many astronomers believe that the correct value is close to $80 \, \mathrm{km \, s^{-1} \, Mpc^{-1}}$, while Sandage and Tammann (whom

I believe to be closer to the mark) believe in a value close to $50\,\mathrm{km\,s^{-1}\,Mpc^{-1}}$. For the simplest (flat) model of the Universe, the larger value of H_0 gives an age of 8.5×10^9 years, while for $H_0 = 50\,\mathrm{km\,s^{-1}\,Mpc^{-1}}$, the age is 13×10^9 years. Those who believe that the larger value is correct have therefore gone back to a model involving a cosmological constant (see Peebles 1995). But even with the smaller value for H_0 there may be a problem with the so-called standard model.

Returning to the earlier period, it is clear from the writings of the 1930s and 1940s that cosmologists in general accepted the idea of an evolving Universe with a beginning, without paying a great deal of attention to the problem of the origin of the galaxies and their contents. There are, of course, one or two notable exceptions. Jeans (1929), in his famous book, tried to understand the origin of galaxies, and in doing this he came up with a very long time scale of $\sim 10^{12}$–10^{13} years for the Universe, incompatible with the accepted cosmology.

In 1948 Hoyle, and independently Bondi and Gold, proposed a Steady State cosmological theory in which there is no beginning and matter is continuously created. For about 15 years after this, many attempts were made to test this theory using the observations, but it is clear to anyone who studies the literature of that period or who listened to the many arguments, that there was a deep-seated dislike of the theory by a majority of observers and theoreticians alike.[†] Many of the objections were based on observations which turned out to be spurious, e.g. the Stebbins–Whitford effect, but they were used for many years as evidence against the theory. Hoyle (1968) gave a summary of all of the objections that had been raised and showed that most of these had not stood up.

The strongest objections until the detection of the cosmic microwave background, in 1965 (Penzias and Wilson 1965), were made on the basis of radio source counts by Ryle and his colleagues (a review was given by Ryle 1968). Then the argument was made that no Steady State model could explain the largely blackbody shape of the background radiation. This, for many people, is still the strongest argument against a Steady State model, but it can be overcome. We shall return to this later.

† For example H. Dingle when he was President of the Royal Astronomical Society devoted his presidential address to an attack on the theory (Dingle 1953), arguing that it was not even science.

In this early period, little attention was paid to the physics of the initial stages of the Universe – the Big Bang itself.

It was Tolman (1934) and Gamow and his associates, Alpher, Bethe, and Herman (Gamow 1948; Alpher *et al.* 1948; Alpher and Herman 1950), who began the study of the physics of the very early condensed phase, which must have been present if there was indeed a Big Bang. Tolman (1934) discussed the properties of the radiation field as the Universe expands, but it was Gamow and his associates who worked on the physics in the early phase and who attempted to estimate the temperature of the radiation field.

A realistic time scale for the evolution of the stars only came after the energy sources had been shown to be the proton–proton chain and the CN cycle and the basic ideas of stellar evolution had been worked out (Hoyle and Schwarzschild 1955; see also the monograph by Schwarzschild 1958). The realization that age determinations for globular clusters and galactic clusters could be obtained from observations of the colour–magnitude diagram led, in the 1950s, to fairly accurate age determinations (see Sandage 1958) of the order of 10×10^9 years, with the oldest clusters having ages of $\sim 15 \times 10^9$ years.

Similarly, the age of the elements was worked out using Rutherford's method (Burbidge *et al.* 1957; Fowler and Hoyle 1960), giving an age for the elements of $\sim 10 \times 10^9$ years.

Over the period 1930–60 the value of H_0 obtained from calibration of the distances of nearby galaxies was corrected downwards, as various errors in earlier work were discovered. As has been mentioned, it progressed from $550 \, \text{km s}^{-1} \, \text{Mpc}^{-1}$ to 180 to $75 \, \text{km s}^{-1} \, \text{Mpc}^{-1}$, which was the value obtained by Sandage in 1957 (Sandage 1958). Since then, Sandage and Tammann have provided strong evidence that the correct value is very close to $50 \, \text{km s}^{-1} \, \text{Mpc}^{-1}$, while others (see Jacoby *et al.* 1992) argue for a value closer to $80 \, \text{km s}^{-1} \, \text{Mpc}^{-1}$. I strongly believe that the correct value lies in the range $50–60 \, \text{km s}^{-1} \, \text{Mpc}^{-1}$, so that H_0^{-1} is close to $18–20 \times 10^9$ years.

To summarize the situation, up to about 1960 there was a general consensus with few exceptions that we live in an evolving Universe which began in a hot Big Bang. Soon after the beginning, galaxies were formed (though there was no understanding of how). It was believed that all formation processes are determined by gravitational forces leading to tidal effects and merging etc. It was also realized that accretion is an important process. It was assumed that galaxies evolve very slowly, with the time scale determined by stellar evolution. In a review written in 1971 (Burbidge 1971), I pointed out that probably the strongest evidence in favour of the Big Bang is the rough agreement between the value of H_0^{-1} and the ages of the stars determined from stellar evolution and from the age of the elements, while the strongest argument against the idea was that we had no understanding of the origin of discrete objects – galaxies etc. If, for the simple flat Universe model, the cosmological age of $\approx 2/3H_0$ turned out to be less than the ages of the stars, one can always go to a Lemaître-type evolving Universe, as is now being suggested (Peebles 1995).

In this period, the only serious alternative suggested was the Steady State model, which was generally in disfavour. It was felt then and is felt now by many people that the observed radio source counts and the microwave background cannot be explained by the simple Steady State model.

Now we turn to the events after about 1960.

2.3 The era of new observations ~1955–90

The new discoveries all followed the opening up to observation of new parts of the electromagnetic spectrum. This started in the 1950s, when the powerful extragalactic radio sources were identified, and it was realized that they were distant galaxies which were generating the radio flux through the incoherent synchrotron process. By some means very large fluxes of relativistic electrons with individual energies $\gtrsim$ GeV and magnetic fields $\lesssim 10^{-5}$ gauss are being generated in regions with dimensions of ~ 100 kpc, outside the main body of the galaxy. Typically, a radio galaxy has two radiating clouds (lobes), which appear symmetrically about the optical galaxy. The galaxies, when they can be identified, are some of the most luminous elliptical galaxies. However, it is found that the sources with large redshifts invariably show strong emission lines, and there is no direct evidence for an underlying stellar galaxy, though it is usually assumed that a galaxy is present.

The most remarkable aspect of these sources is that their energy production is prodigious. From the synchrotron theory, and using the most conservative assumption – that the energies contained in the radio lobes in the form of relativistic particles (electrons and protons) are in rough equipartition with the energy in the magnetic fields (an argument for which there is no good physical basis) – the total energies are $\sim 10^{59}$–10^{60} ergs for the strong sources. If the equipartition argument is not adhered to, the energies go up. The only way of reducing them is to suppose that the sources are closer then their redshifts indicate.

This discovery was the first which indicated that violent energy release must take place in galaxies, thus suggesting that galaxies or parts of at least some of them can rapidly evolve over 10^{10} years. If we also take into account the Seyfert nuclei in spiral galaxies, and much evidence for ejection of gas from the central regions of galaxies, a picture can be built up which suggests that violent events in the nuclei of galaxies are commonplace (Burbidge, Burbidge and Sandage 1963).

Less than ten years after the discovery of radio galaxies, the quasi-stellar objects (QSOs) were discovered, also from identification of radio sources (Sandage 1961; Hazard *et al.* 1963; Matthews and Sandage 1963; Schmidt 1963; see also the monograph by Burbidge and Burbidge 1967). Their very large redshifts, if they are of cosmological origin, mean that these objects are ~ 100 times brighter than bright galaxies ($L \approx 10^{46}$–10^{47} erg s^{-1}), and it is believed that the radiation is non-thermal in origin. Moreover, the rapid variability of these objects meant that the energy sources are exceedingly small, $\lesssim 1$ pc, and there are many indications that they are no larger than the solar system ($\sim 10^{15}$ cm).

In the 1960s there were other observational discoveries which had not been anticipated. The catalogues of peculiar galaxies made by Vorontsov-Velyaminov (1959) and Arp (1966) showed many systems which had every morphological and spectroscopic indication that they are very young compared with H_0^{-1}, i.e. they had ages $\ll 10^{10}$ years. Also, a few compact groups of galaxies were completely studied for the first time. In the cases of the first three of these to be completely measured, each had one galaxy with a redshift very different from the others (see Burbidge and Sargent 1971). Such very compact groups were very hard to understand, since even without the discrepant redshift they could not survive for even 10^9 years without either coalescing or dispersing. Unless the galaxy with a discrepant redshift is a background or foreground object, it must be exploding out of the group, or else its redshift is not due to the cosmological expansion.

Ambartsumian (1958, 1965) brought a completely new approach to cosmology and cosmogony by proposing that groups and clusters of galaxies might very well be systems of positive total energy, i.e. systems similar to expanding associations of stars. This means that they must be comparatively young. He did not develop a detailed theory, but obviously felt that some creation processes must have taken place comparatively recently.

It had been known since the early studies of clusters of galaxies that application of the virial theorem led to the conclusion that the kinetic energy of the visible galaxies is much greater than the potential energy, i.e.

$$2KE + PE > 0. \tag{2.3.1}$$

For the virial condition

$$2KE + PE \simeq 0 \tag{2.3.2}$$

to hold, it is necessary that a large amount of matter, perhaps 90% of the mass of the cluster, must be present in the form of dark matter.

We return briefly to the energy problems raised by the radio sources and the QSOs. All of the possible energy sources were considered, and it was soon concluded that nuclear sources were inadequate, matter–antimatter annihilation was ruled out because there appeared to be no scenario which could allow it to function, and the only likely answers were either that the energy is gravitational in origin or that it is due to the creation of new matter in the nuclear regions of galaxies (for an extensive review see Burbidge 1970). The release of large amounts of gravitational energy requires the collapse of massive objects to dimensions close to the Schwarzschild radius, and this was looked at in detail in the 1960s (see Fowler and Hoyle 1963; Hoyle *et al.* 1964; etc.).

This idea that gravitational energy was released in the collapse of a massive central object soon became the mechanism which nearly everyone accepted as the basic energy source. This theory was extensively discussed by Lynden-Bell, Rees, and Blandford, and others. A scenario was developed in which it is assumed that the central engine (as it came to be called) is a massive black hole surrounded by an accretion disk. Matter falling in, first to the disk and then to the black hole, will give a maximum energy output of about $0.08mc^2$. Various scenarios have been described to show how such a machine can give rise to the non-thermal fluxes that we see. For the radio galaxies, Blandford and Rees developed the idea of jets of relativistic particles which feed energy into the radio lobes (Begelman, Blandford, and Rees 1984; Rees 1984). In the 1970s this turned into *the paradigm* for the energy production in QSOs and radio sources, and, with very few exceptions, it has been blindly accepted by observers.

The problem with this approach is that no real tests of the paradigm are made. All of the observations are interpreted in terms of it, and difficulties with the model, of which there are many, are ignored or glossed over. For example, there is no real understanding of the particle acceleration. Also, the origin of magnetic fields with strengths as large as 10^{-5}-10^{-6} gauss in the extended lobes of radio sources is not understood. Moreover, a very high efficiency of conversion of gravitational energy at the centre to relativistic particle energy in the jets, and then reconversion of this in the lobes, is assumed. This is very problematical.

The discovery of the QSOs brought new problems. In particular, from about 1966 onwards, there were indications that their redshifts might not be due simply to the expansion of the Universe. We shall return to this question in more detail in the later discussion. However, before we do this, it is important to see how the astronomical establishment reacted to the new ideas and discoveries in this period.

2.4 **The establishment view, 1965–present**

As far as cosmological models are concerned, it is clear that there has been no significant evolution of ideas since the 1930s. In that period, it was established that we live in an expanding Universe which, it was felt, could be understood in terms of a standard Friedmann model.

As we have mentioned earlier, there was continuous hostility from the beginning to the idea of a Steady State cosmology, and the approach was always to attack it on the grounds that it did not conform to the observations, or on philosophical grounds.

As far as the new observations in the radio and X-ray wavelengths are concerned, the approach has always been to force-fit the new discoveries into the standard framework, and when they would clearly not fit, to attempt to discredit the observations (or the observer who made them), or just to ignore them. This has gone on up to the present.

The attitudes of the leaders in the field were well established for me by what took place at two conferences of many I attended. The first was the IAU symposium held in Santa Barbara in 1961 (IAU 1962). Three of the most prominent participants were Victor Ambartsumian, Jan Oort, and Fritz Zwicky.

Both Ambartsumian and Zwicky were quite unorthodox in their views (in very different ways), but it was Oort (who was as always quietly magnificent and very sure, but who could change his mind, a very rare quality among the leaders) whose ideas ruled the day. His view was that Ryle's work on radio sources had ruled out the Steady State cosmology. This was a popular view, but I believe it was wrong.[†] He was also quite convinced that all galaxies were old and must have been formed soon after the Big Bang. This meant that Ambartsumian's ideas that some clusters were young and were coming apart must be incorrect. On the other hand, if the virial held, as it must, he felt then that the clusters are bound and a large amount of dark matter must be present. This was then one of the strongest arguments for the presence of dark matter. I remember attempting to argue with Oort that, apart from our own Galaxy and the Magellanic Clouds, we had no good estimate of the ages of the galaxies, but he was adamant that they must all be $\sim 10^{10}$ years old. The idea that young galaxies exist was clearly anathema to everyone, because, among other arguments, the existence of young galaxies would support Steady State cosmology.

At the Vatican conference in 1970 the themes were somewhat different, but the attitudes were the same. Two issues come to mind. The first concerns the microwave background, whose discovery is one of the major pillars of Big Bang cosmology. This meeting came at a time before the background radiation had been shown to be of blackbody form. In fact the observations which had been made above the Earth's atmosphere gave a flux which was considerably higher than that of a blackbody at

---

† Much later Narlikar, Das Gupta, and I demonstrated this in a paper published in *The Astronomical Journal* (Das Gupta, Narlikar and Burbidge 1988).

2.7 K. The approach that was taken to this result by one eminent theoretician was that this result *must* be wrong. He *knew* that the radiation would turn out to have a blackbody shape, and the next experiment being planned at MIT would prove it. In other words, there was no need to wait for proof. In fact he turned out to be correct, but his attitude to any doubter was appalling.

The second issue concerns small groups of galaxies. A paper by M. Burbidge and W. Sargent (Burbidge and Sargent 1971) on compact groups of galaxies was read by W. Sargent. In 1970, only a very small number of compact groups containing four, five, or six members were known. As was mentioned earlier, three of them, Stefan's Quintet, VV 172, and Seyfert's Sextet, each have one member with a redshift very different (by many thousands of $km\,s^{-1}$) from the mean of the others. This means one of the following: (a) the discrepant galaxy is a foreground or background object, (b) part of the redshift is intrinsic, i.e. not of cosmological origin, or (c) the galaxy is literally exploding away from the others in the group. In addition to this, the velocity dispersion of the galaxies within each of these groups (excluding the discrepant ones) is so high that they will disperse in a time $t \ll H_0^{-1}$.

Thus, even in 1970, the compact groups posed a real problem. However, when Sargent presented these results, the establishment position became clear. The discrepant galaxy must simply be a foreground or background galaxy (there had already been a debate in the literature about Stefan's Quintet) and somehow the 'problem' would go away! Sargent was bombarded by critical questions from eminent men until he gave in!

The pattern was clear. In those meetings and others in this period, observations which could not be explained by the classical models and accepted physics were to be questioned, argued away, or ignored.

By 1982, when a conference on cosmology was held at the Vatican, a new approach was taken. The radicals around, such as F. Hoyle, V. Ambartsumian, and this speaker (to mention a few) were not even invited. The conference was confined completely to Big Bang cosmology and its proponents. In fact, in the introduction to the published volume of the proceedings of this meeting (Pontifical Academy of Sciences 1982), it was emphasized that only believers (in the Big Bang) were present, and that this was clearly a deliberate decision of the organizers.

I hope that I have conveyed some of the flavour of the evolution of ideas (or the lack of it) over the last 30 years. In spite of this, I believe that very new developments in theory are to be expected over the next 30 years, and they will stem from the observational evidence that is coming to light, some of which I have anticipated already.

Now we turn to the future and the observations which hopefully will drive us (or more likely the younger generation, if they have the courage) to discard the current view.

2.5 A new approach to cosmology and cosmogony

Let me start on a somewhat pessimistic note. We all know that new ideas and revolutions in science in general come from the younger generation, who look critically at the contemporary schemes, and having absorbed the new evidence, overthrow the old views. This, in general, is the way that science advances. However, in modern astronomy and cosmology, at present, this is emphatically not the case. Over the last decade or more, the vast majority of the younger astronomers have been conformists in the extreme, passionately believing what their leaders have told them, particularly in cosmology. In the modern era the reasons for this are even stronger than they were in the past. To obtain an academic position, to obtain tenure, to be successful in obtaining research funds, and to obtain observing time on major telescopes, it is necessary to conform.

I simply hope that a few will refuse to accept this formula for success, and instead consider the possibility that the new evidence, which has been accumulating over the last 30 years, requires a new approach. In what follows, I shall summarize what I believe should lead us in new directions.

The observations which I shall describe all lead us towards one key concept which, remarkably enough, was first proposed by Sir James Jeans in 1929 (Jeans 1929). In connection with his attempts to understand spiral structure in galaxies he wrote:

> The type of conjecture which presents itself, somewhat
> insistently, is that the centres of the nebulae (galaxies) are of the
> nature of *singular points*, at which matter is poured into our
> Universe from some other, and extraneous dimension, so that, to

a denizen of our Universe they appear as points at which matter is being continuously created (Jeans 1929, p. 352).

This also seems to describe the ideas proposed by Ambartsumian (1958, 1965).

If we accept that the observational evidence supports this concept, then it is fairly clear that this opens the way to a new approach to cosmogony and to cosmology, namely to a version of the Steady State theory in which creation takes place in galactic nuclei and new galaxies are formed by ejection of new material from earlier condensations.

In the remainder of this chapter, I shall discuss in detail the evidence which suggests this. But before doing this, I would like to summarize the attempts of Hoyle, Narlikar, and myself over the last five years to develop a cosmological model along these lines, the so-called 'quasi-Steady State model', as an alternative to the hot Big Bang.

2.5.1 The quasi-Steady State cosmology

The standard Friedmann cosmology, based on Einstein's theory of gravity, starts with an initial explosion. However, Hoyle and Narlikar (1964) argued that, in order to make the classical Einstein theory scale invariant, it is necessary to add a term to Einstein's equation which has the effect of giving rise to the creation of mass from negative energy states. This is then the C-field formulation of the gravitational theory of Hoyle and Narlikar (1964).

Recently, this theory has been used in a series of papers by Hoyle, Narlikar, and Burbidge (Hoyle, Burbidge, and Narlikar 1993, 1994a, b, 1995) to develop the quasi-Steady State cosmology (QSSC), in which creation of matter occurs in a cyclic fashion in the nuclei of active galaxies. Matter is created where there are already mass concentrations. The Universe is slowly expanding on a long time scale of $\sim 10^{12}$ years and smaller time scale oscillations are occurring with a period of $\sim 40 \times 10^9$ years. In terms of this model, the Universe is currently in an expanding phase of the short oscillation. The majority of the creation takes place at the minima of these oscillations. In this theory, there are several free parameters. By using the observed value of the Hubble constant (in our papers we have put $H_0 = 65\,\mathrm{km\,s^{-1}\,Mpc^{-1}}$), we have been able to explain all of the major cosmological results – the abundances of the light elements, the cosmic microwave background, and its blackbody form, etc. We do predict that, at long wavelengths (centimetres to metres), the background radiation will depart from blackbody form. The current measurements are compatible with this.

One interesting prediction of this model is that, since the overall time scale is very long, many generations of stars (and galaxies) will evolve and die. Thus, we naturally expect that a large mass of dark baryonic matter will be present, both in galaxies and in the form of evolved galaxies in clusters. Also, we naturally expect that young galaxies, made of new matter created and ejected from existing galaxies, will be present. In this model there is no necessity to invoke the presence of non-baryonic matter. Further details can be obtained by studying the papers referenced above.

This cosmology fits together naturally with a cosmogony in which the most important property is the creation of new matter from the nuclei of galaxies.

We believe that there is extensive evidence that these processes are going on. It came first from the observations of violent events in galaxies, which we have already mentioned (see Burbidge, Burbidge, and Sandage 1963). Secondly, the observations of the quasi-stellar objects (QSOs) have provided strong evidence that they are physically associated with galaxies, though galaxy and QSO frequently have very different redshifts. Starting in the 1960s, many examples of close pairs and multiple systems of QSOs and bright galaxies were discovered, particularly by Arp (1967, 1987), and the first statistical studies of the brightest radio sample of QSOs and bright galaxies, the 3C QSOs, and the bright galaxies in the Shapley–Ames Catalogue showed that there are far more close pairs than are expected by chance (Burbidge *et al.* 1971).

The implication of these results – that many QSOs have a large non-cosmological component to their redshifts and are comparatively nearby objects – has been largely ignored. Arp's results were widely attacked and he has been denied access to telescopes (see Arp 1987; Burbidge 1988).

Though the observational case for at least some local QSOs could be made early in the discovery phase (see Hoyle and Burbidge 1966), the implications appeared to be so daunting that there has been a refusal for many years to face up to this reality. One of the reasons for this was that there was no easy explanation for the existence of non-cosmological red-shifts.

However, I believe that the evidence by now is very strong indeed and, in what follows, I shall give details taken from several papers (Burbidge *et al.* 1990; Burbidge 1995; Hoyle and Burbidge 1995).

2.5.2 QSOs associated with galaxies

As was just stated, it has been known for some time that there are many high-redshift QSOs which lie so close to bright, comparatively nearby galaxies that probability arguments strongly suggest they are physically associated with the galaxies and lie in the same volumes of space (Arp 1967; Burbidge *et al.* 1971; Burbidge 1979, 1981; Arp 1987; Burbidge *et al.* 1990 and many other references given there).[†] The statistical evidence is supported by a strong inverse correlation between the galaxy redshift (proportional to distance) and the angular separation between galaxy and QSO for a large number of pairs (Burbidge *et al.* 1990, and earlier references).

The statistical evidence is also further supported by morphological evidence. Detailed investigations of several QSO–galaxy pairs show that the galaxy and QSO are connected. For example, the connection is by a luminous bridge for NGC 4319 and Mk 205 (Sulentic and Arp 1987), or by neutral-hydrogen clouds in 3C 232 and NGC 3067 (Carilli van Gorkum, and Stocke 1989; Carilli van Gorkum 1992). There is also evidence of alignment between optical features in the galaxy and the direction of ejection of the QSOs in NGC 3079 (Womble 1992). Other examples are shown in the recent paper by Hoyle and Burbidge (1995).

An ingenious way of explaining the existence of the close pairs of galaxies and QSOs which would still allow them to lie at their respective redshift distances was proposed by Canizares (1981), who argued that it could be due to microlensing and hence amplification of the images of distant QSOs by faint stars in the haloes of the galaxies.

---

† In the lecture version of this chapter I gave many illustrations of the evidence concerning galaxies and QSOs. None of these illustrations is included here, but they can all be found in the literature which is referenced in this section.

However, this argument was demonstrated to fail by Arp (1990) and by Ostriker (1989), because the density on the sky of faint QSOs is far too small to give the required frequency of occurrence of the close pairs. The failure of this argument has also been stressed by Schneider, Ehlers, and Falco (1992) and by Schneider (1994).

Thus, for pairs involving bright QSOs ($m \lesssim 18$) and bright galaxies ($m \lesssim 14.5$) and for separations $\theta \lesssim 3'$ (Burbidge *et al.* 1990) the existence of anomalous (non-cosmological) redshifts is well established, but this is only a very small fraction of the known QSOs. We can either argue by extrapolation that this is true for the whole population of QSOs or work on the assumption that this only applies to a subset of QSOs. In what follows, we shall attempt to explain the whole observed population of QSOs in terms of a model in which they all have intrinsic redshift components.

Because there is now a good deal of new statistical information available, we bring all of these ideas together in this chapter.

We give in Table 2.1 a list of all of the close pairs of QSOs and bright galaxies known to us. There are 46 pairs here, nearly all with separations $\leq 3'$, with the galaxies in 34 of the pairs brighter than $14^{m}.5$. These are prime cases for physical associations. There are many more known cases involving fainter galaxies, but, because the surface density of fainter galaxies is much larger than for the bright galaxies, the possibility that many of these are chance projections cannot be excluded. Also there are many cases involving bright galaxies and QSOs with larger angular separations.

2.5.3 Statistical tests of associations between QSOs and galaxies

Many statistical studies have been made of complete samples of QSOs and galaxies. In the original work of Burbidge *et al.* (1971) we used all of the QSOs in the 3CR catalogue (50) and compared them with the positions of the galaxies in the Shapley–Ames Catalogue (~1200 galaxies). We concluded that four of these are very likely to be physically associated with galaxies. This analysis was confirmed by a study by Kippenhahn and de Vries (1974) using Monte Carlo techniques. The four (and later five) close pairs identified in this analysis are included in Table 2.1. The fifth pair (involving NGC 7413) was not in the original analysis because the radio source had originally been identified with the galaxy and not the QSO. When a correct position was obtained it was shown that the radio source was the QSO (Arp *et al.* 1972).

Seldner and Peebles (1979) found statistically significant evidence for a correlation of the angular position of QSOs taken from the original Hewitt and Burbidge catalogue (Burbidge, Crowne, and Smith 1977) and the galaxies in the Lick catalogue (Shane and Wirtanen 1967). Nieto and Seldner (1982) carried out a further study based on a QSO catalogue of Véron, a portion of the Burbidge *et al.* catalogue and a corrected catalogue of Shane and Wirtanen (1967) and reported that they could not find statistically significant evidence for general QSO–galaxy associations, but they found marginal evidence for associations between *radio* QSOs and galaxies.

Chu *et al.* (1984) carried out a further study using the Hewitt and Burbidge (1980) catalogue of QSOs and the *Second Reference Catalogue of Bright Galaxies* (de Vaucouleurs and Corwin 1976). This catalogue contains 4364 galaxies, all brighter than 16^m, with redshifts $\leq 15\,000\,\mathrm{km\,s^{-1}}$ ($z \leq 0.05$). In that study (Chu *et al.* 1984) we restricted ourselves to objects with $|b| \geq 30°$, so that the number of galaxies fell to 3460. As far as the QSOs were concerned, we removed from the list all of those found originally by Arp and others by searching around bright galaxies. Using the cross-correlation-function technique and the nearest neighbour technique, we found strong statistical evidence for the association of QSOs, at all redshifts, with galaxies with $z \leq 0.05$.

Fugmann (1990) has claimed that the correlations exist only between galaxies and radio-emitting QSOs. However, this may be simply due to observational selection. The reasons are as follows. QSOs were originally identified as radio sources, and the first catalogues of QSOs were dominated by radio-emitting QSOs, while the most recent ones contain a majority of radio-quiet QSOs. This can be seen, for example, if we compare the contents of the 1980 and 1993 Hewitt/Burbidge QSO catalogues (Hewitt and Burbidge 1980, 1993). The radio surveys cover very much larger areas of the sky (many steradians) than do the optical surveys for QSOs, so that large areas of the sky containing many bright galaxies are involved, whereas the optical surveys cover much smaller areas, often only a few square degrees (see the plots in Hewitt and Burbidge 1993), which contain very few bright galaxies.

Table 2.1. *QSOs close to bright galaxies* ($m \leq 15.5$).

Galaxy	m_v	QSO	m_v	z_Q	Sep.$^{(\prime\prime)}$	Remarks
UGC 0439	14.4	PKS 0038−019	16.86	1.674	72	
NGC 470	12.5	(0117+0317g)	19.9	1.875	93	
NGC 470	12.5	(0117+0317g) 68D	18.2	1.533	95	
NGC 622	14.0	0133+004 (UB 1)	18.5	0.91	71	
NGC 622	14.0	0133+004 (UB 1)	20.2	1.46	73	
IC 1746	14.5	0151+048 (PHL 1226)	17.5	0.404	6.4	
NGC 1073	11.3	BSO 1	19.8	1.945	104	
NGC 1073	11.3	BSO 2	18.9	0.599	117	
NGC 1073	11.3	RSO	20.0	1.411	84	
NGC 1087	11.5	0243−007 (UB 1)	19.1	2.147	170	
ZW 0745.1+5543	15.3	0745+557	17.84	0.174	100	
IC 2402	13.5	0844+319 (4C 31.32)	18.87	1.834	30	QSO in direction of radio jet
NGC 2534	14.0	0809+358 (UB 1)	18.7	2.40	121	
NGC 2693	13.1	0853+515 (UB 1)	19.5	2.31	188	
UGC 05340	14.8	0950+080	17.69	1.45	103	
NGC 3067	12.8	0955+326	15.8	0.533	114	21-cm contours connect galaxy to QSO. Absorption in QSO at z of galaxy; active galaxy
NGC 3073	14.1	0958+558 (UB 1)	18.8	1.53	144	
NGC 3079	11.5	0958+559	18.4	1.154	114	Extremely active galaxy. Absorption in QSO at z of galaxy
ZW 1022.0−0036	15.5	PKS 1021−006	18.2	2.547	122	
NGC 3384	10.8	1046+129	20.6	0.497	149	
NGC 3407	15.0	1049+616 (4C 61.20)	16.3	0.422	173	
NGC 3561	14.7	1108+289	20.0	2.192	66	Extremely disturbed galaxy
NGC 3569	14.5	1109+357	18.1	0.91	31	

Table 2.1. (*Cont.*)

Galaxy	m_v	QSO	m_v	z_Q	Sep.$^{('')}$	Remarks
NGC 3842	13.3	QSO 1	18.5	0.335	73	
NGC 3842	13.3	QSO 2	18.5	0.946	59	
NGC 3842	13.3	QSO 3	21.0	2.205	73	
NGC 4138	12.1	3CR 268.4	18.1	1.400	174	
NGC 4319	13.0	Mk 205	14.5	0.070	43	Luminous bridge joining QSO to galaxy. Absorption in QSO at z of galaxy
ZW 1210.9 +7520	15.4	1219+753	18.16	0.645	94	
NGC 4380	12.8	1222+102 (Wdm 6)	17.6	cont.	88	
NGC 4550	12.6	1233+125	17.2	0.728	44	Galaxy in the Virgo cluster
NGC 4651	11.8	3CR 275.1	19.0	0.557	210	Active galaxy with jet and counterjet
NGC 5107	13.8	1319+38	19.5	0.949	40	
ESO 1327−2041	13.2	1327−206	17.0	1.169	38	Jet or bridge pointing to QSO; absorption at z of galaxy
ZW 1338+0350	14.9	1333+0.35	17.98	0.85	41	
NGC 5296	15.0	1342+440 (BSO 1)	19.3	0.963	55	
NGC 5406	13.1	1358+392	17	3.30	95	
NGC 5682	15.1	1432+489	19.2	1.940	95	
ZW 1640.1+3940	15.2	1640+396	18.16	0.54	180	
NGC 5832	13.3	3CR 309.1	16.8	0.905	372	
NGC 5981	13.9	1537+595	19.0	2.132	10.7	
IC 1417	13.6	2158−134	17.8	0.73	76	
Anon	15	2237+0305	17.3	1.41	≤ 0.3	
NGC 7465	13.3	2259+157	19.2	1.66	128	
NGC 7413	15.2	3CR 455	19	0.543	24	
NGC 7714−15	13.1	2333+019 (UB 1)	18.0	2.193	120	Pair of interacting galaxies

Recently, Bartelmann and Schneider (1993, 1994), motivated by the idea that correlations between high-redshift radio QSOs and low-redshift galaxies, which were claimed to exist by Fugmann (1990), could be due to gravitational lensing effects of dark matter associated with the low-redshift galaxies, have studied the correlations between a complete sample

[81]

of high-redshift radio QSOs in the 1 Jansky catalogue of Stickel and Kühr (Kühr *et al.* 1981; Stickel, Fried, and Kühr 1993a, b) and several galaxy samples.

They first looked for correlations between the galaxies in the Lick catalogue (Shane and Wirtancn 1967) and optically identified QSOs in the 1 jansky catalogue. A correlation between 1 Jy QSOs and Lick galaxies, on a 10′ scale, is detected with a significance level of up to 98%. Next, they investigated the same sample of QSOs and looked for correlations with the *IRAS Faint Source Catalogue*. Again, they found highly significant correlations. They found that the 1 Jy QSOs with $z \geq 1.25$ are correlated with *IRAS* galaxies at the 95% confidence level, which increases to more than 99% for QSOs with $z \gtrsim 1.5$.

Most recently, Bartelmann, Schneider, and Hasinger (1994) have looked for correlations between the same sample of QSOs and diffuse extended X-ray sources observed by *ROSAT*. Again, they find correlations with significance levels up to 99.8%. The scale of these correlations is $\lesssim 10′$. They conclude that, for the lower-redshift QSOs in the sample, with $z \approx 0.5$–1.0, the correlations might be due to the fact that the X-ray sources are unidentified galaxy clusters at these redshifts. However, the strong correlations (99.8%) with those QSOs with $z \geq 1.5$ cannot be explained by clusters at those redshifts, since the X-ray luminosities of such clusters would have to be much greater than is normally the case, i.e. the X-ray luminosities would have to be $\gtrsim 10^{46}$ erg s^{-1} (for $H_0 = 50 \, \text{km s}^{-1} \, \text{Mpc}^{-1}$), and there is no independent evidence that such luminous clusters exist.

In summary, very strong correlations of high-redshift radio QSOs have been found successively with:

- The Shapley–Ames catalogue of the brightest galaxies. Here the correlation is with powerful radio QSOs with $S \geq 9$ Jy (0.4 GHz). The result is significant at the 7–10σ level.
- The *Bright Galaxy Catalogue* ($z \leq 0.05$). Here the QSO sample is dominated by radio-emitting QSOs, largely identified from the 3CR, Molonglo, Parkes, and 4C radio catalogues.
- The galaxies in the Lick catalogue ($m \lesssim 17$, $z \lesssim 0.2$). Again, the sample of QSOs is a radio sample.
- The *IRAS* galaxy catalogues, where some fraction of the galaxies may have z up to 0.4, and where a few galaxies may be identical

in position with the QSOs, but where the larger fraction have much smaller redshifts than the QSOs.

- Finally, strong correlations on scales $\lesssim 10'$ have been found between optically bright, high-redshift radio-loud QSOs and the diffuse X-ray emission seen by *ROSAT*. Bartelmann *et al.* (1994) believe that this diffuse X-ray emission is due to galaxy clusters at redshifts significantly less than the observed redshifts of the QSOs.
- In addition to all of this work on radio QSOs, Stocke *et al.* (1987) took a sample of X-ray-emitting QSOs and showed that their associations with moderate-redshift galaxies ($z \leq 0.15$) were statistically significant at a high level of confidence ($> 97.5\%$).

Bartelmann, Schneider, and their colleagues carried out all of their work apparently believing that such strong correlations could only be explained by gravitational lensing due to dark matter underlying the Lick galaxies, the *IRAS* galaxies and the *ROSAT* extended X-ray sources. They make no reference to the earlier statistical work involving bright radio QSOs and brighter galaxies. It is surprising that they nowhere mention the alternative explanation, which is that we are seeing in all of these samples evidence that QSOs with large non-cosmological redshift components are concentrated where the galaxies are concentrated i.e. we *are* seeing at large distances the phenomena discussed in Section 2.1, with examples shown in Table 2.1.

2.5.4 Summary

Close by, far more QSOs with high redshifts are found very close to bright galaxies than are expected by chance. The best statistics come from the 3C sample and the Shapley–Ames galaxies, but some weight must be given to the many remarkable configurations mostly discovered by Arp (see Section 2.5.1). There are also a number of galaxies with two or three QSOs very close to the galaxy (e.g. NGC 622, NGC 1073, NGC 3842 in Table 2.1), and there are many where chains of QSOs, suggesting ejection in specific directions, are found. In addition to this, the luminous connections and other morphological features indicate physical associations.

With the exception of the one early study by Nieto and Seldner (1982), the statistical studies starting with the brightest galaxy catalogue, and then successively the Lick catalogue, the *IRAS* catalogue, and the *ROSAT* survey, all show strong positive correlations between the positions of high-redshift radio QSOs and peaks in the distribution of much lower redshift galaxies, or clusters. While it is possible that some of these effects might be explained by gravitational lensing, involving dark matter underlying the bright galaxies, a strong case can be made for the following interpretation.

The results for the brighter, nearer, close pairs cannot be explained by any form of gravitational lensing, and thus they must be real. This means that we have evidence that low-redshift galaxies are able to eject high-redshift QSOs. While we will only know from which galaxy they came when they are very close to the parent, they will tend to cluster in the regions where galaxies cluster, and thus all of the statistical results are explainable in terms of the non-cosmological redshift hypothesis.

Roughly speaking, the brightest, nearest QSOs will have cosmological redshift components $z_c \lesssim 0.03$, so that, for them, the observed redshifts $z_0 \simeq z_i$, where z_i is the intrinsic redshift component.

As we move out to the QSOs associated with the galaxies in the Lick survey, where $z_c \leq 0.2$, the QSOs will have $z_c \leq 0.2$, and $z_i = (z_0 - z_c)/(1 + z_c)$. Thus already the QSOs will have appreciable components of cosmological redshift and intrinsic redshift.

Thus the debate is no longer about 'local' versus 'cosmological' QSOs. There is a cloud of 'local' QSOs where we interpret *local* to mean distances $\lesssim 200$ Mpc. In this volume $z_0 \simeq z_i$. The remaining QSOs are 'cosmological', but they have significant intrinsic redshift components.

As we move yet further out, the results using the *IRAS* catalogue and the *ROSAT* survey of X-ray clusters show that the cosmological components of the QSOs may be as large as ~ 0.5, so that, if the measured redshift is 2, $z_i \approx 1$.

In the following section we shall discuss, among other nearby systems, QSOs which apparently lie in the Virgo cluster.

2.6 The nearest QSOs

We identify the nearest QSOs as those which are closely associated with nearby galaxies. All of the pairs we know of involving bright galaxies and separations $\lesssim 3'$ are included in Table 2.1. The majority of the NGC galaxies have redshifts $< 10\,000$ km sec^{-1} $(z_c < 0.03)$ so that for them $z_0 \simeq z_i$. Since they are close by it is not surprising that some of the QSOs associated with them are the brightest radio sources in the 3CR catalogue. It is likely that more of the 50 3CR QSOs than the 10% that lie very close to bright galaxies (Burbidge *et al.* 1971) are also at comparable distances.

The QSO–galaxy pairs with separations $\lesssim 3'$ have typical galaxy redshifts $cz \lesssim 10\,000\,\mathrm{km\,s^{-1}}$, or distances $\lesssim 200\,\mathrm{Mpc}$ (for $H_0 = 50\,\mathrm{km\,s^{-1}\,Mpc^{-1}}$). Thus the projected separations are $\lesssim 60\,\mathrm{kpc}$. For much larger separations than this there is no strong evidence for physical associations from probability arguments. At the same time, once evidence is available that QSOs and galaxies are physically associated, and that the QSOs are ejected from the galaxies, it is natural to expect to find QSOs with a wide range of distances from their parent galaxies.

It was pointed out, in the very early days of radio astronomy, that there appeared to be an asymmetry of the bright radio sources on the sky. This was deduced from the source counts by Hanbury-Brown (1962). He suggested that the supergalactic structure, as defined by de Vaucouleurs based on the distribution of the bright galaxies, might be responsible (see also Shaver and Pierre 1989). Arp (1970, 1983) pointed out that there is good correlation between the position of the 3C and the Parkes radio QSOs on the sky and the Shapley–Ames galaxies. These are the bright galaxies, many of which lie in the Virgo cluster, which itself makes up the central part of the local supercluster. Sulentic (1988) has shown that the density on the sky of the bright QSOs in the Palomar Survey is about five times greater in the direction of the Virgo supercluster than elsewhere. This correlation is exactly what we would expect if many of the (nearest) and brightest QSOs have been ejected from galaxies in the Virgo cluster and the local supercluster.

Because of its position and because it is optically the brightest radio-emitting QSO, there has always been a suspicion that 3C 273 might be a member of the Virgo cluster ($d \simeq 21\,\mathrm{Mpc}$). Arp has suggested that both 3C 273 and 3C 279 were ejected from galaxies in the central region of the Virgo cluster. There is considerable evidence from the morphology (see Arp and Burbidge 1990) and from the X-ray emission (Arp 1994) that this may be the case. In addition to this, the very active Virgo cluster galaxy M87 shows evidence of ejection of QSOs in the direction of its jet (Arp 1987), which points directly to, and may have ejected, M84 (Wade 1960).

NGC 4550, which is also a member of the Virgo cluster, has a bright QSO ($17^{m}.2$) only $44''$ from its centre (see Table 2.1).

Apart from the Virgo cluster, and the close pairs listed in Table 2.1, there are a number of groupings of QSOs which have been noted over the years to lie in the vicinity of comparatively nearby galaxies. The major ones are as follows:

    (i) A compact group of four QSOs lie within $10'$ of M82 ($d \simeq 2.5$ Mpc). Three were found serendipitously and a fourth was discovered by Arp (1981). They lie in the direction of the cone of ejection of high-energy matter from M82.

  (ii) A recent *ROSAT* study of X-rays from NGC 4258 ($d \simeq 7$ Mpc) has led to the identification of two compact X-ray sources equally distant from the centre of NGC 4258 (Pietsch *et al.* 1994). They lie along a line passing almost through the nucleus. These two sources are identified with two candidate QSOs which, Pietsch *et al.* suggested, were ejected from NGC 4258. NGC 4258 is clearly an active galaxy. Recently, Burbidge (1995) has shown that both of these sources are genuine QSOs with redshifts of 0.398 and 0.653 respectively.

 (iii) In Table 2.1 one QSO close to NGC 3079 ($d \simeq 16$ Mpc) is listed. There is extensive evidence for explosive ejection of gas, etc. from NGC 3079 (Filippenko and Sargent 1992). Earlier Arp (1974, 1977) identified the QSO listed in Table 2.1, and three more QSOs around NGC 3079.

 (iv) A dense group of QSOs with different redshifts is found in an area ~ 50 arcmin^2 within $2°$ of NGC 3810 ($d \simeq 21$ Mpc) (Hazard, Arp, and Morton 1979; Arp 1983).

  (v) A dense group of five QSOs with different redshifts is found in an area $\lesssim 4.5$ arcmin^2 about $2°$ SW of NGC 450 (Arp 1983).

 (vi) Ten QSOs have been found to surround a faint galaxy close to NGC 2639 ($d \simeq 91$ Mpc) (Arp 1980).

(vii) NGC 1097 ($d = 32$ Mpc) shows evidence of activity in the form of two or more optical jets (Wolstencroft and Zealey 1975; Arp 1976). It is surrounded by a large number of QSOs. There is a dense cluster of them within $24'$ of the centre of NGC 1097 (Wolstencroft *et al.* 1983). This involves at least six QSOs within about 100 arcmin^2. Outside this region there is an extended cluster containing ~ 40 QSOs (Arp, Wolstencroft, and He 1984).

(viii) Five QSOs lie near NGC 2916, which itself is a companion to the bright galaxy NGC 2903 ($d = 12.3$ Mpc) some $40'$ away (Arp 1981).

(ix) Eight QSOs approximately aligned and apparently associated with the triple system NGC 3379, 3389, and 3384 ($d \simeq 17$ Mpc) have been found by Arp, Sulentic, and di Tullio (1979).

(x) Arp and Duhalde (1985) discovered a chain of QSOs apparently ejected from the highly irregular active galaxy NGC 520 (see Stanford 1992). The chain extends over $\sim 7°$ from NGC 520 ($d \simeq 48$ Mpc) (Arp 1987). Because of the large distance involved (~ 6 Mpc), the physical reality of the chain is less likely than in the other groupings described here. Arp (1987) has given an extended discussion of the controversy surrounding this grouping, which he originally pointed out much earlier.

(xi) Recently, Arp (1994) has identified more QSOs close to the pair NGC 4319–Mk 205 (shown in Table 2.1). Again, there is an alignment, suggesting ejection from Mk 205, of the QSOs which are further away.

(xii) Arp and Hazard (1980) discovered two triplet systems of QSOs, each precisely aligned, with redshifts (a) 2.1, 0.51, and 1.7 and (b) 2.1, 0.54, and 1.6. The triplet lines are less than $10'$ apart and lie very roughly parallel to each other ($\Delta\theta \approx 10°$). There is no obvious bright galaxy nearby, but Narlikar and Das (1980) have shown that if one joins together the positions of the pairs with roughly equal redshifts, the three lines intersect at a point (close to the QSO with $z = 1.7$), which may represent an origin.

All of these groups show the same characteristics, namely a large over-density of QSOs in the vicinity of a bright galaxy, and in a number of cases, a preferred direction suggesting that the objects have been ejected in a narrow cone. Very frequently there is independent evidence that the galaxy is active and is giving rise to non-thermal radio emission and the generation of hot gas with high velocities from its central region.

In previous sections, I have made the case for the view that QSOs are being ejected from galaxies and have intrinsic redshift components. Much more could be discussed. For example, we need to understand what is happening in situations in which we see extensive absorption in the spectra of QSOs. This has all been discussed by Burbidge (1995) using the same model, and we refer the reader to that paper for details.

We are now in new territory as far as our detailed understanding is concerned. Hoyle and I (Hoyle and Burbidge 1995) have made a start at explaining the nature of the intrinsic redshifts using the theory underlying the quasi-Steady State cosmology.

There are other observational phenomena which may well be related to the QSO–galaxy associations. We turn briefly to compact groups of galaxies.

2.7 Compact groups

These were mentioned in the earlier part of this chapter when I was discussing the attitudes that were taken to the new results.

In recent years, Hickson (1982) has made a survey and has published a catalogue of 100 compact groups containing four to six members, including those discovered much earlier (see Burbidge and Sargent 1971). All of the redshifts have been measured. Twenty-eight of the systems have one redshift very different from the mean of the others. In Table 2.2 we give a list of these 28 groups. Given these much larger numbers, the fundamental question is, given such a large fraction of the groups with a discrepant galaxy, can they all be interpreted in terms of chance superpositions of background or foreground galaxies? Hickson and his colleagues have argued that at least a fraction of them can be so interpreted (Hickson, Kindl, and Huchra 1988), but Sulentic (1987) has concluded that this is not possible. I believe that the analysis of Sulentic is more likely to be correct. In any case, it appears that at least a fraction are physical systems.

Thus, as was mentioned earlier, either we are seeing galaxies ejected at high speed from the group or an intrinsic spectrum shift is present, as is the case for the QSOs, though in the compact groups the discrepant galaxy can have either a redshift or a blueshift with respect to the mean redshift of the other members. It is also the case that most of the discrepant galaxies in these groups are late-type spirals.

The questions of the ages of the groups and their state of evolution is unclear at present. Recently, strenuous efforts have been made to explain them within the conventional theory of galaxy formation and large-scale structure (see Mendes de Oliveira 1995) and X-ray observations have shown that, in many groups, hot gas is present. From this, attempts have been made to argue that the systems are stable, the virial holds, and large amounts of dark matter are present.

Table 2.2. *Compact groups with at least one discrepant redshift*.

Hickson group no.	Type of discrepant galaxy	Group cz_c (km s^{-1})	z_c (km s^{-1})	$c(z_d - z_c)$ (km s^{-1})	$cz_u = \dfrac{c(z_d - z_c)}{1 + z_c}$
2	SBb	4 320	0.0144	+17 020	+16 780
3	Sd	7 650	0.0255	+3 195	+3 800
4	Sab	8 400	0.0280	+10 080	+9 800
5	Sc	12 300	0.0410	−4 085	−3 920
14	Sd	5 490	0.0183	+2 926	+2 870
18	SOa	4 175	0.0139	+5 844	+5 760
20	SOa	1 420	0.0484	−3 959	−3 780
23	Sm	4 830	0.0161	+5 320	+5 240
28	Sdm	11 400	0.0380	+18 805	+18 120
29	CI	31 410	0.1047	−18 082	−16 370
31	Sdm	4 110	0.0137	+22 790	+22 480
38	SBa	8 760	0.0292	+15 522	+15 080
43	Sc	9 900	0.0330	+9 605	+9 300
52	Sdm	12 900	0.0430	−6 607	−6 330
53	Sc	6 180	0.0206	+2 890	+2 830
55	Sc	15 780	0.0526	+21 100	+20 040
59	Scd	4 056	0.0135	+15 600	+15 390
61	Im	3 900	0.0130	−2 773	−2 740
63	SBbc	9 330	0.0311	−4 102	−3 980
64	Sd	10 800	0.0360	−4 653	−4 490
71	SO	9 030	0.0301	+11 560	+11 220
72	Scd	12 630	0.0421	+11 420	+10 960
78	SO	9 380	0.0313	+8 820	+8 550
79	EO	4 350	0.0145	+15 459	+15 240
84	EO	16 680	0.0556	+15 820	+14 990
92	Sd	6 450	0.0215	−5 664	−5 540
93	Sa	5 040	0.0168	+3 841	+3 780
98	Sc	7 980	0.0266	+6 970	+6 300

z_d is the discrepant redshift.

[89]

For me, none of these arguments is satisfactory. My guess is that these observations, like those of the QSOs, are telling us that we should look in new directions.

2.8 Peaks and periodicities in the redshift distribution

In the discussion following the lecture version of this chapter, I was asked about the redshift periodicities. I therefore thought it worthwhile to add this section, which summarizes what has been going on in these areas.

2.8.1 The Tifft effect

Starting in the 1970s, Tifft (1976, 1980) claimed that ordinary galaxies show quantized differential redshifts with a period $c\Delta z_u = 70\text{-}75\,\mathrm{km\,s^{-1}}$ (here z_u simply means a redshift term of unknown origin). He first found this effect in the differences between the redshifts of members of the Coma cluster and later in the redshift differences between physical pairs of galaxies (Tifft 1980). Also, Holmberg, and later Arp and Sulentic (1985), showed that in small groups of galaxies dominated by a bright galaxy (e.g. the M81 group) the differences are not distributed at random about the redshift of the main galaxy, as would be expected if they were due to satellite motions, or even if they were expanding away from the primary galaxy. It turns out that the mean shift with respect to the central galaxy is displaced to the red. The majority, and in some cases *all*, of the differences relative to the central galaxy are redshifts. Not only that, but the distribution is quantized with $c\Delta z_u \simeq 72.5\,\mathrm{km\,s^{-1}}$. The recent work of Tifft and his associates, on other samples of galaxies, has suggested that the primary value of $c\Delta z_u$ may be 1/2 or 1/3 of the original number, i.e. about $36\,\mathrm{km\,s^{-1}}$ or $24\,\mathrm{km\,s^{-1}}$. Further analyses have been made by Guthrie and Napier (1990, 1992, 1996) of samples of nearby galaxies which have very accurate redshifts, measured using the 21-cm line. Using 89 spiral galaxies, with cz in the range $0\text{-}1000\,\mathrm{km\,s^{-1}}$, whose redshifts were accurately measured ($\sigma \leq 4\,\mathrm{km\,s^{-1}}$), Guthrie and Napier (1992) have shown that when the redshifts are corrected for the optimum solar vector ($v_\odot = 227.9\,\mathrm{km\,s^{-1}}$, $l = 98.7$, $b = -2°.8$) a periodicity is found at $37.22\,\mathrm{km\,s^{-1}}$, with a probability of finding this period by chance of 2.7×10^{-5}. Guthrie and Napier (1996) have extended this result to more galaxies within the supercluster out to $cz = 2600 \, \mathrm{km \, s^{-1}}$ and have confirmed the result. Thus, the Tifft effect is confirmed in a variety of nearby samples of normal galaxies.

It is very important to stress that the result of such high significance as that attained above is only obtained after the correction for our motion with respect to the Galactic Centre; i.e. the periodicity found with $c\Delta z_u = 37.2 \, \mathrm{km \, s^{-1}}$ is associated with the difference in redshifts between the centre of mass of our Galaxy and the other systems.

2.8.2 Periodicity in the redshifts of QSOs and related objects

From the time of the original discovery of the QSOs it was clear that they did not follow a tight Hubble relation. As more and more objects were discovered it became clear that the Hubble diagram has largely the appearance of a scatter diagram (see Hewitt and Burbidge 1993, Fig. 1 for a recent demonstration of this). The conventional interpretation is that they show a very large scatter in their intrinsic luminosities.

Early in the studies of QSOs and related objects, a sharp peak at $z = 1.955$ was reported (Burbidge and Burbidge 1967). Soon after this, it was noticed that if we restrict ourselves to QSOs, and related objects distinguished from normal galaxies by their non-thermal continua and emission-line spectra, which are similar to those of QSOs, the redshifts show a quantized appearance at values $z_u = n \times 0.061$ at least up to $n \simeq 10$. Since the redshifts are mostly very small compared with those of the QSOs, most of the objects in the original survey (70 objects) were those in the second category. In this distribution, a strong peak was seen at $z = 0.061$ and at multiples of this value (Burbidge 1968).

As more QSO redshifts were obtained, several additional peaks in the redshift distribution became apparent, particularly at $z = 0.30$, 0.60, 0.96, and 1.41 (Burbidge 1978). Karlsson (1977) showed that these peaks are periodic with $\Delta \log(1 + z) = 0.089$; i.e. the ratio of successive peaks $(1 + z_0^{n+1})/(1 + z_u^n) = 1.227$. The first peak is at $z_u = 0.061$ and the last discernable peak at $z_u = 1.955$. This analysis, referenced above, was based upon about 600 QSO redshifts, which are mostly comparatively bright radio QSOs. This result was confirmed using larger samples by Fang *et al.* (1982) and by Depaquit, Pecker, and Vigier (1985).

A new catalogue of extragalactic emission-line objects similar to QSOs was compiled recently by Hewitt and Burbidge (1991). It contains 935 objects. More than 700 have redshifts $z \leq 0.2$ and most are Seyfert galaxies, though many emission-line radio galaxies are included with $z \geq 0.2$. A histogram of these redshifts shows a large peak at $z = 0.06$ (Burbidge and Hewitt 1990). There are 89 objects out of about 500 with $z_u < 0.2$, in the very narrow redshift interval $\Delta z = 0.01$ between $z = 0.055$ and $z = 0.065$. Duari, Das Gupta, and Narlikar (1992) did a new analysis, based on all the QSOs which have not been identified by any technique, which determines to some extent the redshift range of the objects being discovered. This meant that they used 2146 objects out of the catalogues, which contain more than 8000 objects (Hewitt and Burbidge 1991, 1993).

In a plot in their paper, the peaks at 0.06, 0.18, 0.24, 0.30, 0.32, 0.36, 0.40, 0.47, 0.55, and 0.62 can easily be seen. Duari *et al.* did a power spectrum analysis similar to that done originally on an earlier sample by Burbidge and O'Dell (1972), who confirmed the original peaks at 0.06 and 1.955. They also carried out the Kolmogoroff–Smirnoff test and the comb-tooth test and found strong evidence for the periodicity at 0.06 (the exact value is 0.0565, and its significance is increased when the redshifts are transformed to the Galactocentric frame). A second period of 0.0128 was also found with high significance. As far as the periodicity at large scales in $\Delta \log(1 + z)$ is concerned, they were more cautious, and the reality of this periodicity has been recently questioned by Scott (1991).

To summarize these investigations, it appears that the peak at 0.06 and the periodicity up to about $n = 10$, which was first noted in 1968, has been shown to exist with something like 30 times as much data as existed then. The peak at 1.955 is also well established, and the larger-scale periodicity may still need more study.

In discussing the reality of such effects, it should always be borne in mind that, if an unknown redshift component z_u is periodic and fairly large, the existence of any significant range of cosmological components z_c with values ~ 0.01 upwards will very easily smear out peaks as far as observation is concerned. Thus, to find peaks at all in the observed data, at multiples of 0.06, or at values up to 1.955, is remarkable. It strongly suggests that the cosmological components of objects in these redshift ranges must be very small ($z_c \ll 0.01$), or that z_c and z_u are related.

2.8.3 Periodicity in z in faint-galaxy surveys

Deep pencil beam surveys of normal galaxies show a periodic redshift effect. The discoverers of this effect do not describe it in this way, but they say that galaxies, mapped in the z dimension, are not distributed randomly but show an excess correlation and apparent regularity in the galaxy distribution, with a characteristic scale of $128\,h\,\mathrm{Mpc}$ for $z \leq 0.2$ (Broadhurst *et al.* 1990; Broadhurst 1994). This corresponds to $cz = 12\,800\,\mathrm{km\,s^{-1}}$ or $cz_{\mathrm{u}} = 0.0426$ for $H_0 = 50\,\mathrm{km\,s^{-1}\,Mpc^{-1}}$.

2.9 **Conclusions**

In this chapter I have tried to give you an extremely compressed but personal view of the way that I believe the major developments have gone in extragalactic astronomy since the 1920s. I have then concentrated on what I think are the indications that the newer evidence obtained since ~1960 will move us away from the standard ideas which were formulated in the earlier period and which have been strongly maintained up to the present.

In my view, it is the newer observational evidence that must lead us to change our views. In taking this position, I am going counter to the conventional method of the time, in which the approach is to fit (and often force-fit) all of the new results into the conventional theory. The underlying belief is that, by now, there is nothing we can learn about physics from astronomy. This is an approach which, for good reason, was not shared by Newton and other giants of the past. The observational data that I have presented in the second part of my lecture, in my view, clearly point to new directions in theory and in model making.

I would like to believe that, by the year 2025, it will be demonstrated that at least some of what I have discussed will have proved this. The actuarial tables suggest that very few, if any, of the contributors to this book will be around to evaluate the situation in 2025, but most of the readership will have the last word.

Acknowledgements

I would like to thank our hosts at the meeting on which this book is based for making it such an exciting event, and for their gracious and generous hospitality.

References

Alpher, R.A., and Herman, R.C.: 1950, Rev Mod Phys, **22**, 153

Alpher, R.A., Bethe, H.A., and Gamow, G.: 1948, Phys Rev, **73**, 803

Ambartsumian, V.A.: 1958, Solvay Conference on *Structure and Evolution of the Universe*, ed. R. Stoops (Brussels) p. 241

Ambartsumian, V.A.: 1965, *Structure and Evolution of Galaxies*, Proc. 13th Conference Physics, University of Brussels (New York: Wiley Interscience)

Arp, H.C.: 1966, Atlas of Pecular Galaxies (Pasadena: California Institute of Technology)

Arp, H.C.: 1967, ApJ, **148**, 231

Arp, H.C.: 1970, AJ, **75**, 1

Arp, H.C.: 1974, IAU Symp. No. 58, ed. J. Shakeshaft (Dordrecht: Reidel), p. 199

Arp, H.C.: 1976, ApJL, **207**, L147

Arp, H.C.: 1977, Coll Int No. 263, 377 (Paris: CNRS)

Arp, H.C.: 1980, ApJ, **236**, 63

Arp, H.C.: 1981, ApJ, **250**, 31

Arp, H.C.: 1983, ApJ, **271**, 479

Arp, H.C.: 1987, *Quasars, Redshifts and Controversies* (Berkeley, CA: Interstellar Media)

Arp, H.C.: 1990, A&A, **229**, 93

Arp, H.C.: 1994, IAU Symp. No. 168, in press

Arp, H.C., and Burbidge, G.: 1990, ApJL, **353**, L1

Arp, H.C., and Duhalde, O.: 1985, PASP, **97**, 1149

Arp, H.C., and Hazard, C.: 1980, ApJ, **240**, 726

Arp, H.C., and Sulentic, J.: 1985, **29**, 88

Arp, H.C., Sulentic, J., and di Tullio, G.: 1979, Nature, **282**, 489

Arp, H.C., Burbidge, E.M., Mackay, C., and Strittmatter, P.: 1972, ApJL, **171**, L41

Arp, H.C., Wolstencroft, R.D., and He, X.T.: 1984, ApJ, **285**, 44

Aston, F.: 1929, Nature, **123**, 313

Bartelmann, M., and Schneider, P.: 1993, A&A, **271**, 421

Bartelmann, M., and Schneider, P.: 1994, A&A, **284**, 1

Bartelmann, M., Schneider, P., and Hasinger, G.: 1994, A&A, **290**, 399

Begelman, M., Blandford, R., and Rees, M.J.: 1984, Rev Mod Phys, **56**, 255

Broadhurst, T.J.: 1994, Proc. of Cambridge Conference, July 1994

Broadhurst, T.J., Ellis, R.S., Koo, D., and Szalay, A.S.: 1990, Nature, **343**, 726

Burbidge, E.M., and Sargent, W.L.W.: 1971, *La Semaine d'Etude sur Les Noyaux des Galaxies* (Rome: Pontifical Academy of Sciences) p. 351

Burbidge, E.M., Burbidge, G.R., Fowler, W.A., and Hoyle, F.: 1957, Rev Mod Phys, **29**, 547

Burbidge, E.M., Burbidge, G.R., Solomon, P.M., and Strittmatter, P. 1971, ApJ, **170**, 223

Burbidge, G.: 1968, ApJL, **154**, L41

Burbidge, G.: 1970, Ann Rev A&A, **8**, 369

Burbidge, G.: 1971, Nature, **233**, 36

Burbidge, G.: 1978, Physica Scripta, **17**, 237

Burbidge, G.: 1979, Nature, **282**, 451

Burbidge, G.: 1981, Ann NY Acad Sciences, **8**, 123

Burbidge, G.: 1988, Sky and Telescope, **75**, 38

Burbidge, G.: 1996, A&A, **309**, 9

Burbidge, G., and Burbidge, E.M.: 1967, *Quasi-Stellar Objects* (San Francisco: Freeman)

Burbidge, G., and Hewitt, A.: 1990, ApJL, **359**, L33

Burbidge, G., and O'Dell, S.: 1972, ApJ, **178**, 583

Burbidge, G., Burbidge, E.M., and Sandage, A.R.: 1963, Rev Mod Phys, **35**, 947

Burbidge, G., Crowne, A.H., and Smith, H.E.: 1977, ApJS, **33**, 113

Burbidge, G., Hewitt, A., Narlikar, J.V., and Das Gupta, P.: 1990, ApJS, **74**, 675

Canizares, C.R.: Nature, **291**, 620

Carilli, C.L., and van Gorkum, J.H.: 1992, ApJ, **399**, 313

Carilli, C.L., van Gorkum, J.H., and Stocke, J.T.: 1989, Nature, **338**, 314

Chu, Y., Zhu, X., Burbidge, G., and Hewitt, A.: 1984, A&A, **138**, 408

Das Gupta, P., Narlikar, J.V., and Burbidge, G.: 1988, AJ, **95**, 5

Depaquit, S., Pecker, J.C., and Vigier, J.P.: 1985, Astron Nachr, **306**, 7

de Vaucouleurs, G.H., de Vaucouleurs, A.P., and Corwin, H.G.: 1976, *Second Reference Catalogue of Bright Galaxies* (Austin: University of Texas)

Dingle, H.: 1953, Observatory, **73**, 42. (This address was given on February 13, 1953)

Duari, D., Das Gupta, P., and Narlikar, J.V.: 1992, ApJ, **384**, 35

Eddington, A.: 1924, MNRAS, **84**, 308

Fang, L.Z., Chu, Y., Liu, Y., and Cao, C.: 1982, A&A, **106**, 287

Filippenko, A., and Sargent, W.L.W.: 1992, AJ, **103**, 28

Fowler, W.A., and Hoyle, F.: 1960, Ann Phys, **10**, 280

Fowler, W.A., and Hoyle, F.: 1963, MNRAS, **125**, 169

Fugmann, W.: 1990, A&A, **240**, 11

Gamow, G.: 1948, Phys Rev, **74**, 505

Guthrie, B., and Napier, W.M.: 1990, MNRAS, **243**, 431

Guthrie, B., and Napier, W.M.: 1990, MNRAS, **243**, 533

Guthrie, B., and Napier, W.M.: 1996, A&A, **310**, 353

Hanbury-Brown, R.: 1962, MNRAS, **124**, 35

Hazard, C., Arp, H.C., and Morton, C.: 1979, Nature, **282**, 271

Hazard, C., Mackey, M.B., and Shimmins, A.T.: 1963, Nature, **197**, 1037

Hewitt, A., and Burbidge, G.: 1980, ApJS, **43**, 57

Hewitt, A., and Burbidge, G.: 1991, ApJS, **75**, 297

Hewitt, A., and Burbidge, G.: 1993, ApJS, **87**, 451

Hickson, P.: 1982, ApJ, **255**, 382

Hickson, P., Kindl, E., and Huchra, J.: 1988, ApJL, **329**, L65

Holmes, A., and Lawson, R.: 1927, Am J Sci, **13**, 327

Hoyle, F.: 1968, Proc Roy Soc A, **308**, 1

Hoyle, F., and Burbidge, G.: 1966, ApJ, **144**, 534

Hoyle, F., and Burbidge, G.: 1996, A&A, **309**, 335

Hoyle, F., and Narlikar, J.V.: 1964, Proc Roy Soc A, **282**, 184, 191

Hoyle, F., and Schwarzschild, M.: 1955, ApJS, **2**, 1

Hoyle, F., Burbidge, G., and Narlikar, J.V.: 1993, ApJ, **410**, 437

Hoyle, F., Burbidge, G., and Narlikar, J.V.: 1994a, MNRAS, **267**, 1007

Hoyle, F., Burbidge, G., and Narlikar, J.V.: 1994b, A&A, **289**, 729

Hoyle, F., Burbidge, G., and Narlikar, J.V.: 1995, Proc Roy Soc A, **448**, 191

Hoyle, F., Fowler, W.A., Burbidge, G., and Burbidge, E.M.: 1964, ApJ, **139**, 909

Hubble, E.: 1929, Proc Nat Acad Sci, **15**, 168

IAU Symp 5: 1962, Santa Barbara (New York: MacMillan)

Jacoby, G., Branch, D., Ciardullo, R., Davies, R., Harris, W., Pierce, M., Pritchet, C., Tonry, J., and Welch, D.: 1992, PASP, **104**, 599

Jeans, J.H.: 1929, *Astronomy and Cosmogony* (Cambridge: Cambridge University Press)

Karlsson, K.G.: 1977, A&A, **106**, 287

Kippenhahn, R., and de Vries, H.L.: 1974, A&AS, **26**, 131

Kühr, H., Witzel, A., Pauliny-Toth, I., and Nauber, U.: 1981, A&AS, **45**, 367

Matthews, T.A., and Sandage, A.R.: 1963, ApJ, **138**, 30

Mendes de Oliveira, C.: 1995, MNRAS, **273**, 139

Narlikar, J.V., and Das, P.K.: 1980, ApJ, **240**, 401

Nieto, J.-L., and Seldner, M.: 1982, A&A, **138**, 408

Ostriker, J.: 1989, in *BL Lac Objects: Lecture Notes*, ed. L. Maraschi (Berlin: Springer-Verlag), Vol. 334

Peebles, P.J.: 1995, preprint

Penzias, A., and Wilson, R.: 1965, ApJ, **152**, 419

Pietsch, W., Vogler, A., Kahabka, P., Jain, A., and Klein, V.: 1994, A&A, **284**, 386

Pontifical Academy of Sciences: 1982 'Astrophysical Cosmology' Proc. Study Week on Cosmology and Fundamental Physics

Rees, M.J.: 1984, Ann Rev A&A, **22**, 471

Rutherford, E.: 1929, Nature, **123**, 314

Ryle, M.: 1968, Ann Rev A&A, **6**, 249

Sandage, A.R.: 1958, ApJ, **127**, 513

Sandage, A.R.: 1961, Sky and Telescope, **21**, 148

Sandage, A.R., and Perelmuter, J.-M.: 1990, ApJ, **350**, 481

Schmidt, M.: 1963, Nature, **197**, 1040

Schneider, P.: 1994, *Gravitational Lenses in the Universe*, eds. J. Surdej *et al.* (Liege: Institut d'Astrophysique), p. 41

Schneider, P., Ehlers, J., and Falco, E.: 1992, Gravitational Lenses (Berlin: Springer)

Schwarzschild, M.: 1958, *Structure and Evolution of the Stars* (Princeton: Princeton University Press)

Scott, D.: 1991, A&A, **242**, 1

Seldner, M., and Peebles, P.J.: 1979, ApJ, **227**, 30

Shane, C.D., and Wirtanen, C.A.: 1967, Publ Lick Obs 22, Part 1

Shaver, P., and Pierre, M.: 1989, A&A, **220**, 35

Stanford, M.: 1992, ApJ, **381**, 409

Stickel, M., Fried, J., and Kühr, H.: 1993a, A&AS, **97**, 483

Stickel, M., Fried, J., and Kühr, H.: 1993b, A&AS, **100**, 395

Stocke, J.T., Schneider, P., Morris, S., Gioia, I., Maccacaro, T., and Schild, R.: 1987, ApJL, **315**, L11

Sulentic, J.W.: 1987, ApJ, **322**, 605

Sulentic, J.W.: 1988, Phys Lett A., **131**, 227

Sulentic, J.W., and Arp, H.C.: 1987, ApJ, **319**, 687

Tifft, W.: 1976, ApJ, **206**, 38

Tifft, W.: 1980, ApJ, **236**, 70

Tolman, R.C.: 1934, *Relativity, Thermodynamics and Cosmology* (Oxford: Oxford University Press)

Vorontsov-Velyaminov, B.A.: 1959, *Atlas and Catalogue of Interacting Galaxies*, Moscow

Wade, C.M.: 1960, Observatory, **80**, 235

Wolstencroft, R.D., and Zealey, W.J.: 1975, MNRAS, **173**, 51P

Wolstencroft, R.D., Ku, W.H., Arp, H.C., and Scarrott, S.M.: 1983, MNRAS, **205**, 67

Womble, D.S.: 1992, PhD Thesis, University of California, San Diego

2.10 Discussion

Question

(Rees): You did not say anything about the projections for the next 30 years as Allan did. We have had 30 years of this debate: I would like to ask what your prognosis is? Would you accept that there is evidence that quite a lot of quasars are a long way away from double lensing, absorption-lines, and all the rest?

Answer

(G. Burbidge): Let me take these one at a time. First of all, the next 30 years? I hope that the next 30 years are not spent convincing people that the evidence that I have presented is real. The fact that many of the cases involve galaxies of modest redshift means that there will be cosmological components and intrinsic components. There is no question that this is a real phenomenon. For example, some of those radio QSOs with measured redshifts of $\sim$0.5 will have cosmological redshift of 0.2, and about 0.2 will be the intrinsic component.

As far as the absorption is concerned, that is a very, very complicated business, and it does not support the cosmological redshift hypothesis, because some of the objects I have shown also show absorption in their spectra at the redshift of the galaxy. If you have those two phenomena present together, you see, the way the game has been played so far is to ignore the statistical evidence, or the connection evidence. In fact, in some of the pictures I did not show (actually, some good ones), people like Jacqueline Bergeron simply say, well, there happens to be a bridge, or a tidal tail pointing in the right direction, but this is an accidental alignment.

In the case of absorption, I believe that what we have to do is think in terms of models in which we have a large halo of gas. It is entirely possible that you can have enough depth in the gas so that the QSO can be embedded in that halo, which gives rise to the absorption. There the QSO is near to the galaxy and not very far behind it. That is the only way that I have been able to think of explaining those phenomena. Where you have got multiple absorptions in the same situation, again, I think you have got to go back and ask the question, which has never been properly dealt with, whether the absorption is intervening or ejected. Sargent has tried to put the lid on this argument by looking at a small sample, demonstrating that the distribution is Gaussian and saying that this shows that the clouds are all intervening. But, for example, if you look at the Ly α absorption in the spectra of 3C 273, which goes all the way from the redshift of 3C 273 down to the redshift of the Virgo cluster, you can find absolutely no evidence for galaxies, or anything else, associated with those absorptions. Why not argue that it could indeed be ejection? It is also the case that, when you have QSOs that show a very large number of absorptions in their spectra, not associated with galaxies, or anything, it is very, very hard – it has always been very hard – to explain those in terms of cloud distributions, because they are not clumped in the way you would expect for clusters. An honest look at that problem suggests that the issue was also not closed, namely whether it is intervening or ejected. What I am trying to say is, when we have cases of this kind, I would argue that you can certainly explain some of the absorption as associated with the galaxy and the galaxy has an appreciable redshift, some intervening clouds may be present; i.e. part of the usual explanation may be correct. Some of the best-known galaxies with QSOs nearby are highly disturbed, and this is very interesting in itself. Work by Carilli, Van Gorkum, and others suggests strongly that much of the gas is being ejected from the galaxies anyway, and it is not simple halo gas.

Question

(Novikov): Geoffrey, you addressed a question to me about the meaning of the problem of events just before the Big Bang. I believe that the question has some meaning because of the following. We know, for example, that events and physics just after the Big Bang led to observable consequences: we can observe them   for example, the huge ratio of number density of photons in the microwave background radiation to the number density of heavy particles. This huge ratio is a direct consequence of the physics of the very early Universe: the first moments, just after the Big Bang. The expansion of the Universe itself is the observable consequence of that. Why can we not take one step back and try to find any consequences of events just before the Big Bang?

Answer

(G. Burbidge): Maybe I am too simple minded, but my problem with this is: it seems to me that if we do not have a Universe, we do not have the laws of physics.

Comment

(Novikov): When we are talking about the Big Bang, it means the beginning of the expansion, not of the Universe itself. Of course, this depends on the definition; but as far as I understand it, you are right:

probably this question has another answer. Probably space and time were born together with matter at this zero point and there was nothing, in this case, before that. Once again, it is a direct question to physics, to astrophysics, and probably to observational cosmology. That is my point.

Answer

(G. Burbidge): I accept your point; I simply have a conceptual difficulty with it, that is all.

Question

(Osterbrock): I am possibly more impressed by the continuity of the active galactic nuclei than you indicated in this talk. Of course, I do not know about all QSOs, but I will show a picture tomorrow, prepared by Morgan and Drysers, showing the continuity of objects from Seyfert galaxies, to what we call QSOs, including the objects which, at one time or another, were called both Seyfert galaxies and QSOs. I have been very much impressed in the last several years that, with improved imaging, improved telescopes, and improved data reduction, many objects, not of course at redshifts of 4, but at redshifts of around 0.5, that were formerly called QSOs turned out to have things that look like galaxies around them. Would you agree that there are some objects like that anyhow?

Answer

(G. Burbidge): Yes, I think so, but I was very struck by what Bahcall demonstrated, that really, when you look carefully at low-redshift QSOs, in some cases you cannot find any serious evidence at all for galaxies, and that is impressive, since we have been bombarded for years with the view that there always must be a host galaxy. Now there is another argument, Don, and I think you are aware of it. Various people have played with it, and that is, if you believe there is a host galaxy and you put a QSO in the middle of it, the activity of the QSO might be so great that it might completely distort or make the galaxy look very different. There has always been that possibility. But the certainty with which people have told me that QSOs are all embedded in galaxies is very great and I think it is fundamental to Martin's theory that you have got to have a galaxy there, or else you are not going to build up your disk and your massive black hole. I believe that it is assumed in the theory that a galaxy must be present.

Question

(Osterbrock): If I could comment on that. First of all, there were some QSOs which Bahcall *et al.* observed, and those were the ones that did not show a galaxy to the limit that they could look. But you agree that there are others that do show one, that were already known before?

Answer

(G. Burbidge): If you require facts to define the galaxy you must get a spectrum of the outer parts and show that stars are present. I only know of one where that was, in my view, quite marginally done, and that was 3C 48. In all other cases hot gas and other material are all that are found. The shapes and so on (as Jerry Kristian first said) were the first evidence proposed to show this continuity.

Question

(Osterbrock): But, for instance, the slide I have shows I Zw 1 and II Zw 1, both of which were called QSOs at one time, and Stockton, I think, has found many others at somewhat larger redshift. I am not saying they all are, but there are some. I think you will agree that, as far as Bahcall's paper is concerned, there is a limit on the surface brightness, and what they expected to see they certainly did not see. I think again that that does not mean that there is not a galaxy there, at a fainter surface brightness, that we are not familiar with.

Answer

(G. Burbidge): That certainly is possible, but in this discussion the onus of proof is on people to find the galaxy, not to tell me it is fainter than they looked. You see what I mean?

Question
(Osterbrock): I agree.

Comment

(M. Burbidge): It is perhaps fair to mention an e-mail I received just before setting off on travels at the beginning of the year, from John Bahcall. It was about something else, but he said: 'By the way, you might like to mention something at the Tenerife meeting. We have now extended our work on looking for faint galaxies around QSOs', the published data that Geoff showed, and he said, 'we have a larger number of galaxies in which we have done this more carefully, with very careful image deconvolution and subtraction of the bright QSO, looking for faint material around'. And this was in a number of quasars where ground-based observations had apparently shown a good galaxy, but Bahcall and colleagues did not find any surrounding nebulosity. In the majority of these cases they found no sign of the galaxies which the ground-based work had shown around the QSO. What they did see was a number of very faint little knots around. I think that he was going to talk about this in Pasadena, but he did say that this would be worth mentioning at the Tenerife meeting.

Question

(Longair): I would like just to confirm this expression of the difficulty of seeing some of the galaxies underneath some of the bright quasars and radio galaxies. I do not want to pre-empt what I am going to say in my own contribution, but some of our *HST* images are showing that what you see in some of these radio galaxies is not the stellar population, but that you are seeing disturbed regions, ionized gas, and all sorts of other things. I think it is very difficult indeed to be absolutely convinced that you see underlying galaxies in the optical observations. I think that, in the infrared, we have got a much better chance of seeing the underlying stellar populations. In my own contribution I am going to be adopting a point of view which Geoff will not particularly like, of maximal conservatism. What I can show there in the samples that I am going to be showing is that indeed this conventional picture can be made to work in a fairly convincing sense. That most, probably all, of the redshifts on the 3CR catalogue do have the sort of cosmological redshift which one nominally would expect. I am not saying that this is correct, I am just saying that I shall try to show that there is a very conventional way of seeing how all this hangs together. But, again, it may all be wrong.

Question

(R. Watson): Everybody has been talking about *Hubble Space Telescope* images, and you show this morphological evidence of associations of quasars with galaxies. Has nobody tested these with the *Hubble Telescope*?

Answer

(G. Burbidge): No, this is part of the problem I mentioned. In fact, at the IAU General Assembly, an amateur from the audience asked why it was that none of these Arp objects had been looked at with the *Space Telescope*. The answer, I think, is clear. As someone who once used to schedule large telescopes at Kitt Peak, I know perfectly well that people do not want to get into this field. Whatever else you say, I think that what happened to Arp was very, very bad and a lot of people know about it. You have got to put that into the equation when you talk about the direction of research. You cannot deny it. Unorthodox things: young people do not want to do them, there is no money in them, no position in them, it is very difficult to get them through. As one of the people who is often appealed to by people who want to do unpopular projects, I know what I am speaking about.

Question

(G. Tenorio-Tagle): I would like to ask why all that ejected matter is always flying away from us. Why is it that we do not see blueshifted quasars?

Answer

(G. Burbidge): This question of why they are not coming towards us was one which was raised a long time ago. In fact, Margaret and I discussed it extensively in our book in 1967. If velocity were the cause of the redshift, you would expect to see very large numbers of blueshifts. So what you conclude from this is that, if indeed they are being ejected, as I am trying to argue here, then the shifts in the spectra are not due to the Doppler shift at all; they are something else. That is a very old argument.

Question

(C. Benn): You mentioned redshift periodicities in passing. These have a long history. Which of them do you think are particularly posing a challenge for the standard model, or for any model? Do you have any particular ones?

Answer

(G. Burbidge): First of all, of course, there is a renewal, I believe, of the idea that the pointed pencil-beam observations of normal galaxies are showing the effect again: the stuff that Broadhurst and his colleagues have been talking about. But as far as the QSO periodicity is concerned, or the objects related to QSOs, I can show you, if the Chairman will let me, one or two more viewgraphs. That is entirely up to you. I have some here, but I do not want to go on more than you expect. However, the best way to do it is just to show you some distributions.

Question

(Rees): Could you also comment about whether you believe Tifft, because one of my worries about all these anomalies is that, as data accumulate, one is bound to find more and more things that look odd. But, the question is, do they hang together in any self-consistent picture? Because, if they do not, they add no cumulative weight to the story. I would like to know if you think that the Tifft anomalies should be taken seriously, and, if so, whether or not they are consistent with the others which you find.

Answer

(G. Burbidge): There are a series of redshift anomalies that you mentioned, and we might as well run through them. There are the Tifft anomalies, which appear in normal galaxies with a periodicity of about $36 \, \text{km s}^{-1}$, something like that. These have been looked at in great detail by Napier and Guthrie, in Edinburgh and Oxford, and there have been two or three papers in *Monthly Notices* which have confirmed these results. I have not seen any papers in the literature which explain why they are wrong, and in fact they have done them with great care. Presumably

Malcolm Longair may have something to say, since he used to direct these people, I believe, but this work is in the literature and, as far as I can tell, a referee found it acceptable; let me put it like that. That basically confirms the Tifft results for galaxies in the supercluster region.

This is a small effect; it is a very important effect, but it is a very small effect; also it will not have any repercussions as far as the results which Allan Sandage has been obtaining, for example. It is a very small ripple on the system and, since it is a differential redshift, this is, as far as I can tell, acceptable, but I have not done any work on this. You have these pointed-beam observations where they find an apparent periodicity. Broadhurst and his colleagues had written another paper, and you probably know more about that than I do. I have only been concerned with the stuff that I know best, and that is the stuff concerned with objects which have been picked out because they have spectra of QSOs, or spectra with characteristics similar to QSOs: for example, broad emision lines, and so on and so forth, so that you are looking at hot gas under these conditions and you are not looking at stars.

I first found periodicity in the data in 1968, which is a long time ago. O'Dell and I did a very detailed analysis of this in the next year and confirmed the effect. Originally these were 47 QSOs and 25 non-QSO emission-line objects out to a redshift of about 0.6. By 1990, when Del Hewitt and I published another catalogue, it had grown into 900 objects and 600 with redshifts out to 0.2. With these we see a gigantic peak at $z = 0.06$. The scale here [on the viewgraph] is not adequate to look at the other bands, but I will show you one other thing.

This larger sample has been subject to a detailed analysis by Narlikar and Das Gupta, who published a paper on this in Astrophysical Journal last January, about a year ago. They indeed confirmed the reality of the periodicity because within one band, between 0.056 and 0.065, in other words 0.01 interval in redshift, there were 90 objects out of a total of 500; I mean 90 objects in one band.

Hanbury-Brown said, 30 years ago, that the bright radio sources tend to be concentrated in the super-Galactic region, and, of course, a lot of them are associated with local systems. Shaver recently wrote a paper in which he said the same thing, and many of those are comparatively local and associated with local systems.

Question

(H. Castañeda, IAC): It is not so strange, when you are observing galaxies and quasars, that, by chance, you would expect there to be some associations between at least some of the quasars and galaxies. So, if we extend the survey, for example, for the next ten years, could you predict by probability how many quasars should associate with galaxies? Secondly, even if one can understand that, by chance, there should be association between at least some quasars and galaxies, I find it very hard to explain, by probability, finding an alignment between, for example, some feature of the galaxy and the position of the quasar. Could you comment on that?

Answer

(G. Burbidge): In answer to your second comment, yes, of course, I agree that, if you see alignments, as I showed, for example, in the case of Mrk 205, that can only, as far as I am concerned, add to the likelihood that you are looking at a real effect.

As far as doing surveys is concerned, for fainter galaxies the problem is, of course, that the best chance you have at finding these effects is when the surface density of the two populations you are comparing are comparatively low. If the surface densities get very high, then you will get a very large number of chance effects, and it is almost impossible, with the present techniques, to dig out the real ones from the chance ones. There are examples of surveys where people have looked and not found the effect, and I suspect that they were comparing a population of QSOs with a population of galaxies, with the two at different average distances.

The Shapley–Ames galaxies are out to a magnitude of about 12.5, which means that they are only out 100 or 200 Mpc away. The *Lick Catalogue* goes out to redshifts of about two-tenths, and there you are looking at somewhat more distant objects, out to whatever a redshift of two-tenths corresponds to, and so on. What I noticed, and what I found so striking, about these recent investigations of Bartelmann and Schneider, is that, as they looked further out, they found the same effects with a comparatively small number of radio-chosen quasars, because it is a good sample (the Stickel and Kühr sample); I think there are 300 radio QSOs, well distributed over the sky. I think, though, that it would be very hard to do unless you chose very carefully to do this. All I would claim at the moment is that we have a good case for the radio QSOs. This may, or may not, extend to the optical QSOs; I simply do not know, but, of course, the radio QSOs only form 1% of the total number of QSOs.

Question

(Rees): Before closing, I would like to say that I think, Geoff, when you said that you felt that young people were somewhat discouraged from working on this subject, it seems to me that they probably do not recoil from the wish to discover something fundamentally new. It is simply that they judge that the prospects of discovering something fundamentally new are slightly greater by adopting other lines of approach. Does not one have to trade off one's estimate of the probability of success against the pay off if one does succeed, and it is really in that trade off that people are making this judgement?

Answer

(G. Burbidge): I would agree, but I would also think that the very large-scale structure is not going to lead a lot of them to great positions, one or two perhaps, but there are very large numbers of people working in a few areas and the distribution is very bad. If I am right, the field is worthy of some effort, not no effort at all. If the other thing is right, it is worthy of 90% of your time, but not 99.5% of your time. That goes particularly if you want to work in a field like non-baryonic matter, or something similar, it seems to me.

Question

(E. Gaztañaga): Just to comment on this discussion. I work on large-scale structure and have been to some of your talks and also some of our talks. The reason why I do not work in your field, just explaining my situation as a young researcher, is that I do not see a clear theoretical model that explains all these strange phenomena. I believe that there are many phenomena in physics that we cannot explain, in many branches of physics. Usually research concentrates on things that make sense when you put it all together and for which you have a priority. I have to complain about this kind of paper, or make a criticism, that you explain all these strange phenomena, or I should say un-explain them, and, even if I believe that these truly represent different phenomena, you do not usually put forward a theory, a paradigm with which to understand this. And this is why I do not ...

Answer

(G. Burbidge): In this particular talk I have been very hasty. Hoyle and I have just completed a paper in which we have put forward a theory. I still have a problem with your method in general, because what it really says is that we have got to have a theory ready to explain the phenomena before we can discover them. We had better get things in the right order. I do not think that Newton did things that way. There are different approaches one can take, but it seems to me, as Allan said in Chapter 1, that there are lots of phenomena in the sky that one cannot understand. Some people work on them, even though they do not understand them. We can perfectly well argue that, well, we should not work on stars for the time being, because we do not know a theory understanding why they shine. If you took that point of view in the 1920s you probably would have made a good engineer, or something, but...

Comment

(Gaztañaga): But we have to have a lead. I agree with you, but as Sir Martin Rees says, we have to have some probability distribution that we have a systematic understanding of what is going on.

Question

(G. Burbidge): Do you think you have a systematic understanding of large-scale structure?

Answer

(Gaztañaga): No, that is why I work on it, but I think I have a framework to work in...

Question

(G. Burbidge): Do you think you have a theory already waiting?

Answer
(Gaztañaga): No, but...

Question
(G. Burbidge): Do you think you will ever find out by your method?

Answer
(Gaztañaga): Yes, I hope so...

Comment
(G. Burbidge): Hope springs eternal!

3 Ω, dark mass, and Galactic history

Donald Lynden-Bell

3.1 Introduction

A crucial question that remains to be answered is whether there is sufficient density to close the Universe. Here I review ways of determining an answer, with emphasis being given to those places where the present consensus may be proved wrong. New work, for instance, on the origin of inertia shows that Mach's principle is satisfied in general relativity, but only if the Universe is closed, i.e. provided $\Omega > 1$.

In addition, speculative work is described, aimed at determining the orbits of past satellites that merged with the Galaxy after its initial formation. Following these precepts, proper motions have been predicted for 22 globular clusters and spheroidal galaxies. These are large enough to be determined using current techniques. Many of the predictions have more than one alternative. When the method is applied to the orbit of the Magellanic Clouds, it gives a motion in agreement with the observed proper motion of the Large Cloud. The possibility that the orbits, sizes, metal abundances, and merger times of the satellites that merged into the Galaxy can still be deduced gives new impetus to the drive to measure the proper motions of the globular clusters.

3.2 A review of methods for determining Ω

The critical density ρ_c is defined in terms of Hubble's constant H_0 by

$$\tfrac{8}{3}\pi G \rho_c \equiv H_0^2 = [100h \ (\text{km s}^{-1} \text{Mpc}^{-1})]^2. \tag{3.1}$$

Ω is defined by

$$\Omega = \rho/\rho_c, \tag{3.2}$$

where ρ is the actual mean density.

A Friedmann Universe is closed provided $\Omega > 1$. Eight considerations, each of which bears on the value of Ω, are listed below.

(i) If gravity has been diminishing the Universe's expansion rate, then the deceleration parameter q_0 should be positive. For a Friedmann Universe $q_0 = \frac{1}{2}\Omega$, but current determinations of q_0 are rather weak and give values in the range between 0 and 1. The main problem is the determination of the intrinsic evolution of the sizes, or the luminosities of the objects measured. These must be taken out before the geometrical factor giving q_0 can be found. No definitive way of using quasars or radio sources to do this has yet been found, though Kellerman's pioneering attempt (Kellerman 1993) shows what may eventually come from these methods.

(ii) Cosmic nucleosynthesis is now a mature field. There are really three important abundances predicted by the hot Big Bang. ^{4}He is not sensitive to the baryon density if Ω_B, i.e. ρ_B/ρ_c is close to one. The predicted helium abundance by mass is then about 27.5%. However, the observed primeval helium abundance is significantly less, with most observers quoting 23.5±1%. This discrepancy is thought to be highly significant. To move the predicted helium abundance down into consistency with the observations, one needs $\Omega_B \simeq 1.5 \times 10^{-2} h^{-2}$.

At such a low value, cosmic nucleosynthesis predicts significant deuterium and ^{3}He abundances. Whereas these can be destroyed in stars, no other source for the observed deuterium in both the solar system and the interstellar gas has been found. Thus, forcing the helium abundance off the high plateau gives a nice explanation for the observed high abundance of deuterium. Furthermore, there is quite good evidence for the low primeval ^{7}Li abundance first suggested by Spite and Spite (1982), and that too is nicely predicted by this Ω_B. This is why, for many years now, this value has been accepted as correctly determined. It was thought that three-figure accuracy might be attainable for the observed helium abundance, but now it has been found that crucial atomic data are not well enough calculated. The LS coupling scheme which was assumed has been found to be an invalid approximation for the relevant transition. This may generate a 5% error, so 23.5 ± 1 should probably be changed to 23.5 ± 2.2. While it is interesting to speculate that another factor of just over 2 in the errors might just allow a primeval helium abundance at the top of the plateau, nevertheless, most astronomers are not that radical. Furthermore, such a move to Ω_B near one would destroy the beautifully consistent picture of the cosmic synthesis of D and ^{7}Li which provided such good confirmatory evidence.

In conclusion, the value of Ω_B now rests primarily on D + ^{3}He and ^{7}Li, with ^{4}He being consistent. There is evidence that the initial abundance of ^{4}He is widespread throughout the Universe. We cannot yet say the same for the others. There is a crucial test of the low cosmic abundance of ^{7}Li. Because ^{6}Li is more fragile, any burning of ^{7}Li inside stars should totally eliminate the ^{6}Li. The finding of ^{6}Li in the stars that have the low cosmic ^{7}Li abundance would confirm that the low value is not the result of stellar processing.

The main strength of the argument from the high D abundance, in both the solar system and the interstellar medium, comes from the lack of any other viable way of making a good quantity of deuterium, without over-producing other light elements.

So, on current data

$$\Omega_B = 1.5 \times 10^{-2} h^{-2} \tag{3.3}$$

is well supported, but it is well worth watching both the interpretation of the observed ^{4}He lines and the attempts to determine ^{6}Li abundances.

(iii) Since Mach's principle seldom enters in other discussions of Ω, I shall give it some prominence here.

Newton postulated absolute space as the arena in which all dynamics occurs. To Newton, absolute space was not the local coupling of the rest of the Universe to the subsystem under study; it was rather the arena in which all the laws of nature played their roles. The arena was there whether or not matter occupied it, just as a theatre is there whether or not there is a play being performed.

Leibniz, Berkeley, and Huygens were all highly critical of Newton's absolute space. Leibniz wished to regard it as a useful invention, like a Cartesian coordinate frame, brought in to simplify calculations of the multitude of interactions on which the relative motions of bodies depend.

However, although there was no test for an absolute velocity through space, nevertheless, Newton provided good arguments for absolute rotation, which were not adequately answered by those of his time who believed that all motion is relative, i.e. the motion of bodies relative to one another. Newton demonstrated that the concavity of the surface of water, rotating in a bucket, was the same whether or not the bucket itself rotated. The concavity depended only on the rotation of the water; the rotation or non-rotation of the bucket was irrelevant. Generalizing from this, Newton (1686) deduced that rotation was absolute, and this view was not seriously challenged for two centuries. Then, Mach (1883) pointed out that Newton's experiment merely demonstrated the irrelevance of the motion of the bucket. It did not show that the non-rotating inertial axes of dynamics were not determined by the distant stars or galaxies. No one was competent to perform the experiment of removing them to demonstrate that the dynamics would not be changed. More provocatively, Mach asked whether the result of Newton's experiment would have been the same 'had the bucket been massive and many leagues thick'. The prescience of this remark became obvious when Thirring (1918) deduced from Einstein's general relativity that inertial frames within a sphere rotating with angular velocity Ω_s rotate relative to those at infinity at the rate $\omega = \frac{4}{3}(\psi/c^2)\Omega_s$, where ψ is the gravitational potential due to the sphere. Thus, Mach's question was vindicated, but, more importantly, *Newton's experiment does not give the rotation of absolute space.* Furthermore, since $\psi = GM/r$, such effects depend weakly on distance, so it is quite possible that inertial axes are determined cosmologically from masses in the depths of space.

Mach's principle states that the local inertial frame is determined by some weighted average of the distribution and motion of the mass of which the Universe is composed.

Although Einstein was much influenced by Mach and hoped to build a new mechanics in which Mach's principle was automatically fulfilled, nevertheless general relativity is too general to do this. It contains Minkowski space as a solution and also spaces close to Minkowski space which are almost flat but contain some bodies with unmodified dynamics. In Minkowski space, unaccelerated frames can be read directly from the metric as those frames in which the light cones are not bent or twisted – as they would appear relative to accelerated or rotating frames. In Minkowski space these inertial axes do not arise from mass: there is no mass for them to come from; they are clearly part of the metric and more generally may be thought of as being determined by the boundary condition that space should be flat at infinity. In general relativity, local inertial frames can be determined directly from the metric: there is no need of any new field, as in Brans–Dicke theory; but, if inertial effects come from spacetime alone, then Mach's principle is false and some possibly modified form of Newton's absolute space is the source of inertia.

To evade this argument, most Machians today believe that not merely inertial axes but also the metric of spacetime arises from some weighted average of the distribution and motion of mass-energy in the Universe. Einstein's equations must be solved under appropriate boundary conditions – the differential equations do not of themselves contain all the physics: only those solutions with the correct asymptotic behaviour are valid physical solutions; the others are mathematical curiosities, like the solution $\psi = xy$ to Laplace's equation, or the negative-m Schwarzschild solution.

With this viewpoint, Katz, Bičák, and I set out to find out what averages of the distributions and motions of the mass in the Universe should be taken in determining the local inertial frames for perturbed Robertson–Walker universes (Lynden-Bell, Katz, and Bičák 1995). Our results, for $\Lambda = 0$ cosmologies, can best be described for perturbations constant on spheres. Then, we can write our results explicitly, although more powerful, more general results were also derived. The angular velocity of the inertial frames is derived from the dragging potential $\omega(r, t)$ which is related to the angular momentum within r, $\mathbf{J}(\mathbf{r}, t)$, by

$$\omega(r, t) = \frac{2G}{c^4} \left[\frac{\mathbf{J}(\mathbf{r}, t)}{r^3} W + \int_r^\infty \frac{W(r', t)}{r'^3} \frac{\partial \mathbf{J}(r', t)}{\partial r'} dr' \right] + \omega_0. \qquad (3.4)$$

Here, the weighting function $W = 3r^3 \int_r^\infty e^{\lambda+\nu} r^{-4} dr$, where λ and ν are the functions appearing in the unperturbed metric $ds^2 = e^{2\nu} dt^2 - (e^{2\lambda} dr^2 + r^2 d\hat{\mathbf{r}}^2)$ and $d\hat{\mathbf{r}}^2 = d\theta^2 + \sin^2\theta \, d\phi^2$. ω_0 is a constant to be determined by boundary conditions and, for an infinite Universe, the 'natural' choice is to take it as zero, corresponding to the idea that there is no inertial frame dragging at infinity. However, this is *not* a Machian choice, since it depends on spacetime *not* on the distribution of masses.

A remarkable property of the equation is that it contains no time delays. The angular velocity of the inertial frames is determined from the current distribution of angular momentum. This result was first discovered for a single collapsing shell by Lindblom and Brill (1974). It stems from the fact that angular momentum conservation arises from integrals over a super-potential, so laws such as eq. (3.2.4) can be written for different time scales, just as the charge within a closed surface in electricity can be written as $4\pi Q = \int \mathbf{E} \cdot d\mathbf{S}$ in any Lorentz frame. Lynden-Bell, Katz, and Bičák (1995) also give equations relating ω to the rotations $\Omega_s(r, t)$ of the different mass shells. In *closed* universes there is no preferred choice for ω_0, but the relative angular velocities of inertial axes (or rather their dragging potentials) can be written in a truly Machian manner, independent of what axes one chooses:

$$\omega(r, t) - \omega(r_*, t) = \frac{2G}{c^2}\left[\frac{\mathbf{J}(\mathbf{r}, t)W^*}{r^3} + \int_r^{r_*} \frac{W^*(r', t)}{r'^3}\frac{\partial \mathbf{J}(r', t)}{\partial r'}dr'\right],$$

(3.5)

where $W^* = 3r^3 \int_r^{r_*} e^{\lambda+\nu} r^{-4} dr$.

In closed universes there are two antipodal points where $r = 0$ and it is better to use the angle around the Universe, χ, as the independent variable since there are two spheres corresponding to a given r. To avoid divergences in eq. (3.5) when $r = 0$ for a second time, the angular momentum of such a closed Universe must be zero. This is a special case of a more general theorem which may be proved for any non-linearly perturbed closed Universe with hyperspherical topology. Lynden-Bell, Katz, and Bičák (1995) show how this theorem follows directly from the existence of a superpotential and go on to demonstrate that ω itself is directly a weighted average of the rotations of the different mass shells, $\Omega_s(r, t)$, in any closed Universe; but this is not true of open universes where the relationship depends on the boundary condition. Mach (1883), Einstein (1918), Bondi (1952), and others have all drawn attention to the difficulty that all Machians have with Minkowski space, or almost Minkowski space, in which inertial axes are determined by the light cones and inertia is almost unchanged, although there is no mass, or almost none. Einstein invented the cosmological constant in the belief that the cosmological solutions would then be closed, like his static model, and the need for all boundary conditions at spatial infinity would be eliminated. However, de Sitter's solution showed open models were still possible, so boundary conditions still entered for them. However, if we postulate that the only physical solutions of Einstein's equations are those with no spatial boundaries, then the conundrum of inertia in an almost flat, almost empty space is beautifully removed. If a is the maximum radius of a closed Universe and M is the maximum of the mass-energy, then $2GM/(ac^2) = 1$. (The total rest mass $\mathcal{M} = \pi M/6$.) If we compare universes of different M, then as M is decreased, a is likewise decreased, and the minimum mean density $M/(\frac{4}{3}\pi a^3)$ is proportional to M^{-2}. Thus, as matter is 'removed' from a closed Universe, its radius of curvature becomes smaller, its density becomes large, and it becomes less and less like Minkowski space. In the limit as $M \to 0$ the Universe shrivels to a point and the time between the Big Bang and the Big Crunch likewise approaches zero. Thus in the limit there is no space, no time, and no inertia. Mach's principle is thus beautifully fulfilled in the closed universes. If, however, one performs the same thought experiment on the open universes, the period of gravitational retardation of the expansion becomes shorter as the mass is decreased, until in the limit one obtains Milne's Universe which is a re-interpretation of Minkowski space. Here the inertia all arises from space, and so Mach's principle is violated. In conclusion, Mach's principle, which embodies the idea that all motion is relative, is only fulfilled in the closed Robertson–Walker universes, so we deduce that $\Omega > 1$.

(iv) Inflation and the prediction that Ω should be very close to 1 has received much discussion in the literature, so I shall only add a few comments.

Inflation is invoked to solve the flatness problem, that the current curvature of the Universe is so small, or that the kinetic energy of expansion is so close to the potential energy. These problems become more acute as one extrapolates back to near the Big Bang because they must then be equal, to amazing accuracy. Furthermore, we can see today regions of the Universe which on a Friedmann model could not have communicated causally with one another at the look-back time at which they are seen. Nevertheless, we see large-scale homogeneity between such regions. How can such homogeneity be achieved? The singularity theorems of Penrose (1965), extended by Hawking, depend on the energy condition $\rho c^2 + 3p > 0$. While this is true of all normal substances, the early stages of the Universe are not closely related to our normal experience. Furthermore, a massless scalar field, if such exists, will under such conditions have an energy tensor for which $p = -\rho c^2$. Such a substance will disobey the conditions under which the singularity theorems were proved, so it raises again the question of whether there has to be a Big Bang singularity. Nevertheless, with T^μ_ν of that form we have $T^\mu_\nu \propto g_{\mu\nu}$, so the Einstein equations reduce to de Sitter's form during the inflationary phase. With inflation, the Universe expands so rapidly that all length scales, including the spatial curvature of the Universe, are ironed out. Similarly, in such a cosmology causal communication between regions which are now well separated was possible at early times. These are the attractions of inflation, but we must balance them against the extrapolation to an unknown substance with new and esoteric properties and against the fact that, for such a substance, gravity is actually repulsive. On reflection, this last may be an advantage, since something must have caused the observed expansion of the Universe, so what better than a phase with repulsive gravity. Nevertheless, a theory in which inflation is an inevitable consequence of some deep understanding of the world, rather than merely an item on the menu from which the cosmologist can choose his Universe, would constitute a real advance. It is worth remarking that Hoyle's creation field had the properties of an inflationary substance and differs from more modern theory mainly in its interpretation. The main prediction of inflation is that Ω should be very close to 1, since all curvature should have been inflated away. Furthermore, if the Universe started from a fluctuation, it is conceptually far simpler to have that fluctuation finite rather than infinite, in which case $\Omega - 1$ should be very small and positive. Such a conclusion is in conformity with our $\Omega > 1$ conclusion from Mach's principle.

(v) Ω can be obtained from large-scale streaming motions. In the linear adiabatic theory of the growth of small perturbations in the Universe, one finds that the current velocity field, $\mathbf{v}$, is related to the acceleration field, $\mathbf{g}$, by

$$\mathbf{v} = \tfrac{2}{3}\Omega^{-3/7}\mathbf{g}/H, \tag{3.6}$$

where H is Hubble's constant. Notice that apart from details concerned with the growth rate, this really says that the velocities are accelerations multiplied by the time since the Big Bang. However, the acceleration is related to the density field $\rho_0 \delta$ via Poisson's integral,

$$\mathbf{g}(r) = G\rho_0 \int \frac{\delta(\mathbf{r}')(\mathbf{r}' - \mathbf{r})}{|\mathbf{r}' - \mathbf{r}|^3} \mathrm{d}^3 r'. \tag{3.7}$$

It is commonly assumed that galaxy densities and the density of all matter are proportional, but a lesser assumption is that the excess luminosity density $\delta\rho_L$ relative to the mean ρ_L is linearly related to the excess density relative to its mean, so that

$$\frac{\delta\rho_L}{\rho_{0L}} = b\,\delta = b\frac{\delta\rho}{\rho_0}, \tag{3.8}$$

where b is the bias factor, which is unity if light traces mass in the sense that they are proportional. With this assumption,

$$\mathbf{g}(r) = \frac{G\rho_0}{b\rho_{0L}} \int \frac{\delta\rho_L(\mathbf{r}' - \mathbf{r})}{(\mathbf{r}' - \mathbf{r})^3} \mathrm{d}^3 r'. \tag{3.9}$$

Now $\delta\rho_L$ and ρ_{0L} are, in principle, observable, but owing to ambiguities in H, we can only really determine luminosities/D^2, and hence $\delta\rho_L \times D = \delta\rho_L/H$. Furthermore, $G\rho_0 = \Omega H^2/(8/3\pi)$, and so

$$\mathbf{v} = \frac{1}{4\pi}\frac{\Omega^{4/7}}{b}\frac{H}{\rho_{0L}} \int \frac{\delta\rho_L(\mathbf{r}' - \mathbf{r})}{|\mathbf{r}' - \mathbf{r}|^3} \mathrm{d}^3 r'. \tag{3.10}$$

Now, even without any knowledge of Hubble's constant, $\delta\rho_L/\rho_L$ can be found by studying luminosities and redshifts, and at the level of linear theory $\delta\rho_L$ is small, so we may substitute the unperturbed values $\mathbf{v}' - \mathbf{v}$ for $H(\mathbf{r}' - \mathbf{r})$. Since $(\mathrm{d}^3 r')/(|\mathbf{r} - \mathbf{r}'|^3) = (\mathrm{d}^3 v')/(|\mathbf{v} - \mathbf{v}'|^3)$, which may be evaluated from redshifts, the only unknown on the right is $\Omega^{4/7}/b$. Thus this quantity can be measured if peculiar motions $\mathbf{v}$ can be measured. There are two ways of doing this:

1. From secondary distance indicators, such as the Tully–Fisher, or D_n-σ relationships, which give relative distances independent of redshifts.
2. From use of the Sun's motion relative to the Cosmic Microwave Background. This can be corrected to give the motion of the Local Group of galaxies, and in this case the integral in eq. (3.10) can be

evaluated directly since it is the dipole of the extragalactic light distribution over the sky once **r** is taken as the position of the observer. The $1/r^2$ weighting is automatic in the observed light distribution (Lynden-Bell *et al.* 1989).

The above methods, applied to the *IRAS* data, give values of $\Omega/b^{7/4} \approx 0.8 \pm 0.2$. Optical determinations give lower values closer to 0.5 (Hudson 1994), but there is evidence that the bias factor is larger for optically selected galaxies. The methods of determining peculiar velocities get progressively less accurate at larger distances and, nearby, it is hard to get a good statistical sample. For these reasons, another method of analysing redshift data has recently come to the fore. The method involves comparing the spectrum of fluctuations angularly over the sky with those determined in depth from redshift distortions of the power spectrum. It gives values of $\Omega/b^{4/7}$ close to 1.0, without the need for peculiar velocity measurements (Fisher, Scharf, and Lahav 1994; see also Fisher *et al.* 1995).

(vi) Some years ago I pointed out that by generalizing the timing argument for the relative motion of M31 and the Milky Way to include a third body, it was possible to measure both the mass of the Local Group and the time since the Big Bang (Lynden-Bell 1981). Peebles, in a series of papers (e.g. Peebles 1994), has provided a nice method of doing this via his action principle. He gets low values of $\Omega \sim 0.2$, but it is difficult in this method to allow for large haloes of dark matter and their dynamical friction. It now seems clear that these methods give smaller Ω values, so it may be some time before agreement is reached by astronomers using these different methods.

(vii) The spectacular arcs caused by the gravitational lensing in large clusters of galaxies provide a new way of determining their mass distributions, including dark matter. With X-ray emissions and velocity dispersions of the galaxies' motions also available, there is great hope that differences between methods will soon be ironed out, so it is likely that these large clusters will soon be well understood. Currently, White *et al.* (1993) have emphasized an important conundrum. If Ω is near 1, then the X-ray emission from the Coma cluster coupled with the optical galaxy masses gives a baryon fraction by mass in Coma of at least four times the mean baryon fraction determined from cosmic nucleosynthesis. Gravitational instability theory does not naturally give such a segregation of baryonic matter. Thus, it seems likely that many of the important problems can be studied using the large clusters, and both gravitational lensing and the Sunyaev–Zel'dovich effect have the possibility of giving good values for Hubble's constant. If I had to choose one key problem, it would be the problem of dark matter and I would study it in the big clusters of galaxies.

(viii) Finally, estimates of Ω in baryons and the dynamical Ω can be made from the intervening Ly α clouds seen in absorption in the spectra of quasars. At present, these Ω estimates are not very useful, since the ionization corrections have large possible errors. However, most matter in the Universe may once have been in the form of Ly α clouds, so improvements in ionization correction determinations may well lead to interesting Ω estimates in future.

3.3 New probes of Galactic history

The evolution of ideas concerning the origin of the Milky Way has been reviewed elsewhere. Here, I wish to update some speculative work that has been done on streams among the globular clusters. The basic ideas are that, after its initial formation, a number of smaller objects merged with the Galaxy, and these were tidally torn up as they passed pericentre, leaving their torn-off globular clusters and other tidal debris in the planes of their original orbits about the Galactic Centre. Whereas the satellite itself may have long since merged with the Galaxy under the influence of dynamical friction, the debris, being of low mass, will nevertheless be left in the original orbit, like a ghost haunting the past abode of a murdered victim. On a smaller scale, we see something rather like this in the meteor streams that still delineate the orbits that were once occupied by their parent comets in the solar system.

Our aim is to find globular clusters, or dwarf spheroidals, in the outer halo of the Milky Way, which lie in planes through the Galactic Centre and follow the same orbits as one another. By studying their ages, metal abundances, etc. it might then be possible to age date, or to order, the possible late mergers by which the outer halo of the Galaxy was made.

Considering first globular clusters etc. at large distances from the Galactic Centre, their observed line-of-sight velocities (after correction for the motion of the Sun relative to the centre) will be almost equal to the components of their velocities radially from the centre. Objects in the same orbit share the same energy and angular momentum, as well as the same orbital plane, so

$$\epsilon_r = \frac{v_r^2}{2} - \psi(r) = \epsilon - \tfrac{1}{2}h^2 r^{-2}, \tag{3.11}$$

where ϵ is the specific energy of all members of the stream and h is their specific angular momentum; $\psi(r)$ is the Galaxy's gravitational potential which, for an infinite massive halo model, would take the form $\psi = -V_0^2 \ln r$, with $V_0 \simeq 220\,\mathrm{km\,s^{-1}}$. We take a finite halo and find that

$$\psi = -V_0^2 \ln\left\{\left[\sqrt{1 + (r/r_\mathrm{h})^2} - 1\right] r_\mathrm{h}/r\right\}, \tag{3.12}$$

where $r_\mathrm{h} = 80\,\mathrm{kpc}$ fits the data slightly better, so we use this formula. Taking a set of objects that lie on a great circle in the Galactocentric sky, we get v_r from the observed radial velocities and $\psi(r)$ from the observed distances. When ϵ_r is plotted against r^{-2}, those objects that belong to a stream should lie on a straight line of gradient $-\tfrac{1}{2}h^2$ and intercept ϵ. Thus, the angular momentum and energy of the orbit can be deduced from radial velocities and distances. This new knowledge, together with the plane of the sky in which the objects lie, allows us to predict the proper motions $\mathbf{h}/r^2$ relative to the Galactic Centre up to an ambiguity in the sense of motion along the orbit.

Once it is realized that transverse motions can be predicted by this method, it is clear that we can correct the observed radial velocity for that part of the motion that is transverse, as seen from the Galactic Centre, but which still contributes to the radial velocity as observed from here. It is then no longer necessary to restrict these considerations to the very distant halo where the solar line of sight almost coincides with the Galactocentric one. However, the correction has two possible values for each object, depending on the sense of the orbital stream motion along its orbit. Rather than regarding such an ambiguity as a defect, we can turn it to our advantage because, with one sense of motion, the different objects will lie more closely on a straight line in the ϵ_r, r^{-2} plot than they do in the other, for which the corrections will have been made with the 'wrong' sense. This difference enables us to resolve the ambiguity and deduce the sense of motion too. Thus, for a number of cases it is possible to predict **h** and the proper motion without an ambiguity of sign. Unfortunately, in spite of this, there is still considerable ambiguity in choosing clusters that lie together on great circles and obey our other criterion, so there are often several alternative predictions. Those interested in the method should read our paper (Lynden-Bell and Lynden-Bell 1995) where possible streams and predicted proper motions are given in the tables. An interesting test is the Magellanic Stream, because the proper motion of the Large Magellanic Cloud has been measured. The prediction, based on the association of the Magellanic Clouds with Ursa Minor and Draco (and conceivably Carina) gives $\mu_{\alpha\cos\delta} = 1.5$ milliarcsecond year^{-1}, $\mu_\delta = 0.0$ milliarcsecond year^{-1}, while the observations (Jones *et al.* 1994) give $\mu_{\alpha\cos\delta} = 1.2 \pm 0.28$ milliarcsecond year^{-1}, $\mu_\delta = 0.26 \pm 0.27$ milliarcsecond year^{-1}, which are in as good agreement as the errors of measurement permit.

Proper motions are now determinable for globular clusters; we predict measurable ones for 22 of them. Furthermore, as more proper motions become available, the associations with other objects on that great circle become surer and the ambiguities of the method become less for all the other objects. We conclude that there is still a great future in the measurement of proper motions and we hope that this method will add interest to the hard work of determining them.

Acknowledgements

I thank the organizers and the benefactors of the meeting on which this book is based for making it a most pleasant one and Bernard Pagel for his advice on the origin of the observed elements (Pagel 1997).

References

Bondi, H.: 1952, *Cosmology* (Cambridge: Cambridge University Press), p. 29
Einstein, A.: 1918, Ann Phys, **55**, 241
Fisher, K., Scharf, C., and Lahav, O.: 1994, MNRAS, **266**, 219
Fisher, K., Lahav, O., Hoffman, Y., Lynden-Bell, D., and Zaroubi, S.: 1995, MNRAS, **272**, 885
Hudson, M.: 1994, MNRAS, **266**, 475

Jones, B.F., Klemola, A.R., and Lin, D.N.C.: 1994, AJ, **107**, 1333

Kellerman, K.I.: 1993, in *Observational Cosmology*, eds. G. Chincarini, A. Iovino, T. Maccacaro, and D. Moccagni, ASP Conference Series Vol. 51, p. 50

Lindblom, L., and Brill, D.R.: 1974, Phys Rev, D10, 3151

Lynden-Bell, D.: 1981, Observatory, **101**, 111

Lynden-Bell, D., and Lynden-Bell, R.M.: 1995, MNRAS, **275**, 429

Lynden-Bell, D., Katz, J., and Bičák, J.: 1995, MNRAS, **272**, 150

Lynden-Bell, D., Lahav, O., and Burstein, D.: 1989, MNRAS, **241**, 325

Mach, E.: 1883, *The Science of Mechanics*, Open Court 1942, (London: Nelson 1957)

Newton, I.: 1686, *Principia*, ed. Cajori (Los Angeles: UCLA Press 1934), p. 12

Pagel, B.E.J.: 1997, this book, Chap. 9

Peebles, P.J.E.: 1994, ApJ, **429**, 43

Penrose, R.: 1965, Phys Rev Lett, **14**, 57

Spite, M., and Spite, F.: 1982, Nature, **297**, 483

Thirring, H.: 1918, Phys Z, **19**, 33

White, S.D.M., Navarro, J., Evrard, A.E., and Frenk, C.S.: 1993, Nature, **366**, 429

3.4 Discussion

Question

(Reeves): I would like to ask one question: how much is your conclusion that $\Omega = 1$ comes from the Mach requirement based on the hypothesis that the Universe is simply connected? Is it not true that, if you have a multiply connected Universe, it might be different? Have you looked at the possibility that the condition at infinity would disappear?

Answer

(Lynden-Bell): We have not looked at multiply connected space-times. What we have looked at, so far, are perturbed Friedmann–Robertson–Walker universes, where it is assumed that the basic structure is a simple closed thing. I am not quite sure even about the angular momentum condition on a torus. It seems to me that there are even two sorts of angular momenta on the torus, and I am not even sure about that. So, my answer is that we have not studied that. We have started to study non-linear perturbations in a Friedmann–Robertson–Walker Universe, and there I think there is great hope that we can make very much the same progress. We already know that the angular momentum is automatically zero, even non-linearly, for however large a perturbation you like, leaving the topology the same as a Friedmann Universe. But no, we do not know anything about other topologies, I think it is true to say.

Question

(Rees): Do any of these Machian models produce an anisotropy of inertia, if the Universe had some small anisotropy, either now or developing?

Answer

(Lynden-Bell): Let's put it this way. In one sense, there would not be an anisotropy of inertia at one point. On the other hand, you obviously will get a locally rotating inertial frame here and another frame rotating at a different rate somewhere else, and there would be a scale on which that varies, and therefore you can get an anisotropy on the variation of the inertial frame.

Question

(Rees): What I meant really was, would it be easier to push in one direction than in a perpendicular direction?

Answer

(Lynden-Bell): No, that is not the case. It is the case of certain quasi-Newtonian models of inertia, but the particular model I was talking about, Newtonian-wise, does not even have that. The very simplest things you can write down, in which the inertia merely comes from the things along the line along which you look, does have such an anisotropy of inertia. That has been shown with amazing accuracy by Hughes and Drever (Hughes, V.W.: 1960, Phys Rev Lett, 4, 342; Drever, R.W.P.: 1961, Phil Mag, 6, 683) not to be there, and, as a result, we already know that inertia is incredibly isotropic. You can make theories in which that is true, even quasi-Newtonian theories in which that is exactly true.

Question

(Novikov): Do you believe that inertia is a result of the gravitational interaction of a test body with all other masses of the Universe? Is that correct?

Answer

(Lynden-Bell): My real answer is 'yes'. I think there is no difference between gravity and inertia, in the last resort.

Question

(Novikov): But in this case, it means that if you put any additional masses near some test body, the mass of this test body, as a measure of the inertia, should become greater and greater, but we cannot observe it, or is it not so, even in theory?

Answer

(Lynden-Bell): I have actually been looking at that very recently, and I do not yet know the answer sufficiently well that I would be happy to tell you, but I will tell you what I think the result is. If you assemble a whole load of objects around a given object and you measure inertia from within, you have got a sphere of heavy objects, with an object inside it. If you look at inertia down inside the sphere and you look at it with experiments inside that sphere, it is no different at all to how it ever was. It is just the same. However, if you attach a string and try to pull the string from way outside this heavy sphere, I think it is harder to pull the string. But I am not even sure about that, because strings are slightly awkward in general relativity, and I am not yet quite sure that I am dealing with strings the right way. You know, I would rather have some more practice with strings, so I will answer you in maybe six months. It is a good question. I think it is harder to pull the string at present from outside, but not from inside.

Question

(Sandage): Why is it that you need the proper motions of these globular clusters, again?

Answer

(Lynden-Bell): We can predict them, OK, that is fine, so we do not need them. But you know, a prediction is nothing like an observation. Predictions can be as wrong as wrong. As I say, we have several predictions for a number of these objects. Here we have Eridanus, with Palomar 14 and Palomar 15, and Fornax [shows viewgraph]. You could probably find even in this diagram, yes, here is Palomar 15, classed with NGC 7006, Palomar 13, and NGC 1851, with one gradient here. Here is Palomar 15, here on this line, with quite a different gradient. Which of those are right? I do not know! In terms of proper motions, what I think is so important is that, once you have a proper motion, you know the great circle in the sky on which the object is moving. You can get that even from the Galactic Centre, because you can transform them to the Galactic Centre. You know the great circle and you know the gradient of the line that you are meant to be looking for in this diagram [shows viewgraph]. As a result, you can throw away most of these so-called predictions: you have now quite a definite line. You can look for things on that same great circle, with a particular sense of rotation, because you know which way it is going in the sky and you have a gradient on this line on which all the other ones have to lie. If you only had the proper motion of one of these globular clusters, then you have much greater power in detecting what other ones might be associated with it. That is very important to you, because once you have got a strong selection criterion you throw away most of the non-sense. We have probably three, or four, possible associations each for many of these objects and we do not even know which one is right.

Question

(Sandage): It is only a technical question but, because your field of view is only 3 arcminutes with *HST*, it is going to be terribly hard to find any non-cluster stars in that area, unless you are really in a lousy cluster like Palomar 14. Almost all the stars in the frame will be globular cluster stars.

Answer

(Lynden-Bell): Go to the outside of the cluster and look for distant galaxies.

Question

(van Woerden): How accurate would your proper motions have to be?

Answer

(Lynden-Bell): The Large Magellanic Cloud has a proper motion of about 2 milliarcseconds per year, actually rather less than that, and that is quite a big one. So, one is after numbers of the order of a quarter of that, in order to have a lot of objects that you can do. You can do several objects at the level of the Magellanic Cloud proper motions. If you wanted to do around 20 objects, fairly far out in the halo, where these approximations are possibly better, you would like to do not 2 milliarcseconds per year, but 0.5 milliarcseconds per year. Then one would get a good sample of objects. There are at present around 15 globular clusters for which somebody says they have the proper motion, most of them too inaccurate to do anything with, but those are the sorts of numbers we are dealing with today.

Question

(M. Burbidge): That was the question that I was going to ask. What was the order of the magnitude of the proper motions that you were hoping to...?

Answer

(Lynden-Bell): 0.5 milliarcseconds year^{-1}.

Question

(M. Burbidge): ...that should be do-able in the lifetime of the *Hubble Space Telescope*?

Answer

(Lynden-Bell): I do not know; that is a *Space Telescope* question. Maybe it is the lifetime of the next *Hubble Space Telescope*. But I would not be surprised. After all, *Hipparcos* claims that it will be getting to the sort of accuracy that we need for the Large Magellanic Cloud in three years. I cannot imagine that the *Hubble Space Telescope* is much worse than

Hipparcos; I really do not know, but I do not imagine it is much worse, and I hope that *HST* is going to last for more than three years.

Question

(Longair): To change the subject somewhat, what is your current view on the nature of the dark halo of the Galaxy?

Answer

(Lynden-Bell): If you would like me to go out on a limb, I would say that the world is made of baryons; that all these people who have done nuclear cosmochronology or cosmogenesis have got it wrong; that there are large numbers of things that Allan did not like, made of hydrogen, in the outer parts of galaxies; and that there are a large number of things that are quite small. I think this is probably knockable down with gravitational lensing in the not too distant future. I think it is nearly there and I think that I may well say that that is all nonsense in less than a year's time. That is what I would say at present. I would think, in the end, the dark matter has to cluster on the scale of a few degrees, from the microwave background experiment. It has got to be in those clusters, otherwise we are in trouble in quite different ways. So, the bumps in the microwave background have got to have been non-relativistic back then. I think quite soon we will find difficulties with the neutrino hypothesis, which is the other one I have liked. Other people have given up on neutrinos long ago, but I rather like neutrinos: at least they exist. I think that is a lot better than other things that do not seem to exist, but on the other hand, I do not have Newton on my side. He talked of particles so subtle that we never observed them and Newton dreamed of things that I do not dream of.

Question

(Reeves): Your interpretation of the data by Simon White was different to the one that I thought is generally accepted. Generally, the fact that he finds this high ratio of baryon to total mass is given as an indication that Ω may be less than 1 after all. You have a different interpretation in the terms of a collapsing separation of baryons with respect to dark matter, is that not it?

Answer

(Lynden-Bell): I happen to believe that $\Omega > 1$. You said 'equal to 1'; I believe it is greater than 1. I have been extreme on those views for a long time and I do not change that position. I have seen what I think is an extra symmetry of nature, and it takes an awful lot of really hard experiments before anybody who thinks they have seen an extra symmetry in nature gives up that extra symmetry, which is really the extra symmetry between gravity and inertia. I am not going to give that up lightly.

Question

(Rees): I want to ask about your method of determining Ω from large-scale structure. You mention that the redshift-only method required assumptions about the galaxies being clustered only under gravity. Am I right in saying that it does not matter how the galaxies are clustered, provided the velocities are induced only by gravity? And that is a slightly weaker statement?

Answer

(Lynden-Bell): No, that is not true when you are using only the redshift. That is true, quite possibly, when you are using the method based on distances; then you do not have to assume that the mountains that you have made are nothing. But, when you are using redshift only, you really do use the fact that the mountains that are there have been made by gravity. Am I not right?

Question

(Rees): But if you can infer the mountains from the galaxy motions, without assuming that the galaxy...

Answer

(Lynden-Bell): We have not got the galaxy motions unless we have got distances. We have only got the real velocities, not peculiar velocities.

Question

(Rees): If there were intrinsic clumps that were not oriented towards us, then any anisotropy would be due to the peculiar motions, would it not?

Answer

(Lynden-Bell): I do not think that is enough. I think we still need the fact that the clumps are due to gravity, otherwise we do not know what we are measuring them against. I may be wrong, though I have not thought too hard about it. I think that you need the clumps to be gravitationally induced and not due to some large scattering, or something pushing photons at some distant epoch, or something.

Comment

(Rees): That is true of the galaxy motions, certainly. I am not sure that it is...

Answer

(Lynden-Bell): I am just saying that it was not moved there by something and then the motions were induced by gravity. I am not sure that if I sort of set lumps there and then let it go at some late epoch I would get the same answer.

Question

(Sandage): Let me come back to the last two or three sentences of your contribution. It was like we were in church: that you could get everything about the formation of galaxies. Could you give those last few sentences again. What do you hope to get from all of this?

Answer

(Lynden-Bell): What I was saying was that, if you could find from some method based, like this, on the proper motions, so that you could really be sure that they really were streams of stuff going round the Galaxy. If you could then age date, using the abundances, and the HR diagrams of the different objects in here...

Comment

(Sandage): You can age date right now.

Answer

(Lynden-Bell): I know...

Question

(Sandage): So?

Answer

(Lynden-Bell): It is more subtle than that. You want to know which objects go together.

Comment

(Sandage): But they are all the same age right now, from the known colour–magnitude diagrams.

Question

(Lynden-Bell): You say that of all the globular clusters in the Large Magellanic Cloud?

Answer

(Sandage): Yes.

Question

(Lynden-Bell): Fine, OK, then they are all the same age. Fine. Are all the ones in the Large Magellanic Cloud the same age as the ones in the Galaxy?

Answer

(Sandage): Yes sir!

Comment

(Lynden-Bell): I see, all right. In that case, we do not have any age discrimination, so we cannot tell which ones came in first.

Comment

(Sandage): It does not matter what the age of the mergers was, if you are trying to ask about the early evolution of our Galaxy.

Answer

(Lynden-Bell): It is interesting, in principle, to see whether, in fact, you can get a sequence, a sort of geological sequence in different layers. I would rather like to know if this one came in that way at such and such a time and if this one came in that way, either before or after, and this one came in later...

Comment

(Sandage): I agree, but that has nothing to do with the initial formation of the Galaxy.

Answer

(Lynden-Bell): Maybe not, but what I am interested in is the whole history of the making of the outside of our whole Galaxy, in terms of these different streams that I think may have made it. I cannot tell you that any of this is right; the only confirmation we have at all is that it gives essentially the right proper motions for the Large Magellanic Cloud, which is buried here, and it would be nice to show you the numbers...

Comment

(Sandage): What I guess I am saying, Donald, is that your age-dating will tell you about the mergers now, but they have not merged yet. So it is a prediction essentially of the future evolution of the Galaxy, not of the formation of the Galaxy.

Answer

(Lynden-Bell): No, the point is that these are the ghosts of past mergers. The heavy objects suffered dynamical friction and have since gone down into the Galaxy, but in the case of the Large Magellanic Cloud it is still there, and in the case of Fornax, it may still be there.

Comment
(Sandage): OK!

Answer
(Lynden-Bell): ...But it may long since have been lost!

Comment
(Sandage): OK! OK! You win!

Question
(A. Aragón): Could this method perhaps tell you how many of these globular clusters actually came from a merger and could you perhaps tell them apart if some were formed in a non-merging event?

Answer
(Lynden-Bell): There is something that I think I ought to elucidate here and that is that the new blue clusters in Perseus A seem to have been formed in a merging event, right there in Perseus A. If I am going to be totally consistent, then I have to believe that it is not too infrequently that globular clusters get formed in such merging events, because I think that Toomre's idea that ellipticals in some sense come from mergings of spirals may be correct. We know van den Bergh's objection to that, and it is that there were too few globular clusters per unit light in the spirals compared with the ellipticals. This observation of new globular clusters in Perseus A has removed that fundamental objection. Therefore I have to admit that some new globular clusters may be formed in merging events. Now, it is not those globular clusters that I have been primarily concerned with here. What I am concerned with is, really, that the Large Magellanic Cloud will in future have globular clusters torn off it as it goes into the Galaxy, and maybe one or two of the objects that we already see as Sculptor and certainly Draco and Ursa Minor, which are oriented along the line. They do seem to define the same stream as the mean of the Magellanic Clouds defines and the line in the sky which the Magellanic Clouds define. I am taking it that such objects are debris torn off at some past epoch, maybe. If you look at the internal constitution of Ursa Minor, if you look at the HR diagram, it is mighty different to Draco and it is also different to Sculptor.

You may ask, why? They are also very different to the Magellanic Clouds, and you may ask, why? There are two very good reasons why they might be different. First, we are wrong, they have nothing to do with one another: they happen to be oriented along the right line, but that is just chance, and we have been giving you a nice story that has nothing to do with the real world. The other one is to go and look into the Magellanic Clouds themselves and you will find clusters with very different metal abundances within them. You will find the body of the Magellanic Clouds rather different in its stellar constitution than a number of the clusters that are within it. Is there really a very strong reason why everything has to be of the same metal abundance, of the same horizontal branch type? Allan may well tell us that there is, but, at present, I have not taken that very seriously into account. One of these streams has, I think, nothing but so-called anomalous systems in it, that is, second-parameter systems. On the other hand, the one that goes with Fornax [refers to viewgraph] – I think that these are both anomalous and this one is not. I do not know whether I should be restricting efforts to put these in two groups that are the same chemically. I tried that and it did not seem to work very well, so I decided that perhaps it was best to ignore that data and then come back to it, when we knew whether any of the proper motions were right or not. Is that an answer?

Answer
(Aragón): Probably. Yes.

Question
(H. Zinnecker): I wonder if you cannot get these proper motions and astrometry even from the ground and do not need *HST*? I hear that people are trying to get the proper motions or astrometry anyway for the Sculptor and other spheroidal galaxies. They are perhaps using X-ray sources, which turn out to be quasars in the background of these galaxies, as references.

Answer
(Lynden-Bell): I think there is real hope that you can get them – if I were prepared to wait till after I was dead, and we may know everything after we are dead, so maybe that is not too long. If I were prepared to wait

that length of time then we would certainly know a good fraction of these. I am much encouraged by the fact that there are attempts at those systems you talked about. Most of these attempts show you that the proper motion is in this quadrant rather than that quadrant, and frankly that is not good enough for me. I want to know the line in this quadrant and I want to know the value of the proper motion with slightly better accuracy than that. There are a number that can probably knock down some of our streams already. I guess that that is, in a sense, good news: it is good to be at the level at which you can be knocked down. You cannot win a fight unless there is someone who is prepared to try and knock you down.

Question
(Zinnecker): But one milliarcsecond seems to be possible, I guess, from the ground?

Answer
(Lynden-Bell): One milliarcsecond? I believe very much that it is possible, yes. I am only asking for the order of half a milliarcsecond per year. So, yes.

Comment
(Sandage): You will live more than two years, Donald.

Answer
(Lynden-Bell): I do not know. That is determined 'up there', as you know, Allan.

Question
(Longair): Suppose your programme works perfectly? You will get these orbits and find these objects. Then, presumably you would get a set of times when individual events must have happened which added to build up the halo in the Galaxy. But would you get some information about how much mass accompanied them that actually disappeared? How could you then go further and say, the sequence of events went like this…this is how it was built up?

Answer

(Lynden-Bell): I think you are more ambitious than I am. I am sorry. I would love to know such things. One of the sorts of things that you might try to do that way is, if you look at the real miseries of galaxies (very low-mass systems) they do not have globular clusters. Fornax is not much of a system and it has maybe 5. The Small Magellanic Cloud has 10. The Large Magellanic Cloud has around 20, or 30, something like that. If you see all the proper motions, you need not even assume that they all remain on the same plane; all you need to do is make sure that the orbits are of the same class and that all the three integrals are the same. Once you have got the proper motions of these things, and the proper motions of stars, you could start classifying them all by orbit and by metal abundance and by everything else. I think this is still a big feature in the history of the Galaxy, and this is what I am trying to say. Whether it will all work out beautifully I do not know because the galaxies have been shaking around a bit and the orbits become a bit shaky, and eventually they become this very smooth thing which Allan says the halo was, which I am not sure it is. It may actually be like a ball of wool and be formed by lots of rather narrow streams, which we do not see as narrow streams because we do not have the data to discriminate it.

Question

(Longair): I suppose if we were being maximally optimistic, you could imagine that each of the progenitor galaxies had lots of globular clusters and you could get their velocity dispersions, so you might get the initial mass of the galaxies?

Answer

(Lynden-Bell): I think that is a bit much. I would rather work it out by counting than I would by using that method.

4   Observations of QSOs which are critical for cosmology

E. Margaret Burbidge

4.1 Introduction

Despite the time and effort spent on observational and theoretical studies of quasars (QSOs) during the past 30 years, and the conviction with which many 'experts' claim to know what is going on, the basic physics of their source of energy and continuum radiation, the distribution and velocity field of the emission-line clouds, the several classes of absorption lines seen in their spectra, and their relation to galaxies are still little understood. In this talk, I will discuss some of the problems which I would like to see addressed in the next 30 years. As I am an observational astronomer, I stress the observations which I would like to see pursued. Since my work has mainly been carried out with the Lick Observatory 3-m Shane telescope from the ground, and with the *Hubble Space Telescope* (*HST*) from space, slides that I show will be recent data obtained with these telescopes, and some with the Keck 10-m telescope obtained by my colleagues. I shall concentrate on the absorption-line phenomena in QSOs, because several of the key problems in cosmology relate fundamentally to the physics of the absorbing gas and its location with respect to the source of the QSO continuum radiation.

First, however, I make some general remarks. New discoveries in astronomy have always followed the technology that enables hitherto unexplored regions of the electromagnetic spectrum to be studied, from radio frequencies to γ-rays. Advances in instrumentation and detector technology combine as essential components, giving us an increased capability to explore the Universe. The *HST*, operating above the Earth's ozone layer, has, since the repair mission to remedy its faulty optics, succeeded in providing high-resolution imaging that was anticipated in the planning and design, and ultraviolet spectroscopy of faint objects with better resolution and signal to noise than was possible before the repair mission.

However, the primary mirror of the *HST* has a diameter of only 2.4 m, and the aberrations produced during the figuring of that mirror still remain and have to be corrected. This has been done very well so far, by clever design of the principal camera which was installed in the repair mission, to replace the original Wide-Field Planetary Camera. This camera is known as WFPC2. The other instruments – of which I shall show results obtained with the Faint Object Spectrograph (FOS) – have had the primary mirror's distortions largely corrected by the very cleverly designed and constructed corrective optics (COSTAR).

Though WFPC2 and COSTAR give results that are not as perfect as would have been produced with a perfectly figured 2.4-m primary, subarcsecond imaging of QSOs has already provided surprises not yet assimilated in our understanding of QSOs and the nuclei of galaxies. These results, together with milliarcsecond resolution with the VLA and similar radio telescopes, which extends to microarcsecond resolution at millimetre wavelengths and ultimately into the infrared, will surely yield new discoveries, as will use of telescope mirrors utilizing adaptive optics. We can look forward to X-ray spectroscopy with good resolution, and X-ray imaging that will reveal structure in hitherto unresolved X-ray sources.

Thus, a summary of future instrumental advances for the study of the physics of QSOs includes: imaging at all wavelengths – radio, millmetre and submillimetre bands, infrared, optical, UV, far-UV, and X-ray – with subarcsecond and, at some frequencies, milliarcsecond and micro-arcsecond resolution, as perhaps the leading priorities, accompanied by high-resolution spectroscopy at all wavelengths and with high spatial resolution.

I now turn to a specific field of investigation: absorption spectra of QSOs.

4.2 QSO absorption lines and their origins

Observationally, absorption lines in the spectra of QSOs can be divided into the following types:

    (i) Broad absorption troughs (BALs).

    (ii) Associated absorption systems (AASs).

    (iii) Narrow-line systems with heavy elements with $z_a \ll z_e$.

    (iv) 'Damped' Lyα systems, with absorption lines from heavy elements.

    (v) Lyα 'forest' lines, the presence or absence of heavy elements needing careful investigation.

    (vi) Far-UV absorption – the search for the signature of He$^+$ near $\lambda304$ Å.

I have selected (i), (ii), (iii), and (vi) for discussion here, because these are the types of absorption with which I have been most concerned observationally. I shall also say something about the phenomena associated with (v). Type (iv), the 'damped' Lyα systems, would require a much more detailed discussion than the available space will allow.[†]

4.2.1 Broad-absorption-line QSOs

A fraction of the QSOs, perhaps $\sim$10% (Weymann *et al.* 1991), display in their spectra broad absorption troughs, starting at, within, or at points displaced some way to, shorter wavelengths from the emission lines, and these extend a long way shortward of the broad emission lines. This phenomenon, discovered by Lynds (1967), is illustrated in Fig. 4.1, from Tom Barlow's PhD dissertation (Barlow 1993), in which a composite of several BALQSO spectra in the UV is compared with a composite emission-line spectrum constructed from the Large Bright QSO Survey (Francis *et al.* 1991). The broad absorption troughs are assumed to arise in gas ejected from the QSOs at high velocity ($\sim c/10$). This gas must lie close to the energy source, since it has been observed to vary in time scales of months (Barlow *et al.* 1989). Variability in the absorption troughs had been detected in the BALQSOs 1303+308 (Foltz *et al.* 1987), 1413+117 (Turnshek *et al.* 1988), and 1246−057 (Smith and Penston 1988), but the first major change throughout the absorption troughs was detected in UM 232 (0019+011) by Barlow *et al.* (1989). CSO 203 (0842+345) proves to be another extremely interesting case (Barlow *et al.* 1992). Many addi-

---

† I shall concentrate largely on the published and unpublished results obtained at UCSD with the support of NASA grants NAG5-1630 and NAG5-1858.

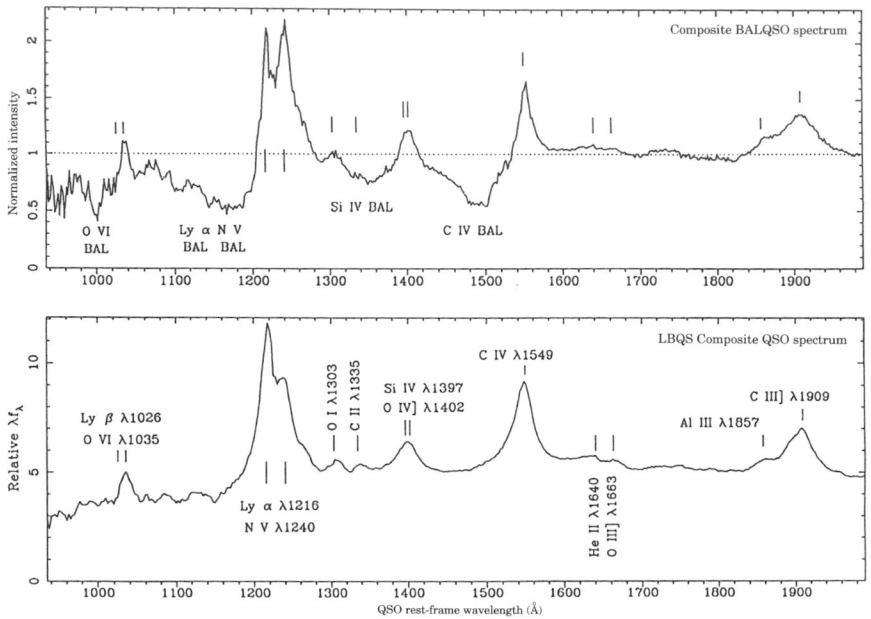

Fig 4.1. Composite BALQSO spectrum averaged over all objects studied by Barlow (1993), the publication in which the original of this figure appears. The non-BAL composite spectrum shown for comparison, also taken from Barlow (1993), was prepared by him from the Large Bright Quasar Survey by Francis *et al.* (1991).

tional references to these and other BALQSOs can be found in Barlow's PhD dissertation previously referenced.

Ionization calculations, using as many ions of different ionization potential as possible, combined with photometric broad-band monitoring and spectroscopic monitoring as frequently as possible, have shown that the variations probably occur in response to changes in the EUV/X-ray ionizing flux (Korista *et al.* 1992; Barlow 1993). Despite the difficulty of obtaining element abundances in the gas producing the broad absorptions, there are several cases where ratios of one element to another have been shown to differ from the solar (or cosmic) ratios. Examples are the C/N ratio, where N appears overabundant.

Major questions to be answered are:

(i) How is the gas ejected, and how much energy is required?

(ii) Where is the absorbing gas relative to the emission-line region?

(iii) What is the geometry of the absorbing gas (cones, jets, clouds?) and what is the covering factor relative to the continuum and emission-line region?

(iv) Is the outflowing gas accelerated or decelerated along the flow?

(v) What produces the apparent abundance anomalies in the outflowing gas (could it be due to extraordinary supernova activity?)?

(vi) What happens to the ejected gas as it departs into the low-density extragalactic medium?

(vii) What is the explanation of the sometimes highly complex multiple absorptions seen at large $\Delta z = z_e - z_a$ and with widths ranging from very broad to narrow?

The observations needed to attack these questions include monitoring for variability with broad-band optical photometry and spectrophotometry, improvement of the scanty data on X-ray emission, high-resolution spectroscopy in the optical and UV, further investigation of element abundances in the absorbing gas relative to those in the broad-emission-line regions and to cosmic abundance ratios, and high-resolution imaging of the field around the QSOs.

Some interesting new data have recently been obtained. Barlow and Junkkarinen (1994) have observed the BALQSO 1246–057 with the High-Resolution Spectrograph, HIRES, on the 10-m telescope at the Keck Observatory. Figure 4.2 shows the C IV $\lambda1549$ absorption as observed both at Keck and with the Lick 3-m Shane telescope. Kwan (1990) suggested that at high resolution the BAL profiles would break up into numerous narrow-line components, but this is not the case. The Keck spectra taken at high dispersion show the same smooth profiles as the spectra obtained at Lick. Further, the profile does not drop to zero intensity, and the amount by which it lies above zero is the same in the Keck and Lick data. If the BAL region is made up of individual clouds, it must be composed of *at least* several thousand components, thus limiting the individual cloud diameters to less than 10^4 km (Barlow and Junkkarinen 1994).

In another case, the BALQSO CSO 755 observed with the Keck 10-m telescope and HIRES spectrograph (Barlow 1995) has a very interesting double BAL trough in C IV $\lambda1549$, detached from the broad C IV emission line; I discuss this object further in Section 4.2.2 where I discuss associated

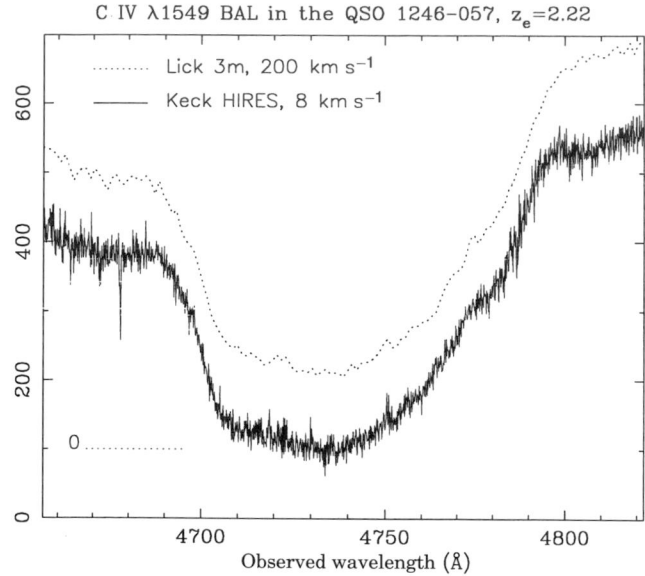

Fig 4.2. Broad absorption-line profile of C IV $\lambda\lambda1548$, 1551 in the radio-quiet BALQSO 1246−057 ($z_e = 2.22$), observed with the Lick Observatory 3-m Shane telescope (dotted line) and with the Keck Observatory 10-m telescope and HIRES spectrograph (full line) (Barlow and Junkkarinen 1994). Note that the HIRES profile is smooth and is not resolved into numerous narrow components, and the low-resolution and high-resolution profiles are very similar. Note especially that the bottom of the profile in each case lies above the zero-intensity line; the base of the diagram represents zero intensity for the Keck profile, and the dotted line indicates zero intensity for the Lick profile.

absorption systems, but I wish to mention here that the Keck HIRES spectra of this BALQSO obtained by Barlow show, as well as the BAL trough and an associated absorption system, some *narrow* C IV $\lambda\lambda1548$, 1551 doublets occurring near the *emission* peak of C IV, and these show interesting examples of C IV doublet line locking.[†]

---

† It is perhaps appropriate to remind readers of the origin of the term *line locking*, because of questions that arose in the discussion that followed the lecture version of this chapter, and to do this I quote from Burbidge and Burbidge (1975) as follows:

Suppose gas has been driven outward (e.g., by a supernova-like explosion) from a central QSO, and that it has a large spread of velocity attaining highly supersonic values. The general

Some low-ionization BALs have been found. Wampler, Chugai, and Petijean (1995) have studied a particularly interesting BALQSO, 0059–2735 ($z_{em} = 1.584$), which has strong low-ionization BALs of Si II, Mg II, Al II, and also narrow lines of singly ionized iron-peak elements (Cr, Co, Ni, Zn) and Fe II absorptions from metastable levels several eV above zero. They have analysed the BAL profiles into nine cloud velocities, and identified more than 1000 narrow lines into four cloud velocities, with evidence for an overabundance Fe/C relative to solar of ~ 10. Evidence for *line locking* is found, i.e. a control of the velocity flow by radiation pressure, as suggested decades ago for hot stars by Lucy and Solomon (1970), and followed up for QSOs by Mushotsky *et al.* (1972), Scargle (1973), and Burbidge and Burbidge (1975). The narrow lines in 0059–2735 must arise only a few pc from the QSO nucleus because of the presence of absorptions from levels several eV above ground level, and a high rate of SN explosions may be responsible.

It is appropriate to add that following the early work the line locking phenomenon has been ignored by most workers on absorption lines who have become obsessed with the idea that all absorptions except in the BALQSOs must be due to intervening galaxies or clouds. I believe that this question is still open.

Supernova activity of massive stars may account for anomalies in the abundances of some elements relative to solar values. There is a particularly interesting case recently found in the low-redshift BALQSO PG 0946+301 (Junkkarinen *et al.* 1995). In *HST*/FOS UV spectra, broad absorptions due to PIV and PV have been identified. Since the cosmic

characteristics of such a flow, even without the probable presence of bursts of relativistic plasma, non-spherical symmetry, and strong magnetic fields will be the onset of instabilities, as in nova shells.

Such outflowing gas can absorb radiation, and hence experience radiation pressure, from the underlying QSOs, whose continuum radiation is modified by emission lines and filtered through absorption by gas closer to the QSO. Ions in the outflow of any particular species will, if they have an appropriate velocity, *see* redshifted radiation from the central source that may have a resonance line or ionization absorption edge falling at the frequency of their particular resonance lines. Such ions will experience less radiation pressure than ions at neighbouring velocities. If an inward force (gravitation) is acting, a balance can be set up, since both scale as $1/r^2$, so that ions will accumulate in velocity space at appropriate velocities at which they experience no net inward or outward acceleration. At the velocities where balance occurs, resonance lines will *lock on* to each other, or on to any point of strong wavelength gradient in the radiation flux such as an ionization edge.

abundance of P is quite low (C/P = 1000), these strong features indicate an overabundance of phosphorus relative to carbon of $\sim$80. Investigation of nuclear reactions, e.g. in supernova explosions of massive stars, that might produce phosphorus is needed.

4.2.2 Associated absorption systems

Among the narrow-line absorption systems with $z_a < z_e$ there is a subset in which z_a is close to z_e and in which high-ionization lines are found. A good example is the radio-quiet UM 675 (0150–202, $z_e = 2.148$) in which Ne VIII absorption was discovered (Beaver *et al.* 1991). Further study of this object (Hamann *et al.* 1995) has revealed that in a time scale < 9 years (2.9 years in the rest frame), the associated absorption lines have strengthened by a factor $\sim$3 (Fig. 4.3). If the variation was caused by changes in the ionization,

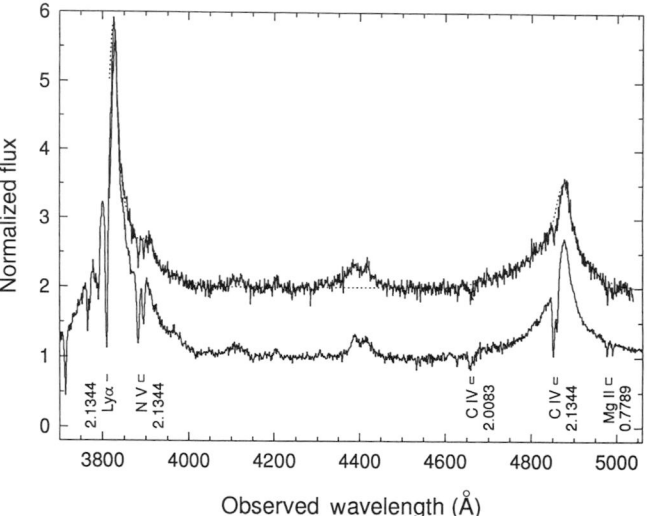

Fig 4.3. Spectrum of UM 675, showing the change that has occurred in the associated absorption system (AAS) at $z_a = 2.1340$. The upper plot shows data obtained in 1981 by Sargent, Boksenberg, and Steidel (1988), which has been normalized to the data obtained in 1990–93 by Hamann *et al.* (1995), from which paper this figure has been taken. The very obvious change in the strength of the associated absorption-system lines by a factor of $\sim$3 occurred some time during this nine-year interval between the 1981 and 1990 observations.

gas densities ≥ 4000 cm^{-3} are required. The presence of Ne^{+7} (ionization potential for Ne VII = 207 eV, and for Ne VIII = 239 eV) requires clouds < 200 pc from the continuum source. Monitoring of this QSO is being continued to check for further variation in the associated absorption system. Future observations to elucidate the nature of such variability are needed, with monitoring with good spectral resolution of many QSOs with associated absorption systems.

More data giving further clues into the nature of the AAS phenomenon are provided by a study mentioned in Section 4.1 of the BALQSO CSO 755 ($z_e = 2.88$) by Barlow (1995). Keck HIRES observations show that in this object the C IV emission peaks near 6000 Å and the absorption shortward of this is very complex, with several narrow C IV doublets in the C IV emission-line peak and in its blue wing, then a strong double detached BAL absorption, and, blueward of this, a very interesting associated

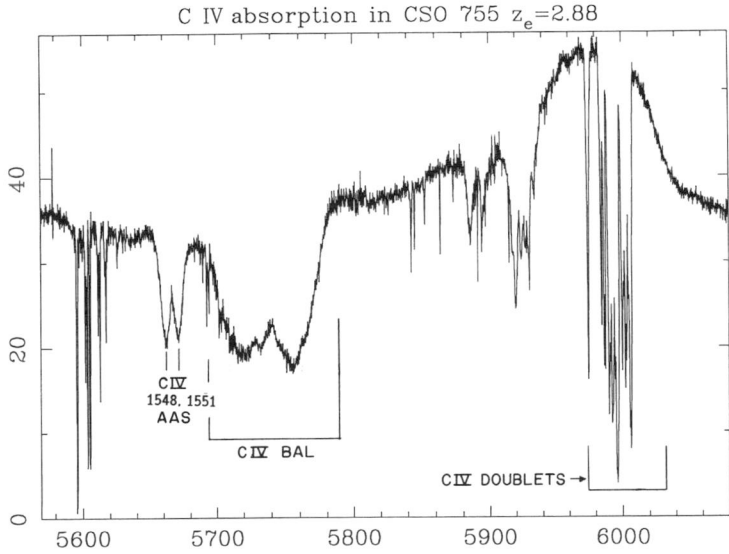

Fig 4.4. A portion of the spectrum of CSO 755 around C IV λλ1548, 1551 observed by Barlow and Junkkarinen at Keck Observatory with the HIRES spectrograph, showing the very complex C IV absorption structure (Barlow 1995). Note that the peak of C IV emission occurs near 6000 Å, that the main (double) BAL trough at 5700–5800 Å is detached from the broad emission, and that numerous narrow lines occur in the peak and blue wing of the C IV emission. Note especially that a *smooth*, resolved, associated absorption system (AAS) of C IV occurs *shortward* of the BAL trough, at ∼5660–5670 Å.

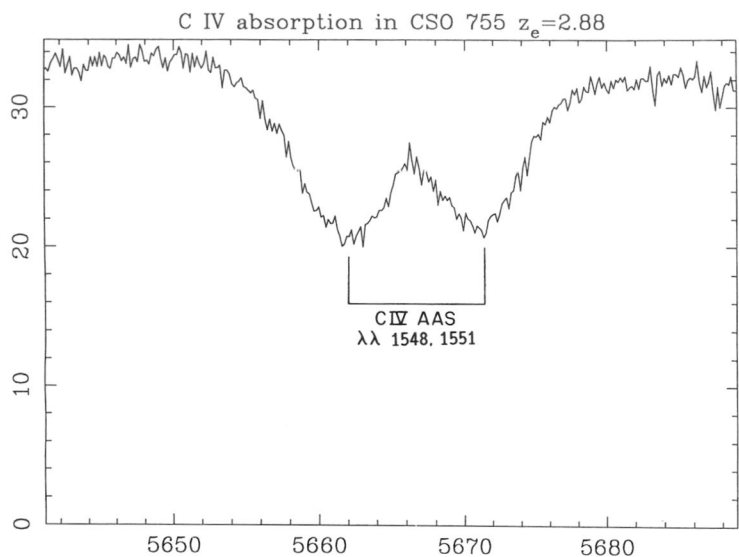

Fig 4.5. Enlargement of the C IV λλ1548, 1551 associated absorption system (AAS) in CSO 755 (Barlow 1995). Note that the doublet is saturated (from the relative intensities which should be 2:1 if unsaturated), that each line of the doublet is smooth and quite broad, and that, although saturated, the bottoms of each component lie well above zero intensity.

absorption system ($\sim$5650–5675 Å) (see Fig. 4.4). The two components of this AAS C IV doublet are well resolved, smooth, quite broad, and do not break up into multiple components at the HIRES resolution of $8\,\mathrm{km\,s^{-1}}$ (Fig. 4.5). They are clearly saturated (they have nearly equal intensity instead of a 2:1 ratio), yet their minima are well above zero intensity, showing that the absorbing gas does not cover the source of light reaching it. This AAS is in sharp contrast to a complex of *narrow* C IV doublets in absorption centred in the peak of C IV emission, which display clear evidence of line locking (Fig. 4.6).

The complexities revealed by high-resolution spectroscopy, such as those seen in this object, and those described by Wampler *et al.* (1995) in Q 0059−2735, may reveal the connection between BALQSOs and AASs and provide an answer to Questions (f) and (g) in Section 4.2.1. Also, we can speculate that they may have some bearing on the origin of narrow-line absorption systems with $z_a \ll z_e$, which are the topic of Section 4.2.3.

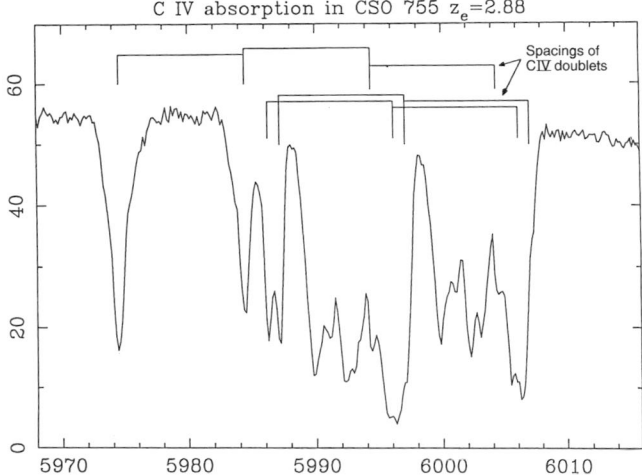

Fig 4.6. Enlargement of the region around the peak of C IV broad emission in CSO 755. Note that numerous *narrow* C IV doublets can be identified (they are marked by the sets of lines along the top of the figure); at the redshift of CSO 755, the C IV doublet separation is ~9 Å. Note especially the evidence for doublet 'line locking'.

4.2.3 Narrow-line systems with heavy elements

The canonical explanation of narrow-line systems with heavy elements is that they are produced by gas in the haloes of intervening galaxies crossed by the line of sight from observer to QSO. An interesting example is provided by recent observations of the QSO 3CR 196 (Cohen *et al.* 1996). The emission-line redshift of 3CR 196 is $z_{em} = 0.871$, and 21-cm absorption was found several years ago by Brown and Mitchell (1983). Foltz, Chaffee, and Wolfe (1988) found low-ionization lines at $z = 0.437$ and also Mg II absorption at $z_{abs} = 0.871$, i.e. an associated absorption system.

The new observations with the *HST* consist of a short-exposure *HST*/FOS 1600–2400 Å spectrum and *HST*/WFPC2 images (Cohen *et al.* 1996). Earlier images (Boissé and Boulade 1990) had shown a diffuse object close to the QSO, but no morphological information on it was available. The post-repair WFPC images show clearly that the diffuse object is a barred spiral galaxy, and the FOS data reveal a strong damped Lyα absorption system at the 21-cm redshift, $z = 0.437$. One of the peculiar 'coincidences' found so

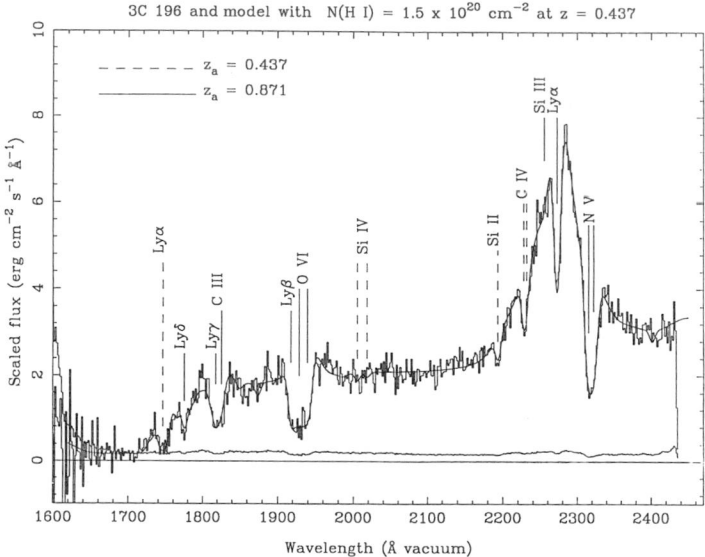

Fig 4.7. Spectrum of the QSO 3CR 196, obtained in 1991 with the pre-COSTAR Faint Object Spectrograph on the *HST*, with Red Digicon and the G160L grating (Cohen *et al.* 1996). The best-fitting model is drawn as the heavy line. Solid lines mark absorption in the $z = 0.871$ associated absorption system, and dashed lines mark identifications in the $z = 0.437$ system which also produces 21-cm absorption in the radio spectrum.

often in QSO research is that Lyα at 0.437 occurs in the region where the AAS Lyman lines converge to the Lyman limit at $z = 0.871$. Figure 4.7 shows the model which provides the best fit to the low-S/N spectral data; the damped Lyα system has an N(H I) column density probably $\approx 10^{20}\,\mathrm{cm}^{-2}$. Details of these observations and the images can be found in Cohen *et al.* (1995).

It has not yet been possible to obtain the redshift of the barred spiral galaxy because of its close proximity to the bright QSO, but this should be possible with the Keck telescope and with future telescopes of the 8–10-m class in good seeing; [O II] $\lambda 3727$ emission and a Ca II H and K $\lambda 4000$ break should be visible. The current hypothesis is that the galaxy will be found to have $z = 0.437$. If so, its luminosity is approximately L_*. However, the 'coincidence' of damped Lyα at $z = 0.437$ and the AAS Lyman limit at

$z = 0.871$ remains intriguing, and the possibility that the barred spiral galaxy has a redshift different from 0.437 should be borne in mind.

The uneven distribution of $z_a \ll z_e$ absorption, in that some QSOs have large numbers of these, in groups (e.g. PKS 0237–23), remains a puzzle. Statistical studies are, at present, difficult to carry out (see Duari and Narlikar 1995), because of the uneven quality of the data in the catalogues, e.g. Hewitt and Burbidge (1993), Junkkarinen, Hewitt, and Burbidge (1991). Key observations would be to randomly select areas of the sky in high galactic latitudes (N and S) and observe all QSOs in the chosen areas with spectral resolution of the same quality. High-resolution imaging is an obvious requirement, and surprises can be expected, such as the recent observations by Bahcall, Kirkakos, and Schneider (1994, 1995) that some QSOs with low-redshift do not appear to be embedded in 'host galaxies', whereas observations made with the Canada–France–Hawaii Telescope (e.g. Stockton and MacKenty 1987 and other papers referenced by Bahcall *et al.*) have detected extended galaxy-like luminosity around QSOs.

Follow up of these observations is very necessary. Adaptive optics on telescopes in the 8–10-m class will make such high-resolution imaging possible. Images obtained with the *HST* require very skilful analysis of the data to extract information from pixels adjacent to the bright image of a QSO, and it must be remembered that the *HST* has only a 2.4-m primary mirror, and it is a mirror that suffers from bad aberration whose effects the clever design of WFPC2 and the corrective optics in COSTAR can minimize, but not fully compensate.

4.2.4 The search for He II $\lambda 304$ absorption in high-redshift QSOs

From the early planning of programmes, before the *HST* was launched, it was realized that in QSOs with $z_e \gtrsim 3.1$, He II $\lambda 304$ would be clear of geocoronal Ly α and observable with the *HST*. The absence of H I continuous absorption shortward of Ly α emission (the Gunn–Peterson effect) has been taken to indicate that any smoothly distributed intergalactic medium must be highly ionized. No dip shortward of the He I resonance line at $\lambda 584$ Å has been detected, e.g. in UM 675 (Beaver *et al.* 1991), so the search for a He II dip, or cut-off, was of high priority with both the FOS and the Faint Object Camera (FOC).

A prime candidate for this search was OQ 172 (1442+101), with $z_e = 3.544$, and with few strong absorption systems observable in the optical region. Short-exposure observations with the FOS prism and with the G160L grating were obtained and showed an absolute cut-off in the spectrum, but this occurs some 20 Å longward of the expected He II cut-off, i.e. at 325 Å (rest) (Lyons *et al.* 1995). If this is due to He$^+$, it would imply an infall of intergalactic gas into the QSO, or a blueshift of the QSO relative to its environment. An alternative possibility is that the cut-off is a Lyman-limit system at $z = 0.621$. However, the spectrum (Fig. 4.8) shows no corresponding Ly α absorption line at $z = 0.621$. *HST* observing time with the FOS to examine the region 1600–2400 Å with adequate S/N to check on a possible Ly α absorption at 1971 Å has been requested, but not so far

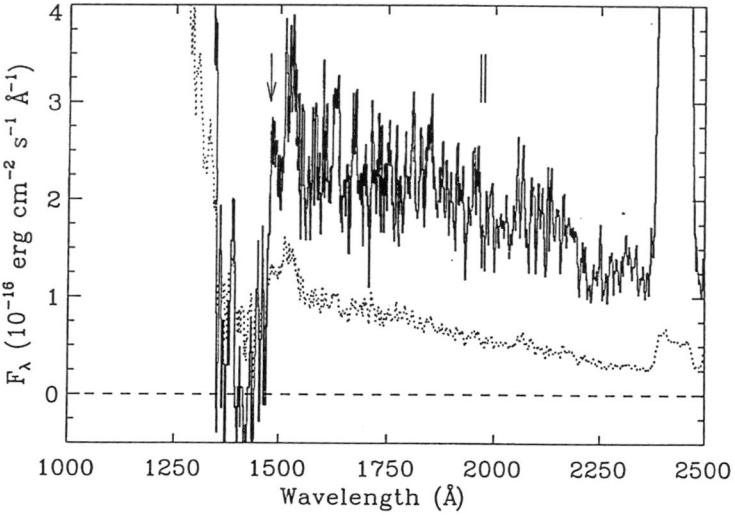

Fig 4.8. *HST* spectrum of OQ 172, from Lyons *et al.* (1995). The spectrum was obtained with the FOS Blue Digicon and the G160L grating. The steep rise at the left end of the spectrum represents the edge of geocoronal Ly α, which appears in second-order at 2432 Å at the right end of the spectrum. The arrow at $\lambda \approx 1475$ Å marks the cut-off which may be due to He II λ 304 Å. Alternatively, if this cut-off is due to a Ly limit system at $z = 0.621$, Ly α absorption should appear at $\lambda_{obs} = 1971$ Å, between the vertical lines marked in the figure; the resolution and S/N are too low to see whether it is present. The dotted line below the spectrum indicates the 1σ errors.

allocated. Some Lyman-limit systems show Mg II $\lambda 2800$ absorption; at $z = 0.621$, this feature would fall in the midst of the Lyα forest of OQ 172. High-resolution optical spectroscopy, e.g. with HIRES on the Keck 10-m telescope, could be used to search for Mg II, which, if present, should be distinguishable from Lyα forest lines because the thermal broadening of Mg is less than for H. However, absence of the Mg II doublet at the expected place would still be inconclusive, and the only sure check would be adequate *HST* resolution covering the 1900–2000 Å region.

The *HST*/Faint Object Camera team, led by P. Jakobsen, has made a dedicated search for He$^+$ absorption, by examining 25 high-z QSOs, of which only 3 showed light at the required wavelength. Most high-z QSOs have strong cumulative absorption from multiple Lyman-limit systems (Möller and Jakobsen 1990). This group has, however, achieved success with the object Q 0302–003, at $z_e = 3.286$ (Jakobsen *et al.* 1994).

Lyons *et al.* (1994) have recently observed another high-z QSO with the *HST*/FOS, and the results are very similar to those obtained for OQ 172. The object is UM 670 ($v = 17.5, z_e = 3.16$), and it was chosen for *HST*/GTO observations because its optical spectrum, like that of OQ 172, is not cut off by any Lyman-limit systems. Again, there is a sharp cut-off in the UV, but, as in OQ 172, this occurs at a wavelength of about 20 Å (rest) *longward* of the expected He II cut-off. Again, the observed cut-off might be due to a low-redshift Lyman-limit system, and, again, UV/*HST* observations with higher resolution are needed to check this possibility.

A different result from these and the Jakobsen *et al.* result has been obtained by Tytler and Zuo (1995). They observed the QSO PKS 1935–692 with the *HST*/FOS and discovered that there is *no* sharp cut-off at the position of the expected He II cut-off. The intensity drops here, but then recovers, implying that there is no smoothly distributed intergalactic He$^+$ along the line of sight to this QSO. This, and recent results reported by the *ASTRO 2* team, obviously impacts current cosmological models.

The nature of the cut-offs in OQ 172 and UM 670 is a challenging problem. Higher-resolution UV observations should indicate whether it will be necessary to tackle the theoretical task of understanding He II cut-offs at velocities several thousand km s^{-1} different from the velocity of the QSOs.

4.2.5 The Ly α forest

The forest of Ly α lines at z_a starting adjacent to Ly α emission was an early discovery in QSO research (Lynds 1971). The gas clouds producing the absorptions extend right down to Virgo Cluster redshifts (Bahcall *et al.* 1991), but searches for low-luminosity galaxies that might be responsible have not been successful. The detection of heavy-element absorption associated with Ly α lines is rare, unless very high spectral resolution is used, although Ly α/Ly β pairs are frequently observed. Of the heavy elements, C IV $\lambda\lambda$1548, 1551 lines are the best indicators of these and have been observed in some Ly α forest spectra. However, Verner, Tytler, and Barthel (1994) have shown that gas producing H I Lyman lines might be so hot that only lines of highly ionized species would be present.

Various attempts to model the distribution in redshift space of the clouds have resulted in parametric fits of dN/dz without a consistent explanation. An outstanding question posed by these observations is: what produces the absorption – small galaxies, protogalactic material, sheets of gas, tubes of gas, or some distribution of gas not hitherto envisaged?

There are some important and puzzling observations that suggest future observational programmes. One is concerned with attempts to detect Ly α absorption common to close pairs of QSOs, which would imply very large absorbing structures. For example, Dinshaw *et al.* (1995), from *HST*/FOS spectra of the pair Q 0107−025 A, B, have found a number of Ly α absorptions common to both spectra, and the QSO pair have an angular separation of 1.44 arcmin! The velocity differences of the common absorptions are very small (50–150 km s^{-1}), and, using the redshifts of the pair ($z_{em} = 0.956, 0.952$), the separations give *minimum* diameters >300 kpc, and 'most probable' diameters ∼700 kpc. The authors of this research suggest the absorbing gas might be in sheets, filaments, or flattened disks. This study indicates the need for more observational programmes of this sort. Some close groupings of QSOs are quite faint (e.g. those near M82; see Burbidge *et al.* 1980), so UV/*HST* spectroscopy would be difficult, but as the group near M82 has $z \approx 2$, ground-based spectroscopy with a spectrograph with good blue/near-UV sensitivity on a large telescope is an important future research programme.

4.3 Conclusions

I have presented examples from several observational programmes, specifically of the absorption spectra of QSOs. Some preliminary results are in and many more will be made in the coming decades. I have not touched on the accompanying theoretical work which will be needed, and how the answers to some of the questions I have raised will impact cosmological theories.

In Section 4.1, I outlined other observations, at all wavelengths, which could be made in the next 30 years as advances in telescopes and instrumentation come along. While many of these involve advances in X-ray and EUV techniques and in the region between the near-infrared and short-wave radio regions, existing and near-future telescopes of the 8–10-m class, specifically the forthcoming telescope on La Palma and the Keck telescopes, can tackle many of the observations which I have described.

All of the QSO absorption-line classes which I have discussed must be tied in to whatever cosmological model best describes our Universe. I have said nothing about the association between galaxies and QSOs, and this topic will be covered by G. Burbidge. I would like to add that while the redshifts of normal galaxies, as described by A. Sandage's contribution, fit well into the standard expanding-Universe cosmological model, the redshifts of the QSOs are another matter. Also, I have said little about the kind of element-building activity in QSOs that may account for abundances of elements that differ from what is seen in our Galaxy and in our neighbours.

In summary, the next three decades will be exciting years for astronomy, if our political leaders will manage to provide us with a world in which the pursuit of basic research can be continued!

Acknowledgements

First, I want to thank the editors of this book, Professors Francisco Sánchez and Guido Münch and Dr Antonio Mampaso, for the opportunity for 'old maestros' to present to a new generation what we see as key problems in astronomy. I am also very grateful to my colleagues Tom Barlow, Ed Beaver, Ross Cohen, Athanassios Diplas, Fred Hamann, Vesa Junkkarinen, and Ron

Lyons for providing figures and data which I have used in preparing this chapter. I acknowledge support by NASA grant NAG5-1630. I wish especially to thank Betty Travell for preparation of this manuscript through many drafts.

References

Bahcall, J.N., Kirkakos, S., and Schneider, D.P.: 1994, ApJL, **435**, L11

Bahcall, J.N., Kirkakos, S., and Schneider, D.P.: 1996, ApJ, **457**, 557.

Bahcall, J.N., Jannuzi, B., Schneider, D., Hartig, G., Bohlin, R., and Junkkarinen, V.T.: 1991, ApJL, **377**, L5

Barlow, T.A.: 1993, PhD Dissertation, University of California, San Diego

Barlow, T.A.: 1995, BAAS, **27**, 872

Barlow, T.A., and Junkkarinen, V.T.: 1994, BAAS, **26**, 1339

Barlow, T.A., Junkkarinen, V.T., and Burbidge, E.M.: 1989, ApJ, **347**, 674

Barlow, T.A., Junkkarinen, V.T., Burbidge, E.M., Weymann, R.J., Morris, S.L., and Korista, K.T.: 1992, ApJ, **397**, 81

Beaver, E.A., Burbidge, E.M., Cohen, R.D., Junkkarinen, V.T., Lyons, R.W., Rosenblatt, E.I., Hartig, G.F., Margon, B., and Davidsen, A.F.: 1991, ApJL, **377**, L1

Boissé, P., and Boulade, O.: 1990, A&A, **236**, 291

Brown, R.L., and Mitchell, K.J.: 1983, ApJ, **264**, 87

Burbidge, E.M., and Burbidge, G.R.: 1975, ApJ, **202**, 287

Burbidge, E.M., Junkkarinen, V.T., Koski, A.T., Smith, H.E., and Hoag, A.A.: 1980, ApJL, **242**, L55

Cohen, R.D., Beaver, E.A., Diplas, A., Junkkarinen, V.T., Barlow, T.A., and Lyons, R.W.: 1996, ApJ, **456**, 132

Das Gupta, P., Narlikar, J.V., and Burbidge, G.: 1988, AJ, **95**, 5

Dinshaw, N., Foltz, C.B., Impey, C.D., Weymann, R.J., and Morris, S.L.: 1995, Nature, **373**, 223

Duari, D., and Narlikar, J.V.: 1995, Int J Mod Phys, in press

Foltz, C.B., Chaffee, F.H., and Wolfe, A.M.: 1988, ApJ, **335**, 35

Foltz, C.B., Weymann, R.J., Morris, S.L., and Turnshek, D.A.: 1987, ApJ, **317**, 450

Francis, P.J., Hewett, P.C., Foltz, C.B., Chaffee, F.H., Weymann, R.J., and Morris, S.L.: 1991, ApJ, **373**, 465

Hamann, F., Barlow, T.A., Beaver, E.A., Burbidge, E.M., Cohen, R.D., Junkkarinen, V.T., and Lyons, R.: 1995, ApJ, **443**, 606

Hewitt, A., and Burbidge, G.: 1993, ApJS, **87**, 451

Jakobsen, P., Boksenberg, A., Deharveng, J.M., Greenfield, P., Jedrzejewski, R., and Paresce, F.: 1994, Nature, **370**, 35

Junkkarinen, V.T., Hewitt, A., and Burbidge, G.: 1991, ApJS, **77**, 203

Junkkarinen, V., Beaver, E., Burbidge, M., Cohen, R., Hamann, F., Lyons, R., and Barlow, T.: 1995, BAAS, **27**, 872

Korista, K.T., Weymann, R.J., Morris, S.L., Kopko, M., Turnshek, D.A., Hartig, G.F., Foltz, C.B., Burbidge, E.M., and Junkkarinen, V.T.: 1992, ApJ, **401**, 529

Kwan, J.: 1990, ApJ, **353**, 123

Lucy, L.B., and Solomon, P.M.: 1990, ApJ, **159**, 879

Lynds, C.R.: 1967, ApJL, **147**, L396

Lynds, C.R.: 1971, ApJL, **164**, L73

Lyons, R.W., Cohen, R.D., Hamann, F.W., Junkkarinen, V.T., Beaver, R.W., and Burbidge, E.M.: 1994, BAAS, **26**, 1337

Lyons, R.W., Cohen, R.D., Junkkarinen, V.T., Burbidge, E.M., and Beaver, E.A.: 1995, AJ, **110**, 1544

Möller, P., and Jakobsen, P.: 1990, A&A, **228**, 299

Mushotsky, R.F., Solomon, P.M., and Strittmatter, P.A.: 1972, ApJ, **174**, 7

Sargent, W.L.W., Boksenberg, A., and Steidel, C.C.: 1988, ApJS, **68**, 539

Scargle, J.D.: 1973, ApJ, **179**, 705

Smith, L.J., and Penston, M.V.: 1988, MNRAS, **235**, 551

Stockton, A., and MacKenty, J.W.: 1987, ApJ, **316**, 584

Turnshek, D.A., Foltz, C.B., Grillmair, C.J., and Weymann, R.J.: 1988, ApJ, **325**, 651

Tytler, D., and Zuo, L.: 1995, private communication

Verner, D.A., Tytler, D., and Barthel, P.D.: 1994, ApJ, **430**, 186

Wampler, E.J., Chugai, N.N., and Petitjean, P.: 1995, ApJ, **443**, 586

Weymann, R.J., Morris, S.L., Foltz, C.B., and Hewett, P.C.: 1991, ApJ, **373**, 23

4.4 Discussion

Question

(Pagel): Perhaps I could open up the discussion myself. I have a couple of questions. The first one is, could you clear up some confusion in my mind between broad-absorption-line systems (BALs) and associated systems? The way I see it is that you can tell the difference between them because the broad-absorption-line systems have broad absorption lines and, presumably, the associated-absorption-line systems (AASs) have more narrow absorption lines. However, you said that in the associated systems, just as in the broad absorption systems, the profiles are smooth, and this caused me some confusion.

Answer

(M. Burbidge): I have not shown you a detailed profile of the UM675 lines, but in one of my illustrations I showed a feature in the broad-absorption-line object that looks like an associated absorption line, although it is displaced considerably from the emission line. It is smooth and does not break up into numerous narrow components. In UM675, the absorption-line profiles are smooth. They are certainly broader than the instrumental profile, and as to whether they would break up or not, we cannot say at the moment. Let's see what I can say about the comparison between the associated-absorption-line systems and the BALs? They are both high-ionization features. Ne VIII is very strong and, in the one associated absorption-line system where we have been able to look at it, very strong. So, the ionization is high in both the BALs and the AASs.

The profiles? More work needs to be done. We need to pin them down, and actually there is some data that Thomas Barlow has from Keck (and he has not even begun to analyse it yet) on some associated absorption systems. He was concentrating on that BAL object that I showed, because it looked so fascinating. He has some data on associated absorption systems, and I guess it will be seen during the coming year. That goes some way towards answering this question.

Element abundances? I suspect this is another key question. There do seem to be element abundance differences from solar in the BALs; there seems to be accumulating evidence that nitrogen is overabundant with respect to carbon. It is difficult to do anything about the hydrogen ratio unless you can get to the Lyman limit and really quantify that for getting abundances relative to hydrogen.

Other abundance differences? There is a tentative identification of phosphorus, which is a low-cosmic-abundance element. It surprised us when one of our group claimed, in the BAL, to have seen phosphorus absorption in the ultraviolet. He has not published that yet, but he has presented it as a poster paper. It would be fascinating to wonder how phosphorus got made, because it is one of the elements that get rather ignored in standard element-production schemes in stars. That is all I can say: that the profiles, the abundances, the high ionization, I think, are related in the BALs and in the associated-absorption-line systems. Can one evolve into another? We do not know that. It needs monitoring, continuous monitoring.

Question

(Pagel): Yes, you answered a question that I obviously was about to ask, but had not asked yet. Apart from phosphorus, which I am not prepared to comment on at this stage, I have the impression that the abundance anomalies in the broad-absorption-line systems are at least to some extent similar to those that are found by Ferland and people in the emission lines. The same sort of thing: high nitrogen to carbon and high iron abundance, just what people have deduced from the emission lines. So, is there actually any difference from the emission lines?

Answer

(M. Burbidge): I do not know how to quantify this, but [shows viewgraph] in this composite of low-resolution spectra compare the NV emission-line in the Large Bright QSO Survey with the CIV peak. In the composite BAL, here is NV and there is CIV. The extent to which CIV has been eaten into by the broad absorption is not clear, but, after all, if the broad absorption by CIV has an exact counterpart in NV, why has it not eaten its way into the side of this emission line? That emission line, I would say [points to NV], to the eye at least (although it needs a lot more quantitative work), looks stronger than this one [points to CIV].

Another person in our group is Fred Hamann, who is going to look into the emission-line abundance ratios; he is more interested in those than in absorption lines. I have not given you a very good answer, but the question of departures from solar abundance ratios needs looking at in more detail.

Question

(Rees): I wanted to ask later about the $Ly\alpha$ forest, but could I first ask a question about BALs? There was, I think, an anticorrelation between BALs and radio properties of quasars. That relates to what seems to be quite an attractive idea, which is that the BALs are due to little cloudlets that are being ejected in a jet in some way. This also relates to Bernard's question, in that, if the BAL material is in cloudlets, which are small compared to the continuum, then, of course, one has to be careful about line ratios because, as was pointed out by White and Morrison and their group, you would not get completely saturated lines, even when the optical depth in the cloudlet is small, if they do not cover the continuum source. So, if the absorption is due to cloudlets in a jet, each of them very small, one has to be careful about interpreting line ratios. I wonder if you would comment on that interpretation problem and also on the radio–optical correlation.

Answer

(M. Burbidge): It is true that there appears to be only one BALQSO which has been found to be a weak radio source, so there definitely seems to be an anticorrelation here. The BALs make up about 10% of the QSOs, so you can ask yourself if the BAL phenomenon is purely an orientation effect, and whether every QSO has a jet, or a cone, or whatever, oriented so that we do not see it, in front of the emission source. I guess that we do not really know that yet, but I am maintaining that there is a difference in the absorption lines themselves from the emission lines. Bernard has raised this question and, of course, it needs a lot more study. The BAL and AAS troughs do not break up into numerous narrow components at Keck HIRES resolution, and are not black in the centre, yet are saturated, from the doublet ratios. Clearly the absorbing material does not fully cover the source.

What about the correlations with X-rays? As far as I am aware, there has not been much work yet done. We have asked for *ROSAT* time but, again, it has not been performed. We have selected some BALs to ask for *ROSAT* observations, to see if they were X-ray sources, but I think if they were X-ray sources, they would be too weak to make it profitable to spend *ROSAT* time on them. However, future X-ray instruments should be able to pin this down. Again, I think that more work needs to be done, looking into anything that you can see with respect to this correlation, or anticorrelation, of radio and BAL phenomena. PHL 5200, the first BAL that was discovered, is a weak radio source, but it has not had high-resolution spectroscopy yet. It does have a very broad trough, which has not changed since it was first observed in 1967, but we do not know what details there are, and we do not know very much about the presence of narrow lines as well in that system. Again, more work is needed, with telescopes of the 8 or 10-m class.

Question

(Osterbrock): About the high-redshift QSO which you talked about, and the attempt to see, or the hope to see, a He II series limit, and you see a cut-off at a somewhat longer wavelength – do you think it is possible that the He^+ gets ionized to He^{++} close to the object, and that is why it does not extend all the way to the redshift of the object?

Answer

(M. Burbidge): We wondered about that. We thought that perhaps this 20 Å difference could be due to the ionized helium being close to the QSO, and while one has a bit of uncertainty in the wavelength scale down in that region, it is not that bad that it accounts for the 20 Å difference, and there are now two objects which have shown this. Jacobsen's object did not seem to show this shift. He seemed to see a drop right where He II 304 Å was, but his FOC observations are not of very high precision. What we would like to do is to look at some $z = 4$ QSOs, but then you are getting to much fainter objects, and if you already have the flux declining as you go to those short wavelengths, it takes a long time to get a decent spectrum with the Faint Object Spectrograph of $z = 4$ objects. We need a better space telescope to answer this.

Comment

(Osterbrock): Yes, I think we all agree that a 25-m telescope in orbit is what we really need!

Question

(G. Burbidge): I have two questions. The first one I think you have really dealt with, in the sense that, about this business about the helium, if I understand you correctly, none of the observations, neither Jacobsen's nor yours, have yet established that the edge is where it should be. Is that correct? Is that a fair statement or not?

Answer

(M. Burbidge): Jacobsen claims that it is in the right place ...

Comment

(G. Burbidge): I understand that.

Answer

(M. Burbidge): ... but the published paper leaves room for some doubt, in my mind at least, because it is a difficult observation with the Faint Object Camera on the *Hubble Space Telescope.*

Question

(G. Burbidge): It was received with great acclaim at the IAU. But the question I wanted to raise is related to what you and Martin Rees were talking about. Namely, first of all, we all agree that radio QSOs are a very small fraction of all the QSOs. Secondly, what most people argue is that the BALs certainly have absorption which is intrinsic to the systems, whereas most people would argue, contrary to the things that I suggested yesterday, that the sharp absorption features, involving heavy elements, are associated with intervening matter, that is, where the redshift of the absorption is significantly different from the redshift of the QSO. My question is the following. If the BALs have this intrinsic absorption, they should also show exactly the same intervening absorption as do all the other QSOs, since they all are out there in space. My question is: can you identify, in all cases in all the BALs, sharp absorptions, as you can in the other QSOs in a similar range of redshifts, because there should clearly be no difference on this common hypothesis?

Answer

(M. Burbidge): I do not know the answer to that. Maybe somebody has done a statistical study of the frequencies of absorption lines of narrow systems at $z_a \ll z_e$ in BALs. I do not know if it has been done, but it should be done. In the object that I showed, which is still being analysed, the object that Tom Barlow is working on, there are a lot of narrow-line systems, but the C IV narrow-line systems are all bunched in the region of the emission lines. They have this very interesting line locking, and I guess we do not understand it yet. I should explain line locking. You can think of it as radiation pressure, which can have some effect in modulating, if you like, ejection velocities, when they fall in a region where there is a strong gradient in the radiation flux. If you have one line almost coinciding, or coinciding, with another different line at a different redshift, you can see that there is a strong gradient in the radiation seen by the gas which is producing the one pair, caused by the profile of the second pair.

In fact, in the object that I mentioned, the low-ionization BAL that Wampler and others at ESO are working on, they found some signs of line locking, and they mentioned this gradient effect in the intensity of flux of photons. It seemed to be caused by what is producing the absorption: if the gas producing the absorption wobbles a bit in velocity in relation to the QSO, it can be caught and put back in place by line locking. It is a thing that I have been interested in for a long time, and got put off from working on it in detail because other workers, particularly Sargent, were very much opposed to it. So that deterred me from spending more time on it.

Question

(G. Burbidge): Could I ask one further question? Considering what you said about this report in Garching, that they had found a $Ly\alpha$ system common to separated QSOs with a very small ΔV associated with them, but a very large distance in space (on the assumption that the redshifts are all what they should be), some of the order of half a million parsecs, or something like that, this has to be a very thin sheet of this kind. As I heard, I think, from you, they told you, 'do not tell Geoff'. I give Martin Rees 30 seconds to come up with an explanation; he probably had one even before I spoke.

Answer

(M. Burbidge): I am not going to attempt an explanation, but perhaps Martin could come up with an answer later.

Comment

(Rees): Could I make some comments on the $Ly\alpha$ forest? It is certainly an important dimension to quasar absorption. I think that, if we look back 15 years, there has been a change of paradigm. Around 1980, there was a paper by Sargent, Young, Tytler, and Boksenberg, who found a lot of data; I think most stands up, but they had a picture where they interpreted the Ly α forest as being due to clouds in pressure balance with a hot intergalactic medium. That was the picture and the paradigm for a number of years. I think we have got a lot more data, particularly on the even lower-column-density systems from Keck. I think this paradigm now changes in one or two ways. There was never any evidence for a hot intergalactic medium, except in clusters of galaxies; it seems to me that there is now evidence against it, because the microwave background distortion, and other parameters, rule it out. So, I think that has gone out of the window.

For that reason, the most natural paradigm now seems to be that the intergalactic medium, except in clusters, is heated only by photoionization, and that means it never gets to a temperature above 50 000 K, or thereabouts. So, the basic picture is that it is gravity which is responsible for the clumping of the gas which you see in the Ly α forest. What we are seeing is a direct reflection of the formation of galaxies on subgalactic scales. If we imagine that the intergalactic medium is photoionized, by a redshift of 4 or 5, by the first quasar, or by pre-quasar activity, then, of course, if it is clumped in any way, the neutral density will go as the square of the total density and so we will see features in it.

So, the most accurate interpretation of the Ly α forest is that it is due to gas that has fallen into bound systems, in the case of high column densities, or is in the process of falling in, in the case of low column densities.

We now see that the lines are almost overlapping in the latest Keck spectra and that tells us that we are seeing, in effect, gas which is filling quite a large fraction of the volume and has an overdensity which is quite modest. So, there is really now a blurred division between the weak lines and the so-called continuum. We have seen gas which is clumpy on scales of 50 or 100 kpc and starting to fall into these systems. That seems to be a very natural interpretation of the line widths and the line densities, and this raises questions which could be addressed by looking at the redshift dependence of the line density.

The redshift dependence of the line width is maybe telling us at what stage helium gets doubly ionized or something like that. Of course, the other new line of evidence is that of the scale of the cloud, which Geoff just mentioned. The tendency would be that the clouds would be of the order of the scale of the systems they are falling into. A single line characterized by a velocity of say $15 \, \mathrm{km \, s^{-1}}$, would have a scale of about 100 kpc. That is something that is going to collapse to form a dwarf galaxy. You would see sheets, or something like that, on that scale, and that is consistent, I think, with the data on the narrow spacings that I got from the double quasars that may be lenses. The data did show that double quasars have a separation which, as Margaret said, is more than arcminutes.

In this scenario we are saying that one is not seeing a single cloud but clumps within something which is going to be a big galaxy or a small group of galaxies. If you took something which was going to be a galaxy with a radius of 0.5 Mpc, then the velocity dispersion would be about $200 \, \mathrm{km \, s^{-1}}$, and so, I think, the separations are consistent with that. One is not seeing single lines; one is seeing a structure within a cloud because where the galaxy starts to collapse one does start to develop a two-phase medium. There is that pressure confinement, not just gravitational confinement, so I think that what one would expect to see is that, at the highest redshifts, there would be subgalactic-scale structure showing the $\mathrm{Ly}\,\alpha$ forest. As we get to smaller redshifts, the scale of structure moves from subgalactic to galactic scales. We would expect, therefore, a change over gravitational confinement toward some kind of pressure confinement, to a two-phase medium in the galactic halo. So, low-redshift systems would be associated with galaxies, the high-redshift systems, not necessarily, because at that stage galaxies have not yet been formed.

I forgot what my question was, but that is really just a comment.

Question

(G. Burbidge): I understand the idea, Martin, but can you tell me, then, what gives rise to the $\mathrm{Ly}\,\alpha$ features in 3C 273? In particular, it seems to me that this is such an awkward situation because there are no faint galaxies (they have looked for them). Why not allow, at least in this case, those $\mathrm{Ly}\,\alpha$ features to be associated with ejection from 3C 273? This does not conflict with anything that I know about.

Answer

(Rees): Two points. First (I think that Don Osterbrock knows more about this), as some work by Simon Morris *et al.* found some correlation between those features and galaxies, or groups of galaxies along the line-of-sight, let me just say that the low-redshift Lyα systems are probably associated with galaxies and groups of galaxies. I would like to make two points there. First, the column density one needs to get a Lyα system is, of course, enormously low compared to anything at 21 cm, and so it would not be at all surprising if any galaxy were to give a Lyα absorption out to a distance of 100 kpc from the nucleus of the galaxy. Just to make that more explicit: everyone is familiar with the so-called high-velocity clouds of neutral hydrogen, in our Galactic halo, which are thought to be neutral-hydrogen clouds, which are in pressure balance, maybe with a hot gaseous halo. If you were to take clouds like that say five times further out in our Galactic halo, the density would go down by a factor of maybe 25, but, more importantly, they would then be ionized all through by the UV background. So, they would then not show up in neutral hydrogen (21 cm), but they would have a large enough neutral column density still to give a Lyα line.

If we were to imagine that a two-phase gaseous halo of the kind manifested within 20 kpc by the high-velocity clouds were to extend out to 100 kpc in galaxies like our own, then out to 100 kpc, you could quite naturally have gas at 10^4 K maybe maintained in pressure balance with a hot medium. That gas at 10^4 K would be mainly photoionized but would have just the parameters needed to show a Lyα cloud. So, I would say at low redshifts most of the Lyα clouds would be associated with the outer parts of galaxies, or maybe with dwarf galaxies. I do not see any evidence against that.

Comment

(G. Burbidge): There is a paper by Simon Morris and others in which they have looked for very faint galaxies and they do not find any. This is my point. There is no real evidence for galaxies associated with those Lyα systems. I will stop because it seems to me that we are ruining the discussion.

Question

(S. di Serego): I have a comment and a question. The comment is: it seems to me that there is an important parameter in trying to understand absorption lines in QSOs: orientation effects. You have hinted at it, but I would like to stress it now. This is obviously very important when you get to imaging the environment, but it is also important, I think, for emission lines. Suppose the velocity distribution is not isotropic. Then, of course, the width of the line depends on the orientation, and this may help you in understanding the relationship between the BAL and associated absorption. The other thing is, of course, when you get the orientation effects, polarization plays an important role because, if the thing is hidden, you get scattered lines being polarized.

I think that it is very interesting that a recent CAT observation has shown that PHL 5200, a BALQSO, is more polarized on the top than it is in the continuum. So, let me come to my question. You have this beautiful spiral galaxy, formed by 3C 196, and you have shown that the arms cover the radio lobes. It seems obvious to me that a very important observation would be to get 21-cm absorptions from the lobes. Have you done that? Maybe this was what the *HST* Time Allocation Committee meant by 'getting ground-based observations first'.

Answer

(M. Burbidge): That might be so. I do not know what the radio astronomers have done, but that is a key object, I think, because you have those very good radio contours, two sets of contours, and the pair that lie in front of the lobes that appear to belong to the galaxy should be examined for polarization, and I hope the radio astronomers will do it. Let me address the radio astronomer on the panel and ask whether or not this is feasible and what could be done about it. Let me turn this to Malcolm Longair.

Comment

(Longair): I think my concern is whether you can make observations at the correct redshift, because it will obviously go through the middle of a very bad waveband which is not protected internationally. So, I think, it will be just luck whether or not you happen to be hitting a wavelength which is not transmitting either radar or commercial TV. One would have to look at it.

Answer

(M. Burbidge): Yes, but it is a direction to go in other objects too. Future observations, which I have not mentioned, are the extension of radio observations.

Comment

(A. Uribe): I would like to make some comments about thermal analysis. Several years ago we began research in Colombia on open clusters. I think that open clusters in our Galaxy can be considered very important, because many astrophysical problems could be addressed in this direction. We are really trying to solve the problem of membership of open clusters using models with many components. You mentioned something about that analysis of these profiles. Maybe some statistical analysis can be done using multiple-component analysis. I think that there is a good future in this field of statistics in astronomy. An astronomer in Canada called James LeMeg is using, with good results, multiple-component analysis, which is a branch of multivariate statistical analysis in astronomy. I would like you to comment about the approach to distinguish different contributions to absorption lines, using multiple-component analysis.

Question

(M. Burbidge): You are talking about taking the profile of an absorption line and deconvolving it?

Answer

(Uribe): Yes.

Answer

(M. Burbidge): In some of these cases we have done that, where there are blended lines, and in the only spectrum that I showed you, the ultraviolet spectrum of 3C 196, you can see that those lines are blended. We have some other Faint Object Spectrograph observations of absorption lines in which the components in one redshift system sit close to another line and the ultraviolet spectrum is quite rich in absorption also. My younger colleagues are doing this sort of thing: working on the profiles to get the separate components. There is a Chinese astronomer called Lin Zuo at UCSD and he is working on this problem.

Question

(Uribe): So you think that this is the right approach?

Answer

(M. Burbidge): I think it is the right approach. Yes, indeed.

Question

(A. Aragon): There seems to be mounting evidence, or at least some people believe so, that the narrow-line metal systems with a lower redshift than the QSO are indeed intervening galaxies. Many people find that it is proved beyond any reasonable doubt that they are intervening. I would like to hear your opinion. What do you think that the situation is at the moment? How sure can we be that they are intervening?

Answer

(M. Burbidge): I think that some are intervening, but I think more work needs to be done on the frequency of absorption-line systems in individual objects. If you find objects that have a great multiplicity of narrow absorption lines over a range of redshifts, then you need to look and see if some of these can be ejected material from the QSO, or whether every one has got to be associated with a galaxy, or the clouds in a galaxy, or whatever. In that BAL object that I showed, those narrow C IV lines were right on top of the emission lines. Somehow those have appeared narrow and yet must be close to the source. I did not mention that, in the low-ionization BAL object that Wampler and the group at ESO have been working on, they found 1000 narrow lines and identified a large number of them as singly ionized iron. Some of these lines were arising from meta-stable levels, a few eV above the ground level. That suggested to them that these really have to be close to the QSO. I mean the absorption-line redshift is close to the emission line, and therefore it is almost like a break up of an associated absorption, except that the associated absorptions seen with higher resolution do not break up like that.

Question

(Aragon): Do you think that in the cases in which the redshift difference is very high, they are all almost certainly intervening?

Answer

(M. Burbidge): I would not say that they are all intervening. It needs more study, it needs a lot more study and statistical analysis over a lot of objects for which there are sufficiently high-resolution observations to detect all the narrow lines that may be present.

Question

(H. Zinnecker): A question regarding the X-rays that you mentioned, the soft X-rays. I understand that, when you go to X-rays of the order of 0.2 keV (which was the range that you were talking about in relation to the Ne VIII ionization), the absorption of X-rays is very crucial owing to heavy elements. I wonder if you can hope to pin down the X-ray emission in this regime, unless you solve, at the same time, the question of what is the absorption of these X-rays: intrinsic or extrinsic. It is a coupled problem: you do not just look there at 0.2-keV X-rays and hope to get an answer. You have to worry about absorption of soft X-rays. Could you comment on that?

Answer

(M. Burbidge): My colleague Fred Hamann is particularly interested in the emission lines. He has suggested that, in the X-ray, you should see a Ne VIII edge. He has already written an abstract for a poster paper for the AAS about this. I think that he has a proposal to work on just that, looking towards edges, particularly Ne VIII. I do not know where the edge of the Ne VIII comes in the X-ray. Do you know?

Question

(Zinnecker): No, I do not by heart. Presumably you meant that we should do X-ray observations, and I was just trying to find out whether or not you think that from these X-ray observations, per se, we can get answers, given the fact that we have additional complications like absorption in soft X-rays. You were showing a different way out: going to emission.

Answer

(M. Burbidge): I am just saying that this is a field that has not been really opened up to observation and is a new observational field. I do not know just how many edges come just where in the soft-X-ray region; I have not really looked at that. The only one I really know about is Ne VIII because of the proposal by Fred Hamann to do just that.

Comment

(Rees): My comment is really on the issue of fitting line profiles with multiple components. It is important that particularly the low-column-density systems will be due to gas so diffuse that it ought to be in a dynamic state, and therefore one would expect that the fit would be not to an exact Voigt profile because there could be some bulk Doppler broadening as well. It is very important to look for that because, if it turns out that part of the line in some of these systems is due to bulk infall motions, that leaves less left over for the thermal broadening, and that is an important constraint on photoionization models and on the question of whether the helium could be doubly ionized or not. It is very important, particularly for the low column densities. One would expect to find non-broad profiles, because there would be bulk Doppler motions contributing to the line profile.

Comment

(Sunyaev): I wish to return to this discussion between Zinnecker and M. Burbidge. I also consider, as M. Burbidge did, that it is extremely important to understand the soft-X-ray flux, or the extreme-ultraviolet flux from quasars and from QSOs. It is now that these high-redshift quasars and *AXAF* [Advanced X-ray Astrophysics Facility] will permit us to go at it from the other side, because we can observe in the ultraviolet and perhaps get information about the extreme-ultraviolet flux. I believe that you are absolutely right and there is absorption, but this absorption is much weaker than the absorption at the Lyman edge, or by helium absorption. At the same time, I believe that the crucial secret here is the question of what the properties of the edge are, in itself.

According to many theories, the bulk of the radiation is in the extreme ultraviolet. You cannot prove this for galactic infrared binaries, but in the case of these distant objects maybe there is a possibility of understanding. Is there emission in the ultraviolet which then drops very strongly with the gap and then with very low luminosity, and then the X-ray component rises again? Or have we something continuous? This is crucial for the theory of accretion, for the systematic observation of the central engine in quasars. This is not only the system, this is the matter surrounding it, which was the main topic of M. Burbidge's talk, but it is just the properties of the central engine: what radiates the power?

5 The nature, structure, refuelling, and evolution of AGNs

Donald E. Osterbrock

5.1 Introduction

Active galactic nuclei (AGNs) are the most luminous objects in the Universe and thus the most distant markers in it which we can observe. They are likewise the most powerful energy sources we know. For both these reasons we must understand them physically. Clearly, AGNs are one of the highest-priority classes of objects for further observational study at all wavelengths, or energy regions, and for theoretical interpretation. No doubt advances in our understanding will continue, but problems and questions will remain for further study. It is relatively easy to see the directions of research likely to prove most profitable in the immediate future, but, as with other topics discussed in this symposium, the further we attempt to extrapolate our predictions, the less likely we are to be complete, or even accurate. Nevertheless, that is the purpose of this book. The attempt must be made, however imperfect the results may be.

This chapter is a continuation and extension of three recent review articles on AGNs which I have written (Osterbrock 1991, 1993a, b). Here I try to emphasize problems for the future, but can only do this in the context of what we know (or think we know) and do not know currently. Hence, the basic ideas are described briefly, but the above reviews should be consulted for further details and for many references to the primary research papers, whose results went into them and into the current chapter.

Let me emphasize that my subject is mostly the observational follow up to many well-known ideas on the nature and structure of AGNs, some of them due originally to several of my colleagues among the invited

[171]

speakers, including Donald Lynden-Bell (1969), Martin Rees (1977), and Igor Novikov.

Active galactic nuclei appear to belong to one family, extending from the highest-luminosity quasars (I continue to use this word to mean quasi-stellar radio sources) and QSOs (quasi-stellar objects, including the radio-quiet ones) through radio and Seyfert galaxies, to the low-luminosity active nuclei studied by Keel (1983), Filippenko and Sargent (1985), and others. This seems the best working hypothesis to me. I am persuaded of this chiefly by the many objects, once called quasars or QSOs, which on more recent, high-resolution images, processed digitally with sky subtraction, reveal faint extensions which may be interpreted as low-surface-brightness, distorted galaxies. But, of course, we cannot be certain that every object that has been called a quasar or a QSO belongs to this family; it is quite possible that some among them are actually nearby objects which mimic much more distant AGNs as has been claimed, for instance, by Arp (1987, 1990) and Burbidge *et al.* (1990). We shall discuss this topic further, with more references, in Section 5.5, and Geoffrey Burbidge treats it in his chapter in this volume.

To say that all AGNs belong to one family does not mean that they all have the same structure, except for scale, any more than to say that all stars belong to one family, including, for example, G dwarfs, giants, and supergiants, means that they all have the same internal structure. Rather it means that the same general physical principles govern their structure. Thus, different types of structure may apply to AGNs in different ranges of mass, luminosity, and age.

On the 'one-family' picture, the only sensible way to distinguish operationally between a QSO and a Seyfert galaxy is on the basis of an arbitrarily chosen absolute magnitude. It is probably most convenient to follow the definitions of the very complete catalogue of 'quasars' and AGNs of Véron-Cetty and Véron (1993), who made the separation at $M_B = -23$, with an adopted Hubble constant $H_0 = 50\,\mathrm{km\,s^{-1}\,Mpc^{-1}}$, $q_0 = 0$, nearly the same as originally proposed and used by Schmidt and Green (1983).

For orientation purposes, Table 5.1 gives very approximate space densities here and now in the Universe (Osterbrock 1989). Note that the Seyfert galaxies make up roughly 1% of the luminous spirals, and the QSOs are only one thousandth as abundant per unit volume of space as the Seyfert

Table 5.1. *Approximate space densities here and now.*

Type	Number Mpc^{-3}
Field galaxies	10^{-1}
Luminous spirals	10^{-2}
Seyfert galaxies	10^{-4}
Radio galaxies	10^{-6}
QSOs	10^{-7}
Quasars	10^{-9}

galaxies. Radio galaxies, which morphologically are mostly N, cD, or D galaxies in the Morgan (1958) classification, further extended by Matthews, Morgan, and Schmidt (1964), Morgan and Lesh (1965), Bautz and Morgan (1970), Morgan (1971), Morgan, Kayser, and White (1975), and Albert, White and Morgan (1977) are only about 1% as abundant as Seyfert galaxies, and quasars are less common (per unit volume of space) than known QSOs by the same factor.

To 'understand' AGNs means, as it does for any other class of astronomical objects, to be able to answer quantitatively the three questions: What are they? How do they work? And how do they evolve? Although ideally we may wish we could fully answer each of the questions in turn, before passing on to the next, in reality for planets, stars, nebulae, galaxies, and every other class of objects our knowledge has been gained by proposing incomplete, trial answers to all three, before even the first is fully answered. It is in this spirit that the present review is written. In it I concentrate on the evidence from optical spectroscopy, with which I am most familiar, but try to bring in as well as I can the data from other wavelength regions, for one of the crucial aspects of AGNs is that they radiate in all energy ranges, from the radio-frequency and far-infrared to the X-ray and γ-ray regions. Most of the discussion is devoted to Seyfert galaxy nuclei, the most common AGNs and hence again the easiest to study. How well the conclusions drawn from them apply also to the more luminous but more distant QSOs is one of the important questions which deserves and needs much more study in the future.

5.2 Spectroscopy and diagnostics

The optical spectra of Seyfert galaxy nuclei, the most common, closest, and therefore easiest type of AGNs to study, have been very thoroughly reviewed in the earlier papers mentioned above and by numerous other authors. To summarize very briefly, Seyfert galaxy nuclei have strong emission lines in their spectra, almost invariably broader than those in starburst galaxies or H II-region galaxies, and covering a considerably larger range in ionization. In the simplest spectral classification, Seyfert 1 nuclei are those with H I, He I, He II, Fe II, and other permitted emission lines considerably broader than [O II], [O III], [Ne V], [N II], [S II], [Fe VII], and other forbidden lines. This leads to the concept of a Seyfert 1 nucleus with a broad-line region (BLR) in which the internal velocities are large, surrounded by a narrow-line region (NLR) in which they are considerably smaller.

In the BLR the electron density must be so high that essentially all forbidden lines ordinarily seen in gaseous nebulae are collisionally de-excited. That condition sets a lower limit to the electron density $N_e \geq 10^8$ cm^{-3}. Diagnostic emission-line ratios give estimates of $N_e \approx 10^4$ cm^{-3} in the NLR and of electron temperature $T \approx 10^4$ K. The electron temperature in the BLR is less well determined observationally but seems to be of the same order of magnitude. The relatively low temperature indicates that the ionization occurs chiefly by photoionization, not by thermal ionization. The wide range of ionization, from [O I] and [N I] through [N II] and [O III] to [Ne V] and [Fe VII], indicates that the ionizing spectrum is hard, extending to high energies, with more high-energy photons than any O star, or collection of O stars, can produce, because of the approximately exponential cutoffs in their continuous spectra. Photoionization models with, for instance, an assumed power-law dependence on frequency for the input radiation give much better representations of the observed narrow-emission-line spectrum than O star model input radiation fields.

The other types of Seyfert galaxy spectrum, in the simplest spectral classification, are those in which the permitted and forbidden emission lines have approximately the same widths, comparable to the widths of the narrow, forbidden emission lines in Seyfert 1 nuclei. These are called Seyfert 2 objects. Their narrow-line spectra are quite similar to the narrow-line spectra of Seyfert 1 nuclei. Thus, the interpretation must be that the Seyfert 2 nuclei either do not have a BLR, or if they do, it is not observed because it is hidden.

Actually, many observed Seyfert galaxy nuclei emission-line spectra are intermediate, to one degree or another, between the two types, Seyfert 1 and 2. Hence, a more detailed classification is possible, namely Seyfert 1.5 for those in which the narrow and broad components of Hα and Hβ have comparable amplitude; Seyfert 1.8, in which strong narrow components, but only weak broad components of Hα and Hβ, are easily detectable; and Seyfert 1.9, in which strong narrow components of Hα and Hβ, and a weak broad component of Hα are easily detectable; but no broad component of Hβ, except with very high signal-to-noise-ratio spectra. These objects form a continuous one-parameter sequence, in decreasing strength of the observed BLR spectrum with respect to the NLR spectrum.

A highly schematic drawing of the 'structure', or overall features of a photoionization model of the gas distribution which will reproduce these observed spectra, is shown in Fig. 5.1. The general idea is that there is a central photoionization source, with the degree of ionization decreasing outwards from it, as the flux of ionizing photons decreases with distance and by absorption. The white areas (not cross-hatched) in this diagram represent the regions which are basically neutral, because most of the ionizing photons have been absorbed before reaching them. This diagram is deceptively simple and should not be taken literally. Undoubtedly the gas in AGNs is not homogeneous, nor does it have a simple, smoothly varying density distribution; if it did, it would be different from any gaseous nebula which is close enough to us to have been observed with good spatial resolution. Rather, all nebulae, envelopes, shells, etc. have strong density fluctuation, 'clumps', 'filaments', 'globules', and the like, and the gas distribution in AGNs must have the same. As in planetary nebulae and H II regions, some of the clumps are undoubtedly optically thick to ionizing radiation, and hence are most highly ionized at their surfaces facing the central ionization source, and neutral at their core, and perhaps at their surfaces facing away from the source (unless the diffuse ionizing radiation field is sufficiently strong). Study of the fine structure in the density distribution in AGNs is a very important problem for the future. We do not yet even understand what the most important parameters are to define it, nor how to study it quantitatively observationally. Understanding it theoretically is even further in the future, but likewise important. Obviously the aim must be to answer all three of the general questions stated in the introduction for the density structure at all scales.

Photoionization by a hard spectrum explains most of the features of the observed emission-line spectra of the NLR qualitatively and, to some degree, quantitatively, but it has not been observationally tested up to high levels of ionization. Observed spectra suggest that it is the main mechanism, for strong [Ne V] $\lambda3426$, a relatively high-ionization line, is well correlated with [Fe VII] $\lambda\lambda5721$, 6087; [Fe VII] with strong [Fe X] $\lambda6375$, [Fe XI] $\lambda7892$; and these two lines with the presence of [Fe XIV] $\lambda5303$ (Grandi 1978; Osterbrock 1981; Osterbrock, Dahari, and Ekberg 1983; Ferland and Osterbrock 1987). Model AGNs with photoionization by hard spectra extending to energies in the keV range can reproduce all these features (except [Fe XIV] $\lambda5303$), with the [Fe X] and [Fe XI] lines emitted within a parsec or two of the central source, as schematically indicated in Fig. 5.1 (Korista and Ferland 1989). However, there are no observed diagnostic line ratios which fix the temperature in the high-

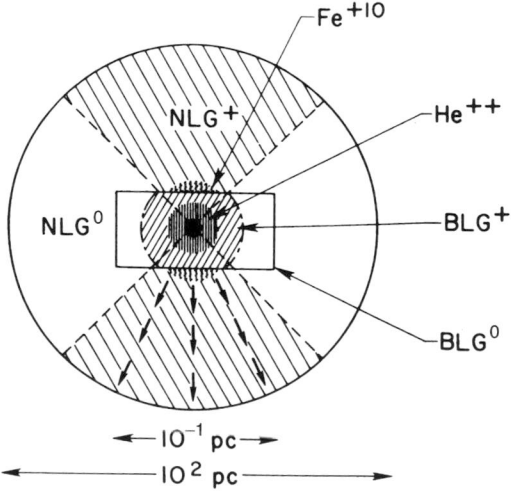

Fig 5.1 A schematic drawing of the ionization structure in a simplified AGN.

ionization zones of the real AGNs. It is quite possible that in fact there are 'coronal', high-temperature regions in them, in which these high-ionization lines are emitted under conditions of thermal rather than radiative ionization, as in the Sun. The process which heats the coronal region in AGNs, possibly dissipation of magnetohydrodynamic waves, or more likely the stopping of jet relativistic plasma in ambient gas, could be correlated with the strength of the high-energy photoionizing flux which produces the [Ne V] and [Fe VII]-emitting zones.

The best diagnostic line ratio which could be measured to determine the temperatures in a high-ionization zone, and hence the mechanism which ionizes it, is [Fe XI] ($\lambda 7892/\lambda 2649$), lines emitted by the $^3P_1$ and $^1D_2$ levels respectively, which differ by 3.1 eV in excitation potential. This separation is ideal for distinguishing between the photoionization and thermal ionization hypotheses, under which the temperature would be expected to be a few eV or a few hundred eV respectively. This is thus an important observational problem which could be carried out in the near future, comparing the ultraviolet and the near-infrared spectral regions, both of which are accessible with the *Hubble Space Telescope*.

To interpret a measured diagnostic line ratio such as this, reasonably accurate collision strengths (or cross sections) are necessary. These can now be calculated using the currently most advanced quantum mechanical methods and computational techniques, as used in the IRON project (Hummer *et al.* 1993). These collision strengths are also needed to calculate the emission rate of the individual lines in photoionization models. It will be important to have and use the best, most recently calculated collision strengths, in place of the previous values, which were based on calculations made at relatively high energies by methods which, though very good for their time, have now been superseded. Thus, for instance, for [Fe X] $\lambda 6375$ the collision strength $\Omega(^2P_{3/2}, \,^2P_{1/2}) = 0.27$, calculated for $kT = 75$ eV (Mason 1975) and used in models at temperatures $kT \approx 2$ eV (Korista and Ferland 1989), differs by more than a factor of ten from the value $\Omega = 3.2$ for $kT = 2$ eV calculated with present, more sophisticated methods (Mohan, Hibbert, and Kingston 1994). The difference is due chiefly to the inclusion of all the detailed resonance structure in the collision strengths (the quoted values are averages over the Maxwellian velocity distribution of the thermal electrons), which could not be handled by the computers of two decades ago. At present no collision strengths of [Fe XI] have been published which were calculated with the best currently available methods and computing power, but cooperation between atomic physicists and observational astrophysicists should make them available soon.

As stated above, only permitted lines are observed from the BLR; evidently all the typical forbidden lines of gaseous nebulae are collisionally de-excited in it. This sets a lower limit of density $N_e \geqslant 10^8 \text{ cm}^{-3}$ in the BLR. In most AGNs observed in the ultraviolet spectral region of [C III] $\lambda 1909$, a semi-forbidden line with transition probability $A = 10^2 \text{ s}^{-1}$ for its strongest component at high density has a broad profile similar to the permitted emission lines. This has been taken to set an upper limit $N_e \leqslant 10^{10} \text{ cm}^{-3}$ in the BLR, but it has turned out to be far too simple an assumption. It is clearly true for the regions which emit the observed [C III] $\lambda 1909$. However, several of the permitted lines, specifically including C IV $\lambda 1549$, evidently come from regions in the BLR with even higher mean N_e. Thus this result again shows the great range of electron density in AGNs, and the importance of density fluctuations, condensations, knots, etc.

The evidence for these statements comes from 'reverberation mapping', or the analysis of the continuum and emission-line variations of AGNs. The idea is simple. If the ionizing radiation from the central source varies, the ionization it produces will vary, and hence so will the emission from the ionized gas. The finite velocity of light will cause a delay in the variation of the ionization at any point; the resulting emission, integrated over the volume of the BLR, will suffer the average delay and also be distorted or 'muffled'. Accurate photometric measurements, closely spaced in time, can be used to find a mean dimension of the emitting region, $R = c\tau$, where τ is the 'observed' delay time, and to eliminate many possible structures which do not predict the observed 'distortion' (actually the convolution of the form of the ionizing radiation variation with the gas distribution). A recent symposium volume (Gondhalekar, Horne, and Peterson 1994) contains a complete up-to-date discussion of this method, from references to the early papers suggesting it, to the most recent results obtained with it.

The best-studied examples, NGC 5548 and NGC 3783, clearly show that the dimensions of the BLRs in these AGNs are smaller (by roughly a factor of ten) than those predicted from the electron-density arguments, and that the mean densities hence must be correspondingly larger. However, these same reverberation-mapping measurements confirm that the density and ionization both decrease outwards from the central radiation source, as earlier found from the line profiles. The [C III]-emitting regions in both these BLRs have larger sizes than the H I-emitting zones (which have different sizes for different lines), which are in turn larger than He II and N V-emitting zones. Thus, within the general distribution there must be extreme smaller-scale density fluctuations. Characterizing them and understanding their physical nature is an important problem for the future.

Reverberation mapping will be important in providing information on the structure within the BLR. The results to date show that simple spherical or cylindrical distributions do not fit the observed data. They do show that in NGC 5548 the structure changed over periods of years; it is certainly not static. This had previously been seen in extreme variations in the forms of the broad-line profiles in some AGNs; for instance, that NGC 4151 has changed back and forth between a Seyfert 1.5 and Seyfert 1.9 more than once in the past 15 years, and has undergone great changes in the shapes of its broad III profiles (Antonucci and Cohen 1983; Penston and Pérez 1984). The sizes of typical NLRs are so large (from their inferred densities and observed total luminosities in Hβ) that no variations are expected to occur in them except over long times, of the order of centuries, and none have been reliably observed to date. Perhaps this may be regarded as a very long-term problem for the future.

5.3 Cylindrical symmetry

Figure 5.1 is intended to show schematically that the BLR has roughly cylindrical symmetry, but is immersed in a much larger NLR, shown with spherical symmetry but more probably simply the central part of the interstellar gas distribution in the galaxy, again with roughly cylindrical symmetry on a very large scale. The main reason for assuming that there is an axis, rather than roughly spherical symmetry, is that the characteristic dimension, particularly of the BLR, is so small that any non-zero angular momentum on the scale of the galaxy corresponds to a very large rotational velocity on the scale of the nucleus. The central source is evidently closely associated with a rotating accretion disk. Very probably this rotational symmetry characterizes the entire BLR (Shields 1977; Osterbrock 1978a, b).

Furthermore, many Seyfert galaxy nuclei have weak radio emission which, with high angular resolution, often shows a jet structure, in some cases with bilateral symmetry, in other cases one sided. However, it is important to note that the jets, which are almost certainly along the axis of the AGNs, are in general not seen in projection perpendicular to the plane of the galaxy, indicating that the axis of the nucleus is not the same as the axis of the galaxy. There is abundant empirical evidence that in many galaxies, including so-called 'normal' ones, there are 'warps' or misalignments of the angular momentum of the central parts or nucleus with the overall angular momentum of the observed galaxy. There is no dynamical reason why this cannot be the case (Tohline and Osterbrock 1982; Rubin 1994). Obviously it is not a stable situation in the very long run, but that it exists thus agrees with the general picture of fuelling and refuelling of the central nucleus as a result of interactions between galaxies (see Section 5.7).

Those ideas were strongly confirmed by the discovery by Antonucci and Miller (1985) that the Seyfert 2 galaxy NGC 1068 shows, in plane-polarized light, the spectrum of a Seyfert 1 galaxy, with broad H I and Fe II emission features, as well as a strong featureless continuum (Miller and Antonucci 1983). Furthermore, the plane of polarization (maximum E vector) is perpendicular to the axis of the radio jet in this nucleus. This is exactly the situation which would result from a BLR 'hidden' within a cylindrical torus whose axis lies nearly in the plane of the sky. No photons can escape directly from the BLR to the observer; they are absorbed or scattered by the torus. Only the photons which escape along the axis and are scattered above and below it can be observed; as a result of the scattering their mean plane of polarization will be perpendicular to the axis. Antonucci and Miller (1985) found the measured degree of polarization to be independent of wavelength, and therefore concluded that the scattering was probably by free electrons.

From Lick Observatory they were able to observe over only a limited wavelength region, approximately $\lambda\lambda 3500$–7000. More recently, measurements with the WUPPE telescopes and spectropolarimeter in the space shuttle *Columbia* in the spectral region $\lambda\lambda 1500$–3300, Code *et al.* (1993), and with the Faint Object Spectrograph polarimeter on the *Hubble Space Telescope* in the region λλ1600–3300, Antonucci, Hurt, and Miller (1994), have shown that the polarization remains independent of wavelength well into the ultraviolet. This clearly establishes electron scattering as the polarization mechanism.

NGC 1068 is not the only Seyfert 2 object which shows a 'hidden BLR' in plane-polarized radiation. The observations are very time consuming, because the degree of polarization is small in all objects studied to date, and the required signal-to-noise ratio is therefore high. This restricts the observations to reasonably bright Seyfert 2 nuclei, but among them eight additional objects with hidden BLRs have been found (Miller and Goodrich 1990; Tran, Miller, and Kay 1992). Hence, this concept must apply to many, and perhaps all, Seyfert galaxies. Detailed spectropolarimetric observations of these additional Seyfert 2 galaxies with hidden BLRs basically confirm the torus model; in particular, the observed plane of polarization is, in all these objects, generally perpendicular to the axis defined by the radio jet or other structure (Tran 1995).

Also, as Fig. 5.1 suggests, the cylindrical symmetry, with ionizing radiation from the central source escaping through the BLR along the axis, may be expected to give rise to 'ionization cones' in the surrounding NLR. As described in my earlier review, such ionization cones were detected in NGC 1068 and several other Seyfert 2 galaxies by long-slit spectroscopy, and by imaging through emission-line filters, especially [O III] and Hα + [N II], by Baldwin, Wilson, and Whittle (1987), Unger *et al.* (1987), Pogge (1988a, b), and others. A very good summary, including a table of ionization cones discussed to date, has recently been published by Wilson and Tsvetanov (1994). Also quite recently, the ionization cone in NGC 1068 has been traced down to very small scales with high-resolution images, taken with the Faint Object Camera on the *HST*, after the COSTAR deployment (Macchetto *et al.* 1994).

Still more recently, Boksenberg *et al.* (1995), also using the Faint Object Camera on the *HST*, obtained an [O III] λ5007 image with very good angular resolution of the central part of NGC 4151, which appears to show both sides of an ionization cone centred on the nucleus but misaligned with the galaxy. According to the interpretation of these authors, the cone opens out enough to include the line-of-sight, consistent with the classification of the nucleus of NGC 4151 as a Seyfert 1.5. Furthermore, in their interpretation the line of sight is close to the edge of the ionization cone, consistent with the fact that NGC 4151 varies in type, sometimes becoming a Seyfert 1.8 or 1.9 (Antonucci and Cohen 1983; Penston and Pérez 1984), no doubt when the dust in the torus or near the BLR blocks the light path from most of that region to the observer for a time.

The ionization cones are distinguished by a relatively high level of ionization, clearly from the central source rather than OB stars. In all cases the position angle of the axis of the ionization cones agrees closely with the position angle of the jet structure. In some Seyfert 2 galaxies double-sided cones are seen, on both sides of the nucleus; in others on only one side, projected on the galaxy. Clearly in the latter objects, the other side of the cone, if it exists, is hidden by extinction in the plane of the disk. Dimensions of the observed ionization cones range from 10^2 pc to 10^4 pc. The (full) opening angles of the cones cover a wide range, centred on about $65°$, and their axes are not perpendicular to the plane of the galaxy. However, Wilson and Tsvetanov (1994) found, for the 11 galaxies with ionization cones measured to date, a good correlation between galaxy type and apparent misalignment of the galaxy and the cone. Further measurements of many more ionization cones would be desirable to confirm or reject this correlation, since, if it is valid, it may have implications for the refuelling processes discussed in Section 5.7.

A good, overall review of 'unified models' with cylindrical symmetry, in which the apparently different types of AGNs are explained in terms of orientation, has recently been published by Antonucci (1993).

5.4 The velocity field

The velocity field in the BLR is surely the subject we understand least well and for which new ideas and new types of measurement are needed. Line-profile studies including comparing Hα , Hβ, He II, and He I agree in showing that the highest velocities occur near the centre of the BLR and decrease outwards. Probably the strongest result of the BLR reverberation-mapping programmes for the velocity field is that it is neither outflow nor inflow primarily, but that both occur and nearly balance (Gondhalekar, Horne, and Peterson 1994, especially Sections 5.1 and 5.3). Evidently the motions are controlled by the gravitational force due to the central object (black hole), and can be regarded as a combination of 'turbulence' and rotation. The structure must be composed of many small, dense clouds, as suggested in Section 5.2. How the clouds are confined, and why they do not expand, merge, and lose their individuality as they are heated and disrupted in collisions with one another at highly supersonic velocities are severe problems.

One possibility, suggested by Kazanas (1989), and further investigated by Begelman and Sikora (1992) and Alexander and Netzer (1994), is that the clouds are giant stars, their outer atmospheres photoionized and heated by the central source, expanding in winds. The gravitational force of the 'bloated star' itself is the 'confinement' (actually restraint) which keeps the 'cloud' from expanding and dissipating too rapidly. Many possibilities are open, in terms of the distribution of stars in the nuclei, their velocity distribution, their mass-loss rates, the physics of the wind, etc. However, the idea is an attractive one, and further investigations are certainly desirable.

In the NLR the velocity field is better understood. Most of the information comes from line profiles. A very good overall interpretation is given by Veilleux (1991), with many references to earlier observational work and theoretical interpretations of it. In the inner region of the NLR (say within 50 pc of the central source), the flow is primarily outwards along the axis of the jet. Evidently the radio-emitting plasma affects the velocity field of the observed NLR gas. The highest densities, ionization, and velocities occur, on average, closest to the central source, but the distribution is decidedly cloud-like, the opposite extreme from homogeneous. There is heavy dust, again in clouds in the central 'plane', perpendicular to the jet. This interpretation is drawn from the observed result that the narrow-line profiles are almost all asymmetric, with a blue wing, indicating more observed gas with velocity approaching the observer than the reverse. In the outer part of the NLR, which merges continuously into the interstellar gas in the galaxy, the velocity field goes over into a rotational flow, indicating the importance of gravitational forces on the larger scale. In general, the axis of symmetry of the outer part (sometimes called the extended NLR or ENLR) goes over continuously to the axis of the galaxy; that is, there is a warp in the gas distribution.

Long-slit, or better, multi-slit, spectra of the nearest Seyfert galaxies with the most favourable angular scale for observing from the Earth, seem to show much of this general, overall picture, but with a great deal of local structure. Some examples are NGC 2110 (Wilson and Baldwin 1985), NGC 5548 (Wilson *et al.* 1989), NGC 5278 (Arribas and Mediavilla 1993), and NGC 3227 (Arribas and Mediavilla 1994). In some regions of all these galaxies the narrow-line profiles are complex, evidently blends, one component apparently representing gas in the main plane of the galaxy, ionized by the OB stars, the other gas in the ionization cone, ionized by the central source. These spectra thus also seem to show the existence of warps. Near the BLR the velocity structure is complicated and difficult to interpret, probably indicating the small-scale warping of the gas distribution close to the central source as the NLR merges with or changes continuously into the BLR. The advantage of the two-dimensional, multi-fibre spectrograph over a one-dimensional, long-slit instrument is evident from comparison of the later two of these four papers with the two earlier ones. This small-scale spectral mapping of the velocity field, and with it the ionization field and temperature field (in various ions, especially O^{++} and N^+) promises to provide important new information and should certainly be continued and extended. The interpretation will be difficult; the hope is to understand the *general* features of the velocity, ionization, and temperature fields by studying several different galaxies, and thus to disentangle them from the specific, unique features in the individual galaxies.

5.5 Continuity

As stated in Section 5.1, QSOs and Seyfert 1 galaxies appear to form a continuum, their names having historical significance only. Better seeing, CCD detectors with their higher quantum efficiencies, and efficient digital methods of reducing images have revealed faint extensions around objects originally classified as QSOs or quasars. Several references are listed in my earlier reviews (Osterbrock 1991, 1993a). In many cases the underlying 'galaxy', to the extent it can be seen, does not seem to fit into a standard morphological classification scheme for 'normal' galaxies but appears to be distorted or otherwise peculiar. This is related to the ideas of evolution, fuelling, and refuelling discussed in Section 5.7.

As mentioned in Section 5.1, the generally adopted, arbitrary definition of the division point between QSOs and Seyfert 1 nuclei is absolute magnitude (in the rest system of the object) $M_B = -23$. This corresponds to approximately $10^{45.6}$ erg s^{-1} $\approx 10^{12} L_\odot$ in the optical, ultraviolet, and mid-infrared spectral regions. However, AGNs radiate over a very wide frequency or energy range, from γ and X-rays to the far-infrared and radio-frequency regions. Hence the overall luminosity of a 'typical' AGN with $M_B = -23$ is approximately $10^{46.3}$ erg s^{-1} $\approx 5 \times 10^{12} L_\odot$. Almost all high-luminosity AGNs and QSOs have Seyfert 1-type spectra; there are practically no high-luminosity Seyfert 2s. However, what is apparently the object with the highest luminosity known in the Universe, an *IRAS* faint source, FSC 10214+4724, at redshift $z = 2.286$, has an emission-line spectrum (observed in the range $\lambda\lambda 1500$–2800 in the rest system of the object) which can be classified as Seyfert 2 (Rowan-Robinson *et al.* 1991; Elston *et al.* 1994; Soifer *et al.* 1995). It has a very high dust content, and will be discussed further in Section 5.7. Except for it, and a very few other, less extreme cases, all the high-optical-luminosity QSOs and quasars have spectra of Seyfert 1 (or similar broad-line radio-galaxy) type. Note that it is possible (but not certain) that FSC 10214+4724 is somewhat amplified by gravitational lensing (Matthews *et al.* 1994; Liu and Graham 1995), so its luminosity is uncertain to some extent.

This suggests that the torus, which blocks the radiation from the BLR from escaping directly, covers a larger range of solid angle in the lower-luminosity objects, and a smaller range in the high-luminosity objects. The 'broad-absorption-line' QSOs or BALQSOs, discussed much more fully by Margaret Burbidge in Chapter 4 of this volume, are QSOs in which broad resonance absorption lines, arising from atoms or ions in their ground levels, appear in the spectrum with blueshifts, indicating a velocity of approach. Evidently they arise in a shell which is being driven off from the BLR; this shell may be the observational manifestation of part or all of the torus. Approximately 10% of optically selected radio-quiet QSOs have this BALQSO property (Weyman *et al.* 1991; Voit, Weymann, and Korista 1993).

This suggests that the torus (or at least the parts of it through which optical continuum radiation can escape, and in which absorption lines can form) covers only about 10% of the total solid angle, 4π, about the BLR in typical QSOs (with M_B brighter than -23). For the lower-luminosity Seyfert galaxies a considerable fraction are Seyfert 2s. Two recent estimates are 52% (Osterbrock and Martel 1993) or 73% (Osterbrock and Shaw 1988), from two different, supposedly complete samples over different regions of space. A rough mean would be 60%, suggesting this figure as an average covering factor for Seyfert galaxy nuclei. This great difference in covering factors between the QSOs and Seyfert galaxy nuclei indicates that the covering factor increases towards lower luminosities. Obviously, however, another effect comes into play as well, namely that the apparent optical luminosity of the central source and BLR is higher as viewed along the axis and lower as viewed through the torus. Disentangling these two effects is an important problem for the future.

At still lower luminosities, LINERs (low-ionization nuclear emission regions) are a class of active galaxies isolated by Heckman (1980). They have relatively small [O III]/Hβ emission-line intensity ratios and relatively large [O I]/Hα and [S II]/Hα ratios. Spectroscopic diagnostics and comparison with models strongly suggest that many of them are extensions of Seyfert 2 galaxies to lower luminosities, photoionized by a weaker, hard AGN spectrum. Detailed studies show that many have very weak, broad Hα emission wings, indicating a weak BLR, detectable only on good signal-to-noise-ratio spectra, with a 'template' galaxy absorption-line spectrum subtracted (Filippenko and Sargent 1985). Other LINERs, however, are objects in which collisions between gas clouds ('shock waves') are the main energy-input mechanism. Many of these are recognizable as objects in which the emission occurs in extended regions, rather than in small nuclei (Heckman 1987).

The weakest known LINERs with nuclear photoionized emission lines go down to the limit of detectability; an early estimate by Keel (1983), based on careful processing of good signal-to-noise-ratio spectra of a well-defined, complete magnitude-limited sample of 'normal' spiral galaxies, was that half showed weak AGN properties in their spectra. No doubt among the other half which did not show these properties, still weaker signs of AGN activity are present and would be detectable with even more sensitive methods. Evidently a very large fraction of spiral galaxies contain AGNs at some level of activity. This is clearly one of the most important future problems of AGN research.

A very complete, recent analysis of 13 LINERs, based on excellent observed spectra covering the range $\lambda\lambda$3400–9800, reduced and analysed with templates to remove the disturbing effects of the underlying starlight of these galaxies, has confirmed that 'compact LINERs' are indeed very probably photoionized and thus members of the 'one family' of AGNs (Ho, Filippenko, and Sargent 1993). The observed spectra can be best fitted with photoionization models, with an assumed hard, power-law-type input spectrum, but with a lower ionization parameter than in Seyfert 2 or NLR models. Available shock-heating models do not fit the observed relative line intensities as well as photoionization models. However, the shock models are greatly simplified, and further theoretical and numerical work in this direction would be desirable. Also, long-slit, or better, multi-slit, spectra would be very useful for disentangling photoionization and shock heating.

Nearly all observed, nearby Seyfert galaxies clearly have star formation going on in them, with many H II regions photoionized by hot stars evident in their images. Generally speaking, the H II regions are widely spread through the galaxy, but many may be near the nucleus, and some within it. The observed emission-line spectrum then depends on the angular size of the slit aperture used to obtain it, but even with a short, narrow slit, some admixture of the H II-region spectrum 'contaminates' the AGN spectrum. For a LINER with a relatively faint AGN-type emission-line spectrum, the H II-region component may have a quite noticeable effect on the overall observed spectrum. This must always be kept in mind and allowed for in comparisons with simplified models, which assume a single type of photo-ionizing spectrum or heating mechanism.

More recent observational data in the X-ray and ultraviolet spectral regions also seem to agree with the idea that many LINERs with condensed nuclei are the low-luminosity extensions of Seyfert galaxies. Koratkar *et al.* (1995) studied the X-ray emission, measured with *ROSAT*, from three low-luminosity Seyfert 1 nuclei and two LINER nuclei. All five have observed broad Hα emission-line components. In all five the strong correlation between soft-X-ray luminosity and broad Hα emission-line luminosity found earlier by Kriss, Canizares, and Ricker (1980), and now confirmed by a larger body of observational data on QSOs and Seyfert galaxy nuclei, appears to continue to low luminosity. The X-ray emission is confined to the nuclei. The X-ray spectra, to the extent they can be measured, fit a power-law ν^{-n} with index $n = 1.6$, quite similar to luminous AGNs ($n = 1.4$) in the same band. All these results are consistent with the picture that LINERs are the low-luminosity members of the one-family of photo-ionized AGNs, although of course they do not completely rule out other, more complicated interpretations.

Likewise, recent *Hubble Space Telescope* measurements confirm that a significant fraction of the LINERs observed with it have a bright ultraviolet continuum (near $\lambda 2270$) confined to their nuclei (Maoz *et al.* 1995). Again, the sample is small, and the observational result does not prove that they are photoionized, but it is consistent with it. Of the sample of 26 LINERs observed with the *HST*, five, approximately 20% of the total number, showed the bright ultraviolet nuclear continuum. Taken at face value, this would imply that the covering factor of the torus is roughly 80% in these low-luminosity LINERs, still larger than in the 'typical' Seyfert 2 nuclei.

An important part of the 'one-family' hypothesis is that QSOs are more luminous than most 'normal' galaxies. Some of them would be expected to be in clusters of galaxies, more luminous than the other cluster members. Because they are so rare, even the nearest QSOs and quasars are distant, and the galaxies in clusters which contain them are expected to be corres-pondingly faint and difficult to detect. More than a decade ago Stockton (1978), examining candidate galaxies within $45''$ of 27 QSOs with redshift $z \leq 0.45$, found 29 galaxies with magnitudes which made them candidates for cluster membership. He obtained redshifts of 25 of them; of these, 13 galaxies in eight fields turned out to have redshifts within $1000 \, \mathrm{km \, s^{-1}}$ of the QSO, far more than expected from chance coincidences. With the improved imaging detectors which have become available in more recent years, tests of this kind have been pushed to fainter magnitude limits, revealing many more galaxies in clusters around some QSOs and quasars, for instance by Ellingson, Yee, and Green (1989, and references given there).

Very recently, using the refurbished *HST* with the COSTAR corrective optics instrument, Bahcall, Kirkhakos, and Schneider (1994, 1995) have obtained images of eight luminous QSOs with redshifts $0.16 \leq z \leq 0.29$. They obtained these images with the Wide-Field Planetary Camera through a filter centred at $\lambda 6060$, similar to the V filter, but slightly redder than it. They attempted to subtract the 'stellar' AGN from the image, to see if an underlying galaxy could be detected. Their general result was that none of these luminous QSOs (their limit was $M_V < -22.9$ for $H_0 = 100\,\mathrm{km\,s^{-1}\,Mpc^{-1}}$) is in a highly luminous galaxy. In three of the objects, residual images which could plausibly be interpreted as galaxies were detected. In the other five, no galaxy was detected, but fairly stringent limits were set to the absolute magnitude of the galaxy which might be present, ranging from approximately $M_V = -20$ down to -19. These limits were obtained using images of individual galaxies taken from similar exposures with the *HST*. Thus, their result contradicts the idea that the more luminous AGNs occur in more luminous galaxies.

That idea was based on the conclusion of McLeod and Rieke (1994a, b), who in two papers surveyed a large number of QSOs and Seyfert 1 galaxies, using a NICMOS array camera working in the near-infrared H band at $1.65\,\mu$m, on the Steward Observatory 2.3-m telescope on Kitt Peak. They obtained images which they used to search for a 'host' galaxy about each nucleus. In the 'high-luminosity' group of 26 QSOs with $M_B = -23.1$ and $z < 0.3$, they reported detections of the host galaxy in at least 23, while in the low-luminosity group of 24 AGNs with $-23.1 \leq M_B \leq -22$, they reported detections of host galaxies in at least 22 (these authors calculated absolute magnitudes assuming $H_0 = 80\,\mathrm{km\,s^{-1}\,Mpc^{-1}}$). The mean absolute magnitude they found for the host galaxies of the high-luminosity QSOs was $\langle M_H \rangle = -24.6$, and for the host galaxies of the low-luminosity group, $\langle M_H \rangle = -23.9$. The difference of 0.7 magnitude is in the sense that the more luminous nuclei are in more luminous galaxies.

However, there is some discrepancy between their results and those of Bahcall *et al.* (1995), whose eight QSOs were all imaged also by McLeod and Rieke (1994a, b). Since one group worked in the H band, and the other in a band near V, their results are not directly comparable. However, Bahcall *et al.* (1995) set upper limits to host galaxies which are inconsistent with magnitudes reported by McLeod and Rieke (1994a, b), unless their $(V - H)$ colour indices are very different from 'normal' galaxies. Since in all cases the galaxies are small, with characteristic scales comparable with the seeing disk of the ground-based images, exact subtraction of the bright nucleus is by no means a straightforward procedure. It seems likely that the *HST* results, in which 'seeing' (actually, apparently due to scattered light) is less of a problem than in the ground-based images, are to be preferred. Obviously, this is a very important question for further observational research.

It has long been known that the host galaxies of AGNs and QSOs are in many cases 'peculiar', or distorted, with more complicated structures than 'normal' galaxies (Miller 1985). In searching for possible faint 'host' galaxies of distant AGNs, some kind of model(s) must be used to represent the galaxy. A wide range of possibilities exist. Models based on exponential disks, King models of ellipticals, or other conventional representations are not likely to apply in all cases. Bahcall *et al.* (1994, 1995) tried to allow for this by using models based on observed galaxies, some of them quite abnormal. Perhaps an even wider range of such 'observational models' must be used.

This is partly confirmed in a third paper by Bahcall *et al.* (1996), based on *HST* images of PKS 2349−014, one of the objects in their sample, at $z = 0.173$, with $M_V = -23.4$. The high-resolution *HST* image clearly shows that it is an interacting system, with the bright QSO nucleus, a large distorted galaxy in which the nucleus is off centre, at least one 'companion' (nearby) galaxy, two 'thin, covered wisps', and possibly a 'host galaxy' centred on the QSO, but not resolved from it even in this image.

Also, the possibility must certainly be kept open that some QSOs may appear in dwarf galaxies. As mentioned previously, AGNs are strongly associated with star formation; as will be discussed further in Section 5.7, it seems quite likely that a small fraction of the total number of starburst galaxies with star formation going on in their nuclei evolve into AGNs. In addition to the well-known starburst galaxies, there are many lower-luminosity dwarf galaxies which have very strong star formation occurring in their nuclei. Salzer, MacAlpine, and Boroson (1989) have isolated several groups of such objects, discovered by their emission lines. In our present state of relative ignorance of the origin, formation, and refuelling of AGNs, it seems not at all unlikely that there may be similar relatively small, low-luminosity galaxies with luminous QSOs in or near their nuclei. Such dwarf galaxies would be very difficult to detect at even moderate redshifts, and may actually surround some of the 'bare' QSOs, for which severe limits have been placed on the luminosity and size of the 'host' galaxy. There may also be low-surface-brightness galaxies, too faint for detection even with the refurbished *HST*.

5.6 Black holes

The best working hypothesis for the energy source in AGNs is the accretion disk around a black hole, as suggested originally by Salpeter (1964). This idea was worked out in considerable detail by Lynden-Bell (1969), and in the intervening years by many subsequent authors. Igor Novikov also discusses it fully in Chapter 7. As discussed in many earlier reviews, including mine mentioned above, the best estimates of the masses of the black holes in Seyfert 1 galaxies appear to be $M \approx 10^{7.5}$–$10^{8.5} M_\odot$, and $M \approx 10^8$–$10^{9.5} M_\odot$ in QSOs. The observed luminosities, in terms of the well-known Eddington luminosity,

$$L \leq L_{\mathrm{E}} = \frac{4\pi cGm_{\mathrm{H}}}{\sigma_T} = 1.3 \times 10^{38} \frac{M}{M_\odot} \mathrm{erg\ s^{-1}}, \tag{5.6.1}$$

the upper limit to the luminosity of a spherical object of mass M held together by radiation pressure, correspond roughly to $L \approx 10^{-1} L_{\mathrm{E}}$ for Seyfert 1 nuclei, but $L \approx L_{\mathrm{E}}$ for QSOs. For a QSO radiating at the Eddington limit, the time scale for its mass to increase by a factor e is

$$\tau = 4 \times 10^7 \left(\frac{\epsilon}{0.1}\right) \mathrm{years}, \tag{5.6.2}$$

where ϵ is the fraction of the mass disappearing into the black hole which escapes as radiation, generally estimated as $\epsilon \leq 0.1$. Thus 10^7–10^8 years are the times expected for a QSO or Seyfert 1 nucleus to exhaust its available fuel, unless the supply is very large. Hence, it would seem that galaxies have relatively short-lived active phases, and that there are probably many currently inactive galaxies with black holes in their nuclei which have run out of fuel.

On this picture, black holes would be expected to be present in all current AGNs and in many inactive galactic nuclei. Very recently Ford *et al.* (1994) obtained high-quality images through narrow-band filters with the new Wide-Field Planetary Camera (WFPC) installed in the *HST* in December 1993, of the nucleus of the nearby radio galaxy M87 = Virgo A. The Hα + [N II] images (with continuum subtracted) show a small disk of ionized gas with spiral arms. The well-known jet appears to be approximately perpendicular to this disk. The radius of the disk is approximately $1''$ or 75 pc. Harms *et al.* (1994), using the Faint Object Spectrograph on the *HST*, with the COSTAR corrective optics in place, obtained resolved spectra of this disk along a slit position perpendicular to the jet. The measured radial velocities appear to show that the disk is rotating with a velocity of 460 km s^{-1} at a radius of 18 pc from the centre. Taking the various projection factors into account, this corresponds to a mass $M = (2.4 \pm 0.7) \times 10^9 M_\odot$ within a radius of 18 pc, which seems (from the measured mass-to-light ratio) to be much too large to be attributed to the stars. Hence, the best interpretation of the observational data is that most of this mass is in a massive black hole and that it provides the energy for the ionizing and other non-thermal radiation from the nucleus. This interpretation is quantitatively in agreement with earlier ground-based and *HST* lower-resolution measurements of M87.

An even more direct apparent confirmation of a massive black hole in a galactic nucleus is the recent radio-frequency measurement of NGC 4258, a mildly active galaxy. Miyoshi *et al.* (1995) measured an H_2O maser in its nucleus with the VLBA of the NRAO, augmented by the VLA operated as a single element. The resulting synthesized beam was 0.6×0.3 milliarcseconds in size. A total of 4096 velocity channels were observed, with strong positive signals in 593 of them. They show a nearly plane structure, with measured velocity decreasing outwards from the centre with distance as $r^{-1/2}$, exactly as expected for a Keplerian velocity field. These velocity features arise in a disk having inner and outer radii ~ 4 and ~ 8 milliarcseconds, corresponding to 0.13 to 0.25 pc. The velocity drops from 1.080×10^3 to 0.770×10^3 km s^{-1} over this range, observed on both sides of the centre. All this corresponds to a nearby edge-on disk, with central mass $3.6 \times 10^7 M_\odot$. The average density derived from this mass, which must be within 0.13 pc of the centre, is $\rho > 4 \times 10^9 M_\odot$ pc^{-3}. Hence, it is almost certainly a black hole.

A review article by Kormendy and Richstone (1995) discusses in detail the observational evidence for central, massive, dark black holes in nearby currently inactive galaxies, based on spectroscopic measurements of rotation curves and of the stellar velocity dispersion very near their centres. Their review summarizes the evidence for 'massive dark objects', presumably black holes, in the nuclei of M31, M32, NGC 3115, NGC 3377, and NGC 4594, ranging from $\sim 2 \times 10^6 M_\odot$ to $1 \times 10^9 M_\odot$. They also discuss the centre of our Galaxy, which has been very thoroughly observed by infrared, millimetre-wave, and radio-frequency techniques for radial velocities of stars and 'gas clouds' (or nebulae). Whole conferences and symposium volumes have been devoted to it. Some of the evidence is contradictory, but the bulk of it favours a central black hole with $M \approx 2 \times 10^6 M_\odot$. Certainly many more observations should be made of it, particularly with techniques which can measure stellar radial velocity as close to the centre as possible.

The nearby spiral M33 is the one well-studied object in which observational evidence rules out a central massive black hole. The upper limit, from the central velocity dispersion, is $\leq 5 \times 10^4 M_\odot$, far too low to be a former AGN (bright enough to be observed) which has exhausted its fuel (Kormendy and McClure 1993).

5.7 Evolution

Empirically, AGNs seem to occur most frequently in interacting galaxies, in galaxies which have recently undergone interactions, or in galaxies which have 'companions', meaning galaxies near them in space and radial velocity (although it is doubtful that many of these are in long-term, stable periodic orbits). The early suggestions along these lines by Adams, Vorontsov-Velyaminov, Gunn, and Simkin have been discussed in my earlier reviews. Also, statistical studies seem to show that there is an excess of Seyferts among galaxies with companions, and an excess of galaxies with companions among the Seyferts, although a single early study reached the opposite conclusion, as also discussed in those reviews.

Most recently Rafanelli, Violato, and Baruffolo (1995), in the largest study of this type to date, have again confirmed that there is an excess of physical companions to Seyfert galaxies. They used two well-defined complete samples, down to apparent magnitude $B = 15.5$, of 99 Seyfert 1 galaxies and 98 Seyfert 2 galaxies. Among the Seyfert 1s they found the lower limit to the fraction with physical companions (with the distance and magnitude difference from the primary which they adopted) to be $12 \pm 4\%$, while among a control sample, treated in exactly the same way, the upper limit to the fraction with companions was between 0% and 5%. Thus, there is a clear surplus of companions to Seyfert galaxies. In a brighter complete sample, limited to $B \leq 14.5$, the CfA sample of Seyfert galaxies, for which published radial velocities were available for all the nearby physical and optical companions, Rafanelli *et al.* found that 19% of the Seyfert 1 galaxies and 18% of the Seyfert 2s do have physical companions (defined as having a redshift difference from the primary $|c\Delta z| \leq 1000 \, \mathrm{km \, s^{-1}}$), well above the lower limit established on statistical grounds.

This observational correlation that AGNs tend to occur preferentially in galaxies with companions suggests that fuelling of inactive black holes can and does occur in gravitational interactions between galaxies. Indeed, this suggestion was first made by Toomre and Toomre (1972) to explain the fact, then observationally known from specific examples, that many 'active galaxies' were also highly distorted. Toomre and Toomre (1972) showed that many of these examples could be understood as the result of gravitational interactions (or 'collisions') between galaxies with specific interaction parameters, on the basis of their model calculations. From subsequent theoretical investigations by several authors, especially Hernquist (1989), it seems highly likely that in many gravitational collisions, interstellar gas in one galaxy may be perturbed into orbits which bring some of it close enough to the nucleus to be captured and, probably as a result of subsequent instabilities, be brought still closer to the centre and ultimately fed into the accretion disk. Thus, it appears that interactions can supply the tidal perturbations which fuel inactive black holes in galactic nuclei, transforming them from an 'inactive' state to an AGN. In many cases, a merger may result, again with the result that gas is 'scattered' into orbits with nearby zero angular momentum about the nucleus, so that it may pass very close to it, and perhaps be captured there.

It can be seen that on such a picture the gas which comes close to the nucleus may have its angular momentum defined more by the orbital parameters of the interaction than by the overall velocity field of the whole galaxy, which is to say that the axis of symmetry near the nucleus may be quite different from the axis of symmetry of the galaxy. As stated above, this is in good agreement with the observational data. The observations show that many starburst galaxies also have companions. Furthermore, the nuclei of most starburst galaxies (in which star formation is going on) are more extended than AGNs. These observational results fit in very well with the theoretical ideas expressed by Hernquist (1989), that a multi-step process which does not go all the way in all cases moves gas down from orbits on the scale of a galaxy to the scale of the AGN, and with the results of Lin, Pringle, and Rees (1988), that perturbations can trigger non-axisymmetric gravitational instabilities which can work from scales of 10^4 pc down to 10^2 pc. Further instabilities, perhaps associated with the occurrence of a burst of star formation, may then further reduce the angular momentum of some of the gas enough for it to reach the potential fuelling region (~1 pc). Probably only a small fraction of the starburst galaxies ultimately become Seyfert galaxies. Understanding the successive processes theoretically is obviously a necessity to understand their evolution.

The most violent perturbations in progress, as shown by the distorted forms of the galaxies, are not observed in Seyferts but in starburst galaxies. This result has been found and confirmed by many authors, as referenced in my earlier review (Osterbrock 1993a). Partly it may be a time effect, that the burst of star formation occurs first and triggers the process, and that the fuelling occurs later, when enough gas has 'fallen' to the fuelling region. Byrd, Sundelius, and Valtonen (1987) have shown that although the fraction of Seyferts reported with close companions is relatively small, if one takes account of more distant companions, of this type of delayed effect, and of mergers in which the companion has disappeared, it is quite possible that all observed Seyferts are the result of fuelling by gravitational interactions. This result should be checked with much more sophisticated model calculations, and by further observational data, particularly on the forms of the distortions which arise in such interactions.

A very important step in this direction is the simulation atlas of tidal features in galaxies produced by Howard *et al.* (1993), based on calculations of a large number of models of specific interactions between a galaxy composed of two components, stars and gas, with an essentially point mass as the perturbing 'companion'. The calculated models cover a wide range of interaction parameters, and the results are presented in video form. The aim is to compare observed galaxies with the atlas and determine the parameters of the interactions which best fit them. One of these parameters is the time (before or after the epoch of closest approach on any other arbitrary zero point), and thus it is possible in principle to date an observed Seyfert galaxy from its form. This gives the hope of studying in detail the evolution of Seyfert galaxies by assigning ages to many individual objects and combining them to trace the evolution of a single subject, just as studies of colour–magnitude diagrams of stars in globular clusters have led to an understanding of stellar evolution.

However, the galactic case is much more complicated, because not just one parameter, stellar mass, comes into play, but several, including the mass of the 'companion' relative to the primary galaxy, the relative velocity at infinity, the impact parameter, and various orientation angles. However, this is the programme which, although difficult, in my opinion is the way in which we shall come to understand the evolution of the spectrum, energy source, and nature of AGNs (Osterbrock 1993a). Some steps in this direction, comparing observed images of galaxies with earlier, less complete numerical models of interactions have already been taken by Hutchings and Neff (1991, 1992), who assigned a 'strength' or a 'mass ratio', and an 'interaction age' to each object in this way. A recent high-resolution study of the Seyfert 1.5 galaxy Mrk 315, based on *HST* WFPC, Mauna Kea, and VLA images, as well as on long-slit spectra, strongly suggests that it is the result of a recent merger, with a remnant of the captured galaxy still visible (MacKenty *et al.* 1994). Detailed studies of this type of many more objects will be extremely valuable in building up an observational picture of AGN evolution, which then can be the basis for theoretical interpretation and understanding.

Heller and Shlosman (1994), in a recent long, theoretical paper, have given a stimulating numerical, smoothed-particle, hydrodynamical treatment of instabilities in a galactic disk, with special emphasis given to the central region, including star formation, possible formation of a massive black hole, and fuelling of a possibly pre-existing black hole. Naturally many simplifications were necessary, particularly in the recipes for star and black-hole formation, but these calculations are very instructive in tracing quantitatively the importance of bars in triggering radial inflow towards the nucleus and in fuelling a growing black hole. Further theoretical work along these lines is clearly highly important.

In a complementary, recent, observationally oriented paper, Moles, Marquez, and Pérez (1995) have discussed the relation between dynamical perturbations, galaxy morphology, and nuclear activity. They emphasize that most Seyfert galaxies are early-type spirals, and that if a gravitational collision is not observed to be in progress, they are often barred spirals, evidently the result of such an interaction in the not too distant past. These authors conclude that 95% of the total, well-defined sample of active spiral galaxies which they studied show signs of organized non-axisymmetric features. From this they conclude that the existence of non-axisymmetric perturbations, preferably in early-type spirals, appears to be a necessary condition for the onset of nuclear activity, in very good agreement with the conclusions above.

Very probably when galaxies first form, the situation is sufficiently turbulent, disorganized, or even chaotic that a black hole may form initially without an external perturbation. This is the idea put forward by Sanders *et al.* (1988) in their study of ultraluminous infrared galaxies, which they interpreted as 'the origin of quasars'. One very good example of these objects is FSC 10214+4724, previously mentioned in Section 5.5. We have observed very few such objects except at quite large redshifts. It seems to me that some at nearer redshifts, perhaps including FSC 10214+4724, may well be objects in which a black hole has recently formed (an 'origin'), while others of them may be objects in which a slow merger with another gas and dust-rich galaxy has only recently occurred, refuelling a pre-existing massive black hole, or injecting a second black hole. Clearly, observational methods for distinguishing between these possibilities are highly desirable. Numerical simulations of mergers of two gas-rich galaxies should be a profitable way to attack this problem.

Most of these 'ultraluminous infrared galaxies' have Seyfert 2 emission-line spectra. Evidently the AGN and BLR are hidden 'below', or 'behind', or 'within' so much dust, not yet fully organized by rotation about the direction of the angular momentum vector of the central region, that there is no clear axis along which the optical radiation can escape. This suggests that for such objects the spectral evolution may be along the path: Seyfert 2 $\rightarrow$ Seyfert 1 $\rightarrow$ Seyfert 2 $\rightarrow$ LINER $\rightarrow$ inactive. Others which 'form' (appear when refuelled) in gravitational collisions with less chaotic resulting dust and gas distributions may evolve as follows: Seyfert 1 $\rightarrow$ Seyfert 2 $\rightarrow$ LINER $\rightarrow$ inactive (Osterbrock 1993a). Testing these ideas observationally, along the lines sketched above in this section, seems to me to be the most important immediately foreseeable direction of research on AGNs.

5.8 Dust

Dust is generally present in AGNs, including in the BLR. Although it may be difficult to understand theoretically how it can survive, there is no doubt that it actually is present. Some of the best evidence is provided by analyses of the light and profile variations of the broad $H\alpha$ and $H\beta$ emission-line components in Seyfert 1.8 and 1.9 galaxies. In the most complete, recent paper on this subject, Goodrich (1995) shows that in most, but not all, of the well-studied objects of these types, the variations in the broad-line profiles and strengths can be understood in terms of variations in the amount of dust along the line of sight to the BLRs, confirming and extending earlier, less complete observational results. These variations probably result from transverse motions of dust-rich clouds within (or just outside) the BLRs. One conclusion which I would draw and emphasize from these results is that the best hypothesis for beginning a theoretical analysis of any region of any AGN is that it does contain dust. The apparently 'simpler' starting hypothesis that it does not contain dust is almost certain to be wrong!

Acknowledgements

I am very grateful to the organizers of the symposium on which this book is based and to our hosts at the Instituto de Astrofísica de Canarias for inviting me to participate in it, and for their many kindnesses during the course of it. I am also most grateful to J. S. Miller, R. W. Goodrich, and A. R. Martel for many helpful discussions on AGNs, while I was preparing this chapter. My research at the University of California on AGNs over the years has been partially supported by the National Science Foundation, most recently under grant AST 91-23547.

References

Albert, C.E., White, R.A., and Morgan, W.W.: 1977, ApJ, **211**, 309

Alexander, T., and Netzer, H.: 1994, MNRAS, **270**, 781

Antonucci, R.: 1993, Ann Rev A&A, **31**, 473

Antonucci, R., and Cohen, R.D.: 1983, ApJ, **271**, 564

Antonucci, R.R.J., and Miller, J.S.: 1985, ApJ, **297**, 621

Antonucci, R., Hurt, T., and Miller, J.S.: 1994, ApJ, **430**, 210

Arp, H.: 1987, *Quasars, Redshifts and Controversies* (Berkeley: Interstellar Media)

Arp, H.: 1990, Ap&SS, **167**, 183

Arribas, S., and Mediavilla, E.: 1993, ApJ, **410**, 552

Arribas, S., and Mediavilla, E.: 1994, ApJ, **437**, 149

Bahcall, J.N., Kirhakos, S., and Schneider, D.P.: 1994, ApJL, **435**, L11

Bahcall, J.N., Kirhakos, S., and Schneider, D.P.: 1995, ApJ, **450**, 486

Bahcall, J.N., Kirhakos, S., and Schneider, D.P.: 1996, ApJ, **457**, 557

Baldwin, J.A., Wilson, A.S., and Whittle, M.: 1987, ApJ, **319**, 84

Bautz, L.P., and Morgan, W.W.: 1970, ApJL, **162**, L149

Begelman, M.C., and Sikora, M.: 1992, in *Testing the AGN Paradigm*, eds. S.S. Holt, S.G. Neff, and C.M. Urry, p. 568

Boksenberg, A., *et al.*: 1995, ApJ, **440**, 151

Burbidge, G., Hewitt, A. Narlikar, J.V., and Das Gupta, P.: 1990, ApJS, **74**, 675

Byrd, A.A., Sundelius, B., and Valtonen, M.: 1987, A&A, **171**, 16

Code, A.D., *et al.*: 1993, ApJL, **403**, L63

Ellingson, E., Yee, H.K.C., and Green, R.F.: 1989, AJ, **97**, 1539

Elston, R., McCarthy, P.J., Eisenhardt, P., Dickinson, M., Spinrad, H., Januzzi, B.T., and Maloney, P.: 1994, AJ, **107**, 910

Ferland, G.J., and Osterbrock, D.E.: 1987, ApJ, **318**, 145

Filippenko, A.V., and Sargent, W.L.W.: 1985, ApJS, **57**, 503

Ford, H.C., Harms R.J., Tsvetanov, Z.I., Hartig, G.F., Dressel, L.L., Kriss, G.A., Bohlin, R., Davidson, A.F., Margon, B., and Kochar, A.K.: 1994, ApJL, **435**, L27

Gondhalekar, P.M., Horne, K., and Peterson, B.M. (eds.): 1994, *Reberberation Mapping of the Broad-Line Region in Active Galactic Nuclei* (San Francisco: Astronomical Society of the Pacific)

Goodrich, R.W.: 1995, ApJ, **440**, 441

Grandi, S.A.: 1978, ApJ, **221**, 501

Harms R.J., Ford, H.C., Tsvetanov, Z.I., Hartig, G.F., Dressel, L.L., Kriss, G.A., Bohlin, R., Davidson, A.F., Margon, B., and Kochar, A.K.: 1994, ApJL, **435**, L35

Heckman, T.: 1980, A&A, **87**, 152

Heckman, T.: 1987, in *Observational Evidence of Activity in Galaxies*, eds. E. Khachikian, K.J. Fricke, and J. Melnick (Dordrecht: Reidel), p. 421

Heller, C.H., and Shlosman, I.: 1994, ApJ, **424**, 84

Hernquist, L.: 1989, Nature, **640**, 687

Ho, L.C., Filippenko, A.V., and Sargent, W.L.W.: 1993, ApJ, **417**, 63

Howard, S., Keel, W.C., Byrd, G.G., and Burkey, J.: 1993, ApJ, **417**, 502

Hummer, D.G., Berrington, K.A., Eissner, W., Pradhan, A.K., Saraph, H.E., and Tully, J.A.: 1993, A&A, **279**, 298

Hutchings, J., and Neff, S.N.: 1991, AJ, **101**, 34

Hutchings, J., and Neff, S.N.: 1992, AJ, **104**, 1

Kazanas, D.: 1989, ApJ, **347**, 74

Keel, W.C.: 1983, ApJ, **269**, 466

Koratkar, A., Destua, S.E., Heckman, T., Filippenko, A.V., Ho, L.C., and Rao, M.: 1995, ApJ, **440**, 132

Korista, K.T., and Ferland, G.J.: 1989, ApJ, **343**, 578

Kormendy, J., and McClure, R.D.: 1993, AJ, **105**, 1793

Kormendy, J., and Richstone, D.: 1995, Ann Rev A&A, **33**, 581

Kriss, G.A., Canizares, C.R., and Ricker, G.R.: 1980, ApJ, **242**, 492

Lin, D.N.C., Pringle, J.E., and Rees, R.J.: 1988, ApJ **328**, 103

Liu, M.C., and Graham, J.R.: 1995, BAAS, **27**, 886

Lynden-Bell, D., 1969, Nature, **223**, 690

Macchetto, F., Capetti, A., Sparks, W.B., Axon, D.A., and Boksenberg, A.: 1994, ApJL, **435**, L15

MacKenty, J.W., Simkin, S.M., Griffiths, R.E., Ulvestad, J.S., and Wilson A.S.: 1994, ApJ, **435**, 71

Maoz, D., Filippenko, A.V., Ho, L.C., Rix, H.-W., Bahcall, J.N., Schneider, D.P., and Macchetto, F.D.: 1995, ApJ, **440**, 91

Mason, H.E.: 1975, MNRAS, **170**, 651

Matthews, K. *et al.*: 1994, ApJ, **420**, L13

Matthews, T.A., Morgan, W.W., and Schmidt, M.: 1964, ApJ, **140**, 35

McLeod, K.K., and Rieke, G.H.: 1994a, ApJ, **420**, 58

McLeod, K.K., and Rieke, G.H.: 1994b, ApJ, **431**, 137

Miller, J.S.: 1985, in *Astrophysics of Active Galaxies and Quasi-stellar Objects*, ed. J.S. Miller (Mill Valley: University Science Books), p. 367

Miller, J.S., and Antonucci, R.R.J.: 1983, ApJL, **271**, L7

Miller, J.S., and Goodrich, R.W.: 1990, ApJ, **397**, 452

Miyoshi, M., Moran, J., Herrnstein, J., Greenhill, L., Nakal, N., Diamond, P., and Inoue, M.: 1995, Nature, **373**, 127

Mohan, M., Hibbert, A., and Kingston, A.E.: 1994, ApJ, **434**, 389

Moles, M., Marquez, I., and Pérez, E.: 1995, ApJ, **438**, 604

Morgan, W.W.: 1958, PASP, **70**, 364

Morgan, W.W.: 1971, AJ, **76**, 1000

Morgan, W.W., and Lesh, J.R.: 1965, ApJ, **142**, 1364

Morgan, W.W., Kayser, S., and White, R.A.: 1975, ApJ, **199**, 545

Osterbrock, D.E.: 1978a, Proc Nat Acad Sci, **75**, 540

Osterbrock, D.E.: 1978b, *Astronomical Papers Dedicated to Bengt Strömgren*, eds. A. Reiz and T. Anderson (Copenhagen: Copenhagen University Observatory), p. 299

Osterbrock, D.E.: 1981, ApJ, **246**, 696

Osterbrock, D.E.: 1989, in *Astrophysics of Gaseous Nebulae and Active Galactic Nuclei* (Mill Valley: University Science Books)

Osterbrock, D.E.: 1991, Rep Prog Phys, **54**, 579

Osterbrock, D.E.: 1993a, ApJ, **404**, 551

Osterbrock, D.E.: 1993b, Rev Mexicana A&A, **26**, 65

Osterbrock, D.E., and Martel, A.: 1993, ApJ, **414**, 552

Osterbrock, D.E., and Shaw, R.A.: 1988, ApJ, **327**, 89

Osterbrock, D.E., Dahari, O., and Ekberg, J.O.: 1983, ApJL, **273**, L31

Penston, M.V., and Pérez, E.: 1984, MNRAS, **211**, 33P

Pogge, R.W.: 1988a, ApJ, **328**, 519

Pogge, R.W.: 1988b, ApJ, **332**, 702

Rafanelli, P., Violato, M., and Baruffolo, A.: 1995, AJ, **109**, 1995

Rees, M.J.: 1977, QJRAS, **18**, 429

Rees, M.J.: 1984, Ann Rev A&A, **22**, 471

Rowan-Robinson, M. *et al.*: 1991, Nature, **351**, 719

Rubin, V.C.: 1994, AJ, **108**, 456

Salpeter, E.E.: 1964, ApJ, **140**, 796

Salzer, J.J., MacAlpine, G.M., and Boroson, T.A.: 1989, ApJS, **70**, 479

Sanders, D.B., Soifer, B.T., Elias, J.H., Madore, B.F., Matthews, K., Neugebauer, G., and Scoville, N.Z.: 1988, ApJ, **325**, 74

Schmidt, M., and Green, R.F.: 1983, ApJ, **269**, 352

Shields, G.A.: 1977, ApL, **18**, 119

Soifer, B.T., Cohen, J.G., Armus, L., Matthews, K., Neugebauer, G., and Oke, J.B.: 1995, ApJL, **443**, L65

Stockton, A.: 1978, ApJ, **223**, 747

Tohline, J.E., and Osterbrock, D.E.: 1982, ApJL, **252**, L49

Toomre, A., and Toomre, J.: 1972, ApJ, **178**, 623

Tran, H.D.: 1995, ApJ, **440**, 597

Tran, H.D., Miller, J.S., and Kay, L.E.: 1992, ApJ, **397**, 452

Unger, S.W., Pedlar, A., Axon, D.J., Whittle, M., Meurs, E.J.A., and Ward, M.J.: 1987, MNRAS, **228**, 67

Veilleux, S.: 1991, ApJ, **369**, 331

Véron-Cetty, M.-P., and Véron, P.: 1993, *A Catalogue of Quasars and Active Nuclei*, 6th edn (Garching: ESO)

Voit, G.M., Weymann, R.J., and Korista, K.T.: 1993, ApJ, **413**, 95

Weymann, R.J., Morris, S.L., Foltz, C.B., and Hewitt, P.C.: 1991, ApJ, **373**, 23

Wilson, A.S., and Baldwin, J.A.: 1985, ApJ, **289**, 124

Wilson, A.S., and Tsvetanov, Z.I.: 1994, AJ, **107**, 1227

Wilson, A.S., Wu, X., Heckman, T.M., Baldwin, J.A., and Balick, B.: 1989, ApJ, **339**, 729

5.9 **Discussion**

Question

(Pagel): Perhaps I could kick off the questioning by asking for your comments on the suggestions, mainly by Terlevich and Melnick, that a lot of the high-energy ionization phenomena can be explained by the later stages of a starburst. When the more massive stars have lost their outer layers and become Wolf–Rayet stars, with very high effective temperatures, they produce collectively something like a power-law spectrum, which Terlevich and Melnick refer to as 'warmers'. How far along the sequence do you think this kind of explanation can go, as opposed to, say, the black-hole phenomena?

Answer

(Osterbrock): My own expertise is much more in what I have been telling you about. I have never studied the 'warmer' picture in great detail. I have heard talks about it, and I have read some papers about it. I would say that I am very impressed by the evidence for photoionization. We observe a lot of X-rays. I think that if you are willing to postulate the existence of stars of any temperature then surely you could take a spectrum of stars and get any radiation field you want. What we see, though, is not the kind of star which they are talking about, as far as I am aware. I think that the X-rays observed from the Seyfert 1 galaxies are very hard to get from stars alone. But I think each person has to work on what seems more plausible to him, or her, and so I am careful as I am not an expert on that subject.

Comment

(Pagel): I think that, in fairness, I should say that (although perhaps you do not actually see them very much) it is based on specific stellar-evolution models. It is not just pulled out of a hat, in order to get the right sort of spectrum.

Answer

(Osterbrock): No, I do not mean to imply it is pulled out of a hat. And I believe that there are specific models, but we do not seem to see in nature many of those stars. We do see many starburst galaxies, which show starbursts like we see in our Galaxy.

Question

(Pagel): I also have a supplementary question. Do you think that IRAS 10214 is some kind of a Rosetta stone for the sort of effects that you are talking about?

Answer

(Osterbrock): I am completely incapable of answering that question: I do not know what the object is, or anything about it!

Comment

(Pagel): It is a dusty Seyfert 2 galaxy, at a redshift of 2.4, which is quite possibly the most luminous object that has ever been found, and has these great starbursts and lots of CO and also a somewhat Seyfert 2-like spectrum, but probably basically a very powerful starburst caught at a rather interesting stage.

Answer

(Osterbrock): This is the object known to me as FSC 10214+4724, which I discussed in some detail in Section 5.5. It is a good representative of the objects that Sanders *et al.* (1988) studied. Those are the objects that I think are the births of QSOs, or the rebirths of them if they have been refuelled in an interaction. I think that, if there is enough dust, you do not see what is coming out from down inside. I also think that what we are talking about is the early stages of a Seyfert 2, and so it is in the direction of what you are talking about.

Question

(M. Burbidge): In the Antonucci and Miller model, the electrons that are scattering the Seyfert 1 spectrum – are they just from ionized gas that is flowing out, or is there a high-energy electron component needed?

Answer

(Osterbrock): I think that it is ionized gas flowing out. The widths of the lines are somewhat larger than in these single clouds that they see, which are also ionized, where we are not seeing gas flowing but more or less fixed. On the other hand, they have done analysis in the last few years, with Mathews, and almost any specific model they take does not agree with what they see. I would say that is another problem for the future.

Comment

(Sunyaev): I want to return to the subject which was raised by our chairman (B. Pagel): this is about the stars. I myself am very impressed by the discovery, by Genzel, of seven helium stars, in the vicinity of the centre of Sagittarius A. This, I believe, is an extremely important discovery: that you really have stars which radiate a great deal, that maybe they make the main contribution to the ultraviolet luminosity of the centre of our Galaxy. But, at the same time, I believe that such a starburst model cannot give us the full explanation of what we are observing in both Seyfert 1s and Seyfert 2s. To me, the strongest evidence is the very strong variability in the continuum and in the lines. As you know, NGC 4151 changed from having broad and strong lines to a state where the lines are very weak. You remember this. I believe it is fully impossible to understand this change when the lines are due to the photoionization of a starburst. This shows us that we have these accretion disks. I believe myself that only a black hole can produce such a hot continuum, which can give you all variety of ions that we are observing in these broad-line regions.

Answer

(Osterbrock): Yes, I agree with that; the variability is a very important point.

Question

(Sandage): On the question of whether or not companions are necessary to create Seyferts, is it true that all luminous spirals have AGNs in the centre. I know the Vérons at one time, and, I think, Sargent and Filippenko, looked for low-level Seyfert activity in many field galaxies, and almost always found, in every single galaxy, a small low-level AGN in the centre. If that is true, then it does not require companions and the AGN phenomenon is part of the formation process itself, as Martin Rees said privately earlier. As soon as a bulge forms, you will get a black hole in the middle.

Answer

(Osterbrock): First of all, do all spiral galaxies have AGNs in the centre? Looking at them optically, Keel, I think, did the most complete study of this, some years ago, with a quite large sample of well-defined spirals, and about 50% showed signs of being an AGN down to some very weak level.

Question

(Sandage): Broad Hα too?

Answer

(Osterbrock): No. Strong N II, strong S II, that kind of thing. The broad Hα, I think, is much less than that (at a detectable level). I think that Sargent and Filippenko found it in many cases, but not in all, by any means.

Question

(Sandage): And the Vérons, earlier than Sargent and Filippenko, did a similar thing by subtracting off the continuum; I thought most of their galaxies showed Hα.

Answer

(Osterbrock): I will not disagree with you, but I think it is 50% in the work by Keel. I would say that, although I did not talk about it, as you know there are several groups looking for signs of inactive black holes in nearby galaxies, including Kormendy, Dressler, and others. I would say that the evidence keeps changing one way or the other, but it seems likely that there are in a lot of them.

Question

(Sandage): Those are ellipticals: the ones that they are doing?

Question
(Unidentified): Is Andromeda an elliptical?

Answer
(Osterbrock): Yes, M31 is one of the sample.

Answer
(Sandage): Andromeda is elliptical. In the centre, yes!

Answer
(Osterbrock): That there is some weak activity going on all the time would not be surprising, if there is a central nucleus there. There are stars near it, and some of those must be losing mass. There must be ways! But to get a lot of mass in, I think, requires refuelling.

Question
(Sandage): So, the strong ones need companions, while there are weak ones without companions?

Answer
(Osterbrock): Yes, it is quite possible that a different fuel source, namely mass from individual stars, works at very low AGN luminosities. As I say, companion is the wrong word. When observers say companion, what they mean is another galaxy close by. In many cases that nearby galaxy may be merged, and in other cases it goes past, probably, but it leaves this refuelling, which dies out in a time of the order of a few times 10^7 years. It does not die out. It dies out going down.

Question
(Lynden-Bell): I would just like to say that I think that this discussion is very apposite, but I think that we would all agree that, probably, near the beginning there is lots of fuel, and so, right at the beginning of things, you probably do not need a companion at all. It is only when you are thoroughly exhausted, like most of us will become at the end of this long session, that you need some of the young people to come round and restimulate you to think about these matters. Then, of course, there is the

possibility of the real interaction, with some of the gas and dust that have previously been around the edges of this coming down into the middle and causing fireworks yet again.

Answer

(Osterbrock): I agree completely. You realize that what I am talking about are Seyfert galaxies where you see the galaxy; there are lots of them to observe, at redshifts of a few tenths.

Question

(Sandage): What was so surprising last year when Jerry Kristian and I were down at Las Campanas and we had superb seeing, and we had a list of 50 ScI galaxies with CCD detectors, was that every single one of those had an unresolved bright thing in the centre. Amazing!

Answer

(Osterbrock): Well, Keel would say 50%! You go ahead!

Question

(Rees): A comment and a question. The comment following Allan Sandage's point is that Andromeda is perhaps a good example of a galaxy that is thought to have a black hole in the centre but is extremely inactive, and of course these guys can be inactive if swept clean of gas. There is though a distinct possibility that they cannot be kept completely free of gas, because of the possibility of a star getting on the radial orbit and getting tidally disrupted. This is something which will happen about once every 10^4 years in a nucleus like that of Andromeda. The kind of flaring events that would follow would cause these quiescent nuclei to flare up to quasar-level activity for over a year. This is the kind of phenomenon that ought to be looked for and ought to be expected, if indeed those black holes are present in the nuclei of these quiescent galaxies. That is the comment.

The question I wanted to ask is a different one. I am going back to the broad-line region. All these models for broad-line regions involve little clouds, with a small volume-filling factor. There is the issue then of whether or not they have to be confined, and if so, what confines them?

I get the impression that there is no entirely accepted mechanism: a hot gas does not seem to work too well; there is work by some of your colleagues on this. My opinion is that maybe magnetic fields may be the best bet, for lack of anything better. I wonder if you would like to comment on the structure of the clouds and what confines them.

Answer

(Osterbrock): I think that you are quite right. The picture certainly has to be of small clouds, and what keeps them together I have no idea. We keep observing that situation. What you say about magnetic fields sounds plausible to me. I did not have time or space to talk about it, but in the end, we have very little information on the velocity field in the broad-line region. How you are to arrange these clumps so they do not completely dissipate seems very difficult. There is no preferred symmetry in the broad-line region as there is in the narrow, so that suggests that maybe rotation is part of it. People have tried the time delay studies that I talked about. They look for variability, comparing the red wing of the line with the blue wing, to see whether there is generally expansion, or contraction. The first-order result is that there is no difference. The second-order result is that there is some at very low level. Some observers find there is a slight tendency towards outflow and some find there is a slight tendency towards inflow. So, there are lots of problems for the future, I would say, but it is apparently not at all well determined now.

Question

(Longair): Just before this point, I have personally been very impressed by the reverberation-mapping techniques for trying to understand the structure of the clouds. I just wondered what your view is on how far one can actually go to understand what is happening within the central region with these techniques. I say this because it is very time consuming on telescopes to do it in a systematic way, but if it is the most important thing to do, perhaps we should be setting aside large amounts of time to do this properly.

Answer

(Osterbrock): I think it is very important, and I certainly will not say that you should not give time to it. I think it is an important problem. Whether it will lead to the answer, I cannot tell you. The methods I am familiar with have been based on very simplified models, eliminating all the models I could, and then keeping the result from the simplest one. There are certainly more sophisticated methods than those which are available. There are many measurements which have been made in the past. If I were a director, which I am not, luckily, I would say 'why don't you try your hand at these data first and see what already exists and show me what you can get out of it, or tell me why you have never done that'.

Comment

(Longair): I have this grand vision that, in principle, it might be possible to determine all the dynamics by using these techniques. Inflows, outflows, rotations, differential phenomena – in principle it might be worked out by these techniques. I have this great problem: that it would take up so much telescope time to do even one object, one optically violent variable object, that it will never be done.

Answer

(Osterbrock): I hope it will be done. I hope that AGNs, asymptotic giant branches, etc. will be studied. I hope lots of things will be done!

Question

(Sandage): You also are a superb historian, and I would like to make a comment, as I know it, as to how Seyferts were named. You probably know this as well. One of the greatest advantages of being old is that one was around when giants walked the Earth. When I went to Mount Wilson, in 1952, two of those giants were still alive: they were Milton Humason and Rudolph Minkowski. You know how Seyferts were named?

Answer

(Osterbrock): You tell me!

[209]

Comment

(Sandage): Starting in the 1930s, Humason began the redshift determinations of field galaxies and, lo and behold, he found in his Cassegrain spectra, at the 100-inch reflector, sharp nuclei with broad Hα. He had about six or seven of these and Minkowski had a few. Carl Seyfert, perhaps then not even a PhD, came out as a summer student to Mt Wilson and wanted a project. So, Humason and Minkowski laid the spectra on Seyfert's desk and asked him to trace them. He got broad lines and then he wanted to publish a paper. He asked, 'will you sign it?', and neither Minkowski nor Humason signed it. So, in the *Astrophysical Journal* 1938 there was Seyfert's first paper. They could have been called Humasons, or Minkowskis, as well as Seyferts.

Answer

(Osterbrock): If I could respond, as a great historian, to my dear friend Allan, Seyfert was actually a PhD at that time, from Harvard, but, apart from that, I think that it is quite true!

Comment

(Sandage): Carl Seyfert was one of the first post-docs in the nascent US education system of the time. But Curtis at that time probably had...

Answer

(Osterbrock): Actually, it was E. Fath, who was a Lick student, who took some spectra of these galaxies, partly at Lick and partly, afterwards, at Mt Wilson. He was very confused by NGC 1068, because he realized, from all the other spirals he observed, that they were collections of stars: he saw the absorption lines. But this darned NGC 1068 had emission lines and absorption lines, and so it was a sort of a cross between a planetary nebula and a star cloud in his view. He did not want to publish it, or his bosses would not let him, one or the other.

Comment

(Lynden-Bell): I am sorry, I am not a historian and I did not exist at the right time, but William Herschel noticed the extreme brightness of the nucleus of NGC 4151 and remarked that it had an extremely bright nucleus. V. M. Slipher, in a paper of 1917, pointed out that there was a class of galaxies with strong emission lines, and I think he even mentioned the broadness of them. This, I think, pre-dates what we have talked about here.

Answer

(Osterbrock): No! No! It does not pre-date Fath! Slipher certainly did pre-date Seyfert! That is right! But let's get back to reality.

Comment

(S. di Serego): I have a comment on the question by M. Burbidge on whether the electrons doing the scattering in NGC 1068 are from hot gas or from ionized gas. You can tell from the fact that you see broad polarized lines that have about the right width that you expect from broad lines that they are not broadened by the Doppler effect. So the gas, the electrons due to scattering, cannot be warmer than say about 10^5 degrees because, otherwise, you would see the lines unbroadened and probably hidden in the continuum.

The other comment I wanted to make refers to this question of whether every galaxy might have a black hole in the centre or an active nucleus: a nucleus which had some activity at some time. It relates to what M. Rees said about possibly observing flares in galaxies. We happen to have observed a flare using *HST*, in an elliptical galaxy in the Virgo cluster. This flare occurred right in the nucleus and has the right luminosity and time scale to be explained by a star falling into a black hole. I am not completely convinced that this is the only explanation and am considering also the possibility of supernovae, or maybe microlensing, but it is interesting that these things are truly observed as predicted by M. Rees. A paper by Alvio Renzini, various colleagues, and myself has been submitted to *Nature* on this.

Question

(A. Díaz): I have, many times, talked about this dichotomy, about the morphological types of spiral galaxies in which starbursts and activity have taken place. If you look at late-type galaxies you will find preferentially starbursts. If you look at early-type spirals you will find preferentially Seyferts, or LINERS, on looking at the emission-line spectra. This has been related to metallicity, and I think this was one of the reasons why Roberto Terlevich and Jorge Melnick started to think that, since late-type galaxies have low-metallicity nuclei and early-type spirals have high-metallicity environments, maybe star-formation processes were different in different metallicity environments. Still, it is puzzling why this dichotomy exists. I am sure you have thought about this many times. Maybe you have an idea why we observe starburst in late-type spirals and active nuclei in early-type spirals?

Answer

(Osterbrock): I think that, in active galactic nuclei, you also observe starbursts. There might be more or less, but there are always some there. No, I have no further interpretation.

Question

(H. Zinnecker): You have mentioned, but not elaborated very much on, the ultraluminous *IRAS* galaxy. You only mentioned Sanders's name and the word *IRAS*. Could you comment a little more about the relation between the ultraluminous *IRAS* galaxies and the AGN phenomenon?

Answer

(Osterbrock): I had a slide from Sanders's paper. The objects they observe are very luminous and they have lots of dust and they are all mixed up and look to me like a recent, strong interaction in my terms. There appears that there is something deep down inside that is putting out lots of energy. The spectrum in both cases I have not observed myself but, as published, it seems to be Seyfert 2. I would say that these objects look to me like the formation of a new AGN, or the refuelling by very strong interaction where a lot of gas is transferred.

Question

(Zinnecker): Can I make another comment on this? I guess one of the arguments at that time when this was discussed was the space density of the ultraluminous *IRAS* galaxies and the active nuclei. It was thought that these space densities were similar, so that gave rise to this explanation, or this hypothesis. I guess the real question is, that there probably will be a mixed phenomenon in there...AGN and starburst. The problem is, how does one disentangle the two?

Answer

(Osterbrock): I guess that is a problem for the future! I think that, in many cases, you see objects that are not as dusty, and not as messed up, as those in which these high-ionization lines do not come out. It seems to me that the spectra, as described in these papers, show signs of Seyfert 2, which requires high-energy photons to ionize it. So I tend to think of those as having a black hole deep down inside, but that the direct radiation from it does not come out. But, as I said, this is a working hypothesis. I would like to do more work on it.

Question

(Zinnecker): Another further comment. Do you really expect that, for those dusty galaxies, the optical wavelength regime would provide the answers? Of course, I am thinking of the impact that we will soon have if the *ISO* (*Infrared Space Observatory*) flies. There are so many lines, mid-infrared lines, that pertain to the state of ionization and excitation, that presumably the progress of disentangling these two phenomena will come from infrared observations, not from optical observations. Do you agree with that?

Answer

(Osterbrock): Yes, I could not agree with you more! I thought that I had said that! I said at one point that, with dusty objects, you have to observe in the near infrared and the far infrared. That is one reason why I know nothing about these objects, because I cannot do it, but certainly the way to go is to observe the entire wavelength region. In the infrared there had been some so-called Seyfert 2s, in which you do not observe any broad Hα, but it has been possible to detect broad wings in Paschen α or Brackett lines by getting more space penetration, and that is certainly the direction to go. There are also diagnostic lines of high ionization and low ionization in those regions. Yes, I strongly agree with you.

Question

(G. Tenorio-Tagle): Regarding the debate about 'black holes versus warmers', although the black-hole model should be continued and investigated in that particular way, as it has been done so far, you should also allow other possibilities, and this other possibility should be mentioned in a forum like this one.

Answer

(Osterbrock): I could not agree with you more. I would not want to stop any research; I think that I can best talk about the things that I know best myself.

Question

(A. González): If you need a companion in order to create a Seyfert, I would expect that the number of AGNs, or Seyfert galaxies, would change with redshift. Have you done any analysis of this?

Answer

(Osterbrock): No, I have not, but others have tried to do some. There is a recent paper that I do not have the notes for, by Roger Blandford and someone else. This was within the last two years or so. Several scientists are trying to do that kind of thing. I would say that, to me, trying to understand physically what is going on, I do not want to do such studies myself. I may be wrong, but the number-versus-magnitude counts have so little physical information that I am not sure that it will be possible to distinguish between models on that basis. Certainly some have worked in the direction of trying to include the decreasing density as a result of the expansion. I do not know any results yet though.

Question

(Rees): Just to follow up on that. It is not as straightforward as one might guess to understand how an interaction can actually do what you want it to do, because, whereas interaction can have tidal effects that will stimulate star formation on the outer parts of the galaxy, the tidal effects will not then directly, of course, be significant in the inner kiloparsec. Thus, any effect that has to be communicated by viscosity in a disk in the outer part is also too slow. So it is not at all obvious. I think that the best bet might be an indirect effect via a gaseous halo, whereby star formation is triggered in the outer part of the galaxy disk, and that generates a denser gaseous halo than there was otherwise. Then, on the dynamical time scale, that effect could be communicated to the centre. I am not saying that is the best idea, but there is the non-trivial problem of understanding how the core, which does not directly feel the tidal effects, can be influenced fast enough by the interactions.

Answer

(Osterbrock): I agree. From the observational point of view, if you try to classify galaxies in terms of how disrupted they are, or how strong the interaction was or is, whatever it might be, the ones that are the most extremely messed up are not the ones that are AGNs, but almost always those that have starbursts going on in them. Evidently, an AGN is either a later stage in the interaction, when it has died down somewhat, or an interaction which is not this extreme. Trying to make a quantitative comparison between observed structures and models is important. What the parameters are that you should use to classify an interaction, I do not know, but the direction of trying to compare simulations with actual galaxies to get information about which ones have AGNs and which ones do not, and what the spectral type is, I think, is the direction to go to try to understand the evolution, rather than trying to think it up entirely.

Question

(Lynden-Bell): I just wanted to say that – I think it was even as early as in the Toomre and Toomre paper, but it may have been a bit later – Alar was certainly interested in this problem, and he suggested that the spiral wave generated from outside the galaxy actually concentrated things down towards the nucleus and therefore had a much larger amplitude when it went down towards the middle. This caused a sufficient asymmetry to push things down into the nucleus. Hummel wrote a thesis at Gröningen, in 1980, about radio activity, and I think he deduced that there was no significant extra activity produced on the outside, but there was very significant nuclear activity from interacting galaxies.

Comment

(G. Burbidge): What Martin said reminded me of a basic point that really has not been stressed here. Somehow this discussion has all gone in the sense that things from outside trigger the inside. I thought there was an argument going on for some time, and many people in this field have felt that it was violent events going on in the nucleus that triggered the starburst events, and not necessarily going the other way. I am not sure whether this argument exists today, but it always sounded to me a much more plausible way of looking at things.

Question

(Sandage): In that regard do you not, Guido, find in the very centre of M31 outflow in the O II λ3727 emission lines?

Answer

(Münch): It was very hard to see the emission lines.

Question

(Sandage): But you did...

Answer

(Münch): ...and even harder to decide whether it was an outflow. It looks like it, but the evidence is a little bit weak.

Answer

(Osterbrock): There are some more recent high-resolution spectra of the nucleus by Vera Rubin. Again, I would say that I have not studied it carefully, but it does not seem to me that there is neither outflow nor inflow, but there are some different phenomena, perhaps starbursts quite close to the nucleus. Here only very small amounts of gas have been observed.

Comment

(Zinnecker): I wanted to comment on this discussion about whether the starbursts are in the innermost part of the galaxies or in the outer parts. It was one question at least. Martin Rees mentioned that the starburst may do something on the outside of the galaxy which then has subsequent effects. However, the observations do show that starbursts are in the middle. Infrared observations say basically that starbursts are in the middle. And for whatever reason (I do not know if it is a triggering from inside out or from outside in), the fact is that the starbursts are in the middle. There is one case which kind of argues in your favour, and that is NGC 7469, which is an active nucleus and has a ring of star formation. A ring of HII regions both seen in optical speckle and in infrared speckle work. It is a beautiful example which looks at least like something from the inner part that has triggered something in the surroundings. That does not violate the other idea that the tidal interaction somehow manages, in combination perhaps with a bar, or whatever there is in the galaxy, to shovel material also to the inner part. But then, of course, there is a time-scale problem. To get something to one kiloparsec is kind of easy, but then to get it down to the nucleus, that is much, much harder.

6   The high-energy radiation of active galactic nuclei

Malcolm S. Longair

6.1 Introduction

My task is to survey the 'non-thermal' aspects of the radiation of active galactic nuclei. This is an enormous subject and results in many 'key' astrophysical problems. Donald Osterbrock has set much of the scene for my review and described how optical spectroscopic observations tell us a great deal about the disposition and properties of the clouds which radiate the narrow and broad-line emission observed from different types of active galactic nuclei. I will adopt a complementary approach and describe a wide range of observations from different wavebands, which indicate what the essential ingredients of any successful model of active galactic nuclei must be. I will approach the problem from the outside, gradually moving inwards towards the supermassive black holes in the nucleus. I will cover briefly the following topics:

   (i) Unification of different types of active galactic nuclei.
   (ii) Extragalactic radio-source asymmetries and unification.
   (iii) *Hubble Space Telescope* observations of 3CR radio galaxies.
   (iv) Superluminal radio sources.
   (v) γ-ray sources.
   (vi) X-ray observations of active galactic nuclei.
   (vii) How to put it all together.
   (viii) A selection of key problems.

These are enormous areas of current research and I hope that omissions and personal biases will be neutralized by contributions in the discussion section.

6.2 Unification of active galactic nuclei

One of the more remarkable developments over the last ten years has been the attempt to 'unify' different classes of active galactic nuclei. The concept of unification resulted from the realization that projection effects must be important in the observation of certain types of active galactic nuclei. A number of workers therefore proposed that much of the diversity of active galactic nuclei can be accounted for in terms of a model in which a single class of active galaxy is viewed at different angles to the line of sight. It is beyond question that such projection effects are important, but the intriguing question is the extent to which *all* types of active galactic nuclei can be unified within a single scheme. An amusing aspect of these endeavours is that the authors of some of the most important articles consider this hypothesis to be an issue of faith and they identify with some precision the dates of their 'conversion' to strong unification scenarios.

The idea that projection effects are important in distinguishing between Seyfert 1 and Seyfert 2 galaxies was first convincingly demonstrated by Antonucci and Miller (1985) in their now classic study of the nearby Seyfert 2 galaxy NGC 1068. This galaxy is a typical Seyfert 2 galaxy in that its nucleus exhibits a narrow-line spectrum. When observed spectropolarimetrically, however, the polarized line emission is as broad as the broad permitted lines observed in Seyfert 1 galaxies. Antonucci and Miller interpreted these observations in terms of what has become the standard model for the unification of Seyfert 1 and Seyfert 2 galaxies. In this model, there is some form of 'obscuring torus' about the nucleus so that, when the galaxy is observed at a small angle to the axis of the torus, the nuclear regions, containing the active nucleus itself and the broad-line-emitting regions, are observed and the galaxy is classified as a Seyfert 1 galaxy. If the axis of the torus is observed at a large angle to the line of sight, the nuclear regions are obscured and so only the narrow-line regions, which are located further from the nucleus, are observed. The central regions can, however, be observed in the reflected light of the nucleus, and this is what is observed spectropolarimetrically. Radiation from the nuclear regions is reflected by electrons or dust in the vicinity of the nucleus and, in the process of reflection, the scattered light is polarized.

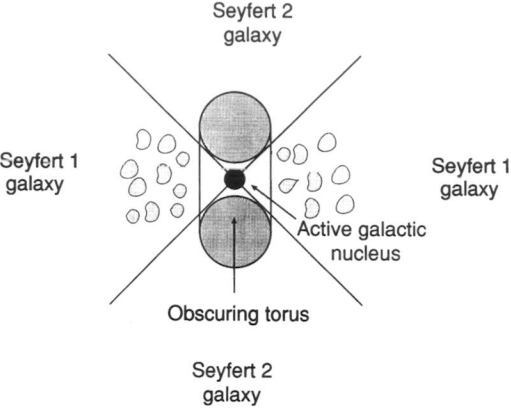

Fig 6.1. A schematic diagram illustrating the unified picture for Seyfert 1 and Seyfert 2 galaxies. The nuclear regions contain the source of continuum radiation and the broad-line regions which can be observed in scattered light, if the axis of the torus is observed at a large angle to the line of sight.

This unified scheme is illustrated in Fig. 6.1, which is a highly schematic representation of a cross section through the nuclear regions of a Seyfert galaxy. The active nucleus itself contains the source of continuum emission and the broad-line regions, whereas the narrow-line regions originate from outside the vicinity of the torus, but are illuminated by the beamed radiation from the nucleus. Excellent (and amusing) up-to-date surveys of the unification hypothesis for Seyfert 1 and Seyfert 2 galaxies are provided by Antonucci (1993) and Miller (1994). The key observations which support this picture are:

(i) The spectropolarimetry of Seyfert 2 galaxies which reveal the broad-line regions in scattered light. This observation has been successfully repeated for a number of other Seyfert 2 galaxies, in addition to NGC 1068. This is a difficult observational technique, since the percentage polarizations are normally only a few per cent.

(ii) Further evidence for the obscuration of the ionizing continuum emission from the nucleus comes from comparing the flux of ionizing photons observed from the nucleus with the number required to excite the emission-line regions. In a number of cases, the observed flux of ionizing radiation is less than that required to account for the ionization and excitation of the gas clouds, suggesting that they are exposed to a more intense radiation field than that which we observe.

(iii) This suggests that the radiation from the nucleus is emitted in an 'ionization cone', and examples of these have been observed by the *Hubble Space Telescope*. Figure 6.2 shows two examples in which the radiation from the nucleus is believed to take place in an 'ionization cone'. In the case of NGC 1068, the cone structure is entirely consistent with the type of geometry inferred by Antonucci and Miller (1985) and with photon-counting analyses (Kinney *et al.* 1991). In the case of the radio galaxy NGC 4261, 3C 270, there seems to be a cone-like structure emanating from the nuclear regions of the galaxy and there is also a remarkable ring about the nucleus (Stockton 1994). The elongated large-scale radio structure lies along the axis of the ring. Although NGC 4261 is a radio galaxy rather than a Seyfert galaxy, it shows the type of geometry expected in unified pictures of Seyfert 1 and Seyfert 2 galaxies.

(iv) There is a difference in the X-ray absorption properties of Seyfert 1 and Seyfert 2 galaxies in the sense that the nuclear X-ray-emitting regions of Seyfert 1 galaxies are observed through much greater column depths of cold gas as compared with the Seyfert 2 galaxies. We will return to this point in Section 6.7.

    Another major unification industry concerns the strong radio sources. The importance of projection effects in determining the observed properties of extragalactic radio sources has been advocated by Barthel (1989, 1994), who dates his conversion to unification to 1989. Barthel's analysis concerned the unification of radio galaxies and radio quasars. By the terms *radio galaxy* and *radio quasar* are meant those extremely luminous radio sources which appear in catalogues of bright radio sources, such as those found in the 3CR catalogue (Laing, Riley and Longair 1983) and other bright-radio-source catalogues (for example, Wall and Peacock 1985). These are very much rarer objects than the Seyfert galaxies, and the typical radio sources in these catalogues have redshifts in the range $0.1 < z < 2$. They have, however, the advantage that the double structures of the intense radio emission of these radio galaxies and quasars delineate well-defined axes, and so projection effects can be studied in a straightforward way. The arguments for the importance of projection effects and unification advanced by Barthel (1994) included the following:

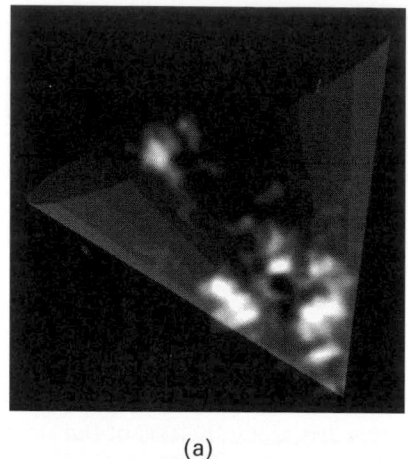

(a)

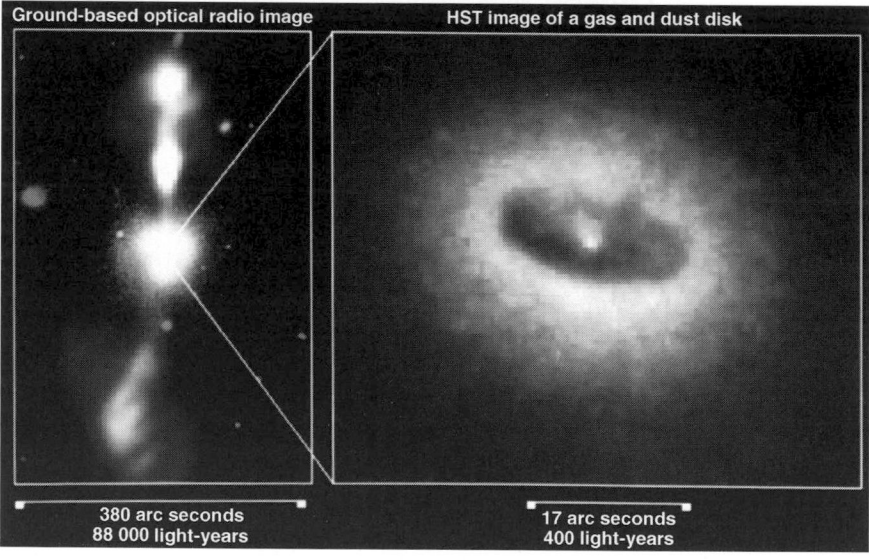

(b)

Fig 6.2. (a) The structure of the line-emitting regions close to the nucleus of the Seyfert 2 galaxy NGC 1068. Photon-counting analyses suggest that the ionizing radiation is emitted within the cone shown in the diagram. (b) The central regions of the radio galaxy NGC 4261 showing the central ring and evidence for a cone-shaped emission region, the axis of which is aligned with the radio jet. In the left-hand panel, the jets to the north and south of the galaxy show the distribution of the radio emission from the radio source.

(i) Superluminal motions in compact radio quasars in bright source samples indicate that these sources are observed at a small angle to the line of sight, and so projection effects are certainly important for these classes of radio source. We will return to these observations in Section 6.5.

(ii) In some radio quasars, which have symmetric large-scale radio structures, asymmetric radio jets are observed emanating from their nuclei, suggesting that Doppler enhancement of the radio emission is important.

(iii) Depolarization asymmetries are strongly correlated with the presence of asymmetric radio jets, as expected if the jets were beamed towards the observer.

(iv) The average projected linear sizes of the radio structures of the radio galaxies are greater than those of the radio quasars in samples which span similar redshift ranges, as would be expected if the radio quasars are observed at a smaller angle to the line of sight as compared to the radio galaxies.

(v) The optical emission observed about a number of radio galaxies is linearly polarized, as expected if the radiation of an obscured nucleus were being scattered into the line of sight, just as in the case of the Seyfert 2 galaxies.

I will return to a number of these observations in the next section. The upshot of these considerations is that a unified picture for radio galaxies and radio quasars, illustrated in Fig. 6.3, has become popular.

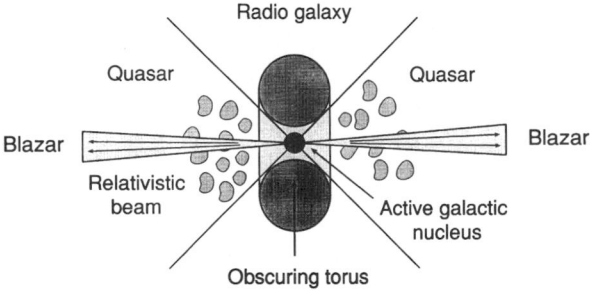

Fig 6.3. The unified model for the radio galaxies and radio quasars observed in bright samples of radio sources. Radio quasars are observed when the axis of the radio source lies within about 45° of the line of sight. When observations are made almost along the axis of the radio jet, superluminal radio sources and blazars are observed.

This picture bears a strong resemblance to the unification scenario for Seyfert galaxies, but now radio quasars are observed when the nucleus is observed within a cone of half-angle roughly 45° with respect to the axis of the radio source and a radio galaxy observed when the nuclear regions are hidden by the obscuring torus. Highly collimated radio jets are assumed to be emitted from the nucleus along the axis of the torus, and these are responsible for powering the outer radio hot spots and extended radio lobes. If the radio jet is observed at an angle close to the line of sight, the emission of the relativistic jet is strongly enhanced by the Doppler effect and superluminal motions may be observed. The class of objects observed close to the direction of the beam is referred to generically as 'blazars'. This is the orthodoxy advocated by Barthel, and we have recently come across some remarkable results which would be consistent with this picture.

6.3 Asymmetries in double radio sources and the velocities of hot spots

Radio maps of extragalactic radio sources of superb quality are now available in large numbers thanks to observations with the VLA and, in particular, radio maps of the brightest radio sources in the sky, for example, those in the 3CR sample, have improved enormously. Philip Best, Daniela Bailer, Julia Riley, and I have used these data to study asymmetries in the structures of the Fanaroff–Riley class-2 3CR radio sources, that is, sources in which there are hot spots towards the outer edges of the radio-source components (Best, Bailer, Longair, and Riley 1995). An example of the structure of these sources is shown in Fig. 6.4.

A convincing case can be made that the 'hot spots' observed towards the outer edges of these radio structures are continuously supplied with energy by highly collimated jets originating in the active galactic nucleus. The consequent ram pressure of the jet causes the hot spots to 'burn' their way out through the intergalactic medium at a high speed. Thus, asymmetries in the disposition of the hot spots and the fact that they are generally not perfectly lined up provides information about the dynamics of the jets and their interaction with the surrounding media. To study these asymmetries, we defined three asymmetry properties of the hot spots using

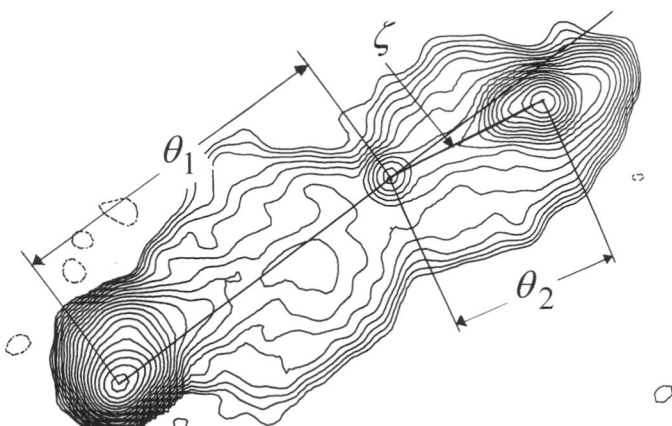

Fig 6.4. Illustrating the typical structure of a Fanaroff–Riley class-2 radio source. The various angles used in the analysis of their structures are shown on the diagram.

the quantities defined in Fig. 6.4 for all sources in which there are hot spots in both radio components. These are the *separation quotient* $Q = \theta_1/\theta_2$, a related quantity, the *fractional separation difference* $x = (\theta_1 - \theta_2)/(\theta_1 + \theta_2) = (Q - 1)/(Q + 1)$, and the *asymmetry angle* ζ. There were 23 quasars and 72 radio galaxies for which the asymmetry parameters could be defined for a complete sample of 3CR radio sources which lie in directions away from the Galactic plane.

First of all, we assumed that the sources were linear, that is, $\zeta = 0$. The surprising result we found was that the distributions of the fractional separation quotients were significantly different for the quasars and the radio galaxies (Fig. 6.5). The radio galaxies have a much narrower distribution of x as compared with the quasars which have a broad distribution and an obvious deficit of sources with $x = 0$. The difference is significant at the 99% significance level.

This difference in the x distributions can be explained naturally within the context of a simple relativistic model in which the hot spots move outwards from the nucleus at a constant relativistic velocity $v = \beta c$. It is a simple calculation to show that, because of the light travel time across the source, if the axis of the source lies at an angle θ to the line of sight, the separation quotient is $Q = (1 + \beta \cos \theta)/(1 - \beta \cos \theta)$ and the fractional

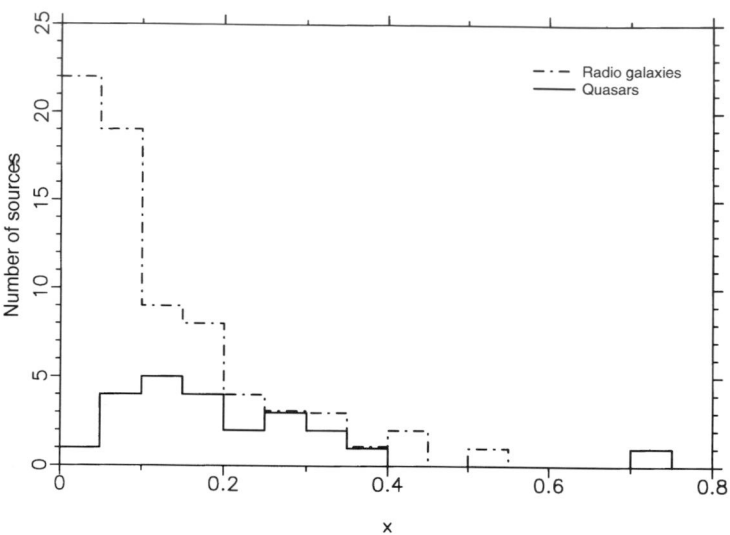

Fig 6.5. The distributions of fractional separation difference x for quasars and radio galaxies in a complete sample of 3CR radio sources (Best *et al.* 1995).

separation difference $x = \beta \cos\theta$ (Ryle and Longair 1967; Longair and Riley 1979). Thus, if quasars are observed at smaller angles to the line of sight than the radio galaxies, they are expected to be more asymmetric and there should be no symmetric quasars at all, unlike the radio galaxies, which should have many values of x close to zero.

To find out if this hypothesis can explain the observed distributions quantitatively, we have followed the procedure recommended by Banhatti (1980). For a sample of sources with a fixed velocity βc, the probability distribution of x is flat from 0 to β. Therefore, if the probability that the velocity of the hot spots lies in the range β to $\beta + d\beta$ is $g(\beta)\, d\beta$, it can be seen that the probability distribution for x is

$$p(x) = \int_{\beta=x}^{1} \frac{g(\beta)}{\beta}\, d\beta. \tag{6.1}$$

The procedure is therefore straightforward. We fit the distribution $p(x)$ by some suitable function and then find $g(\beta)$ by differentiation. This procedure is shown in Fig. 6.6, in which it can be seen that there is a broad distribution of inferred velocities with an average value of $\beta \sim 0.2$ but with a distribution which extends to large velocities.

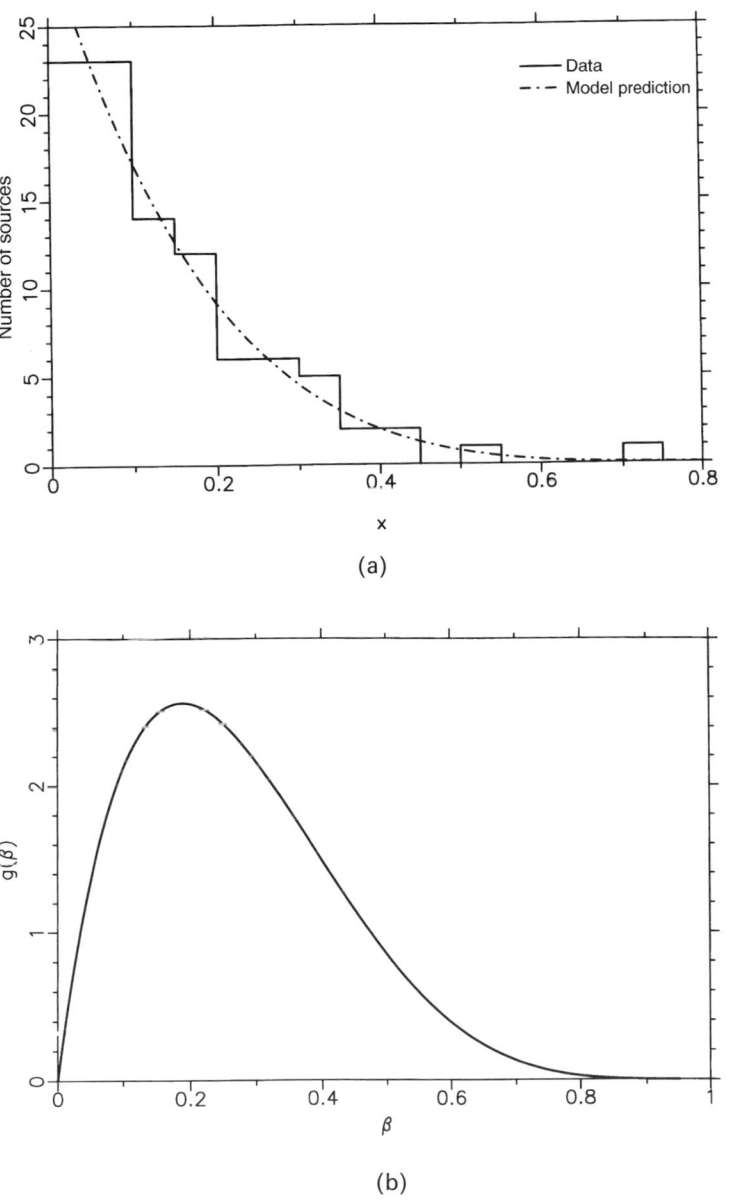

Fig 6.6. (*a*) The fit of the function $p(x)$ to the observed distribution of fractional separation quotients for the complete sample. (*b*) The inferred distribution of hot-spot velocities found by differentiation of the distribution of x shown in (*a*), according to the procedure discussed in the text.

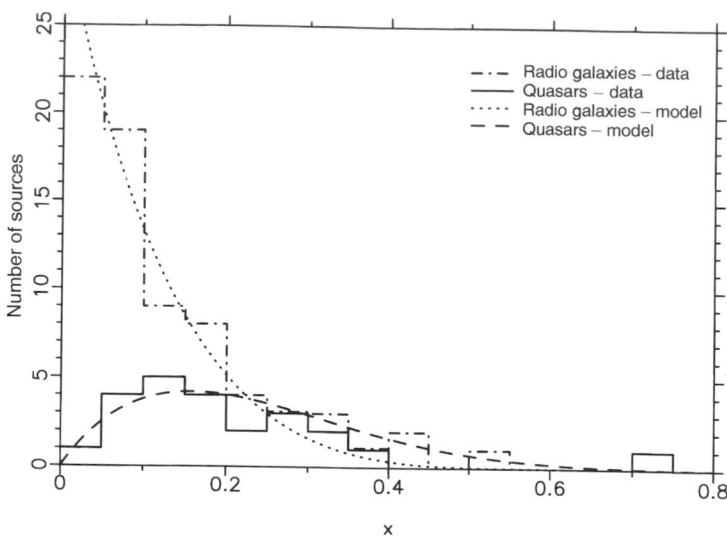

Fig 6.7. Comparison of the observed and predicted distributions of the fractional separation difference for radio galaxies and quasars in the complete 3CR sample of sources.

Then, assuming that a double radio source is classified as a quasar if it is observed within $\theta = 45°$ of the line of sight and as a radio galaxy if $\theta > 45°$, we can derive the expected probability distributions of $p(x)$ for the radio galaxies and quasars. This comparison is shown in Fig. 6.7, in which it can be seen that the observed distributions and the simple model are in good agreement.

In a second analysis, we investigated the distribution of asymmetry angles ζ for the quasars and the radio galaxies. Figure 6.8 shows that there is a significant difference between the probability distributions for the quasars and the radio galaxies in the sense that the quasars have a broader and flatter distribution than the radio galaxies.

Again, this is exactly as expected if the quasars and radio galaxies have the same intrinsic distribution of asymmetry angles but the quasars are observed preferentially within 45° of the line of sight. The projection onto the plane of the sky results in a broader distribution of observed asymmetry angles. To quantify this difference, we have constructed a simple model for the asymmetry in which one component is ejected at an angle ϕ with respect to the axis of the source. We have assumed that the source com-

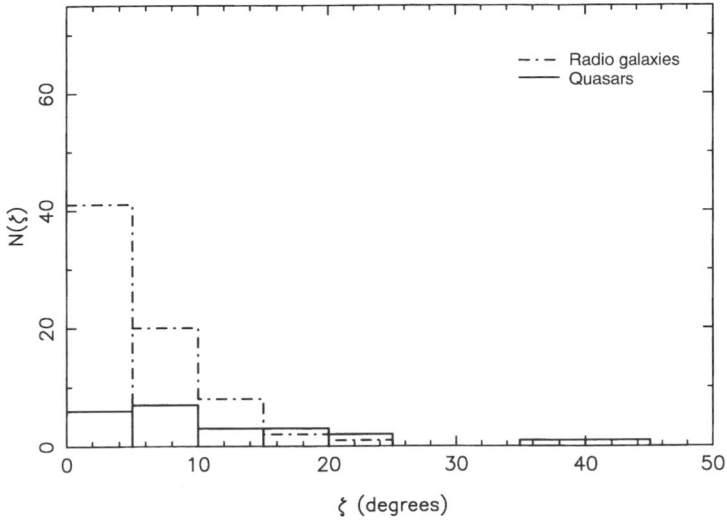

Fig 6.8. The distribution of asymmetry angles ζ for the radio galaxies and radio quasars in the complete 3CR sample.

ponent can be ejected randomly within an angle ϕ_{max} with respect to the axis of the source. The geometry is illustrated in Fig. 6.9.

The predicted distributions for different assumed values of ϕ_{max} are shown in Fig. 6.10, from which it is apparent that typical values of ϕ_{max} of about 10–15° can account for the observed distribution of asymmetry angles.

An important aspect of this analysis is that it is *independent* of any assumption about the velocities of the hot spots – it is simply a piece of three-dimensional geometry. Thus, the fact that the radio structures of the quasars are found to be more asymmetric than the radio galaxies is simply a projection effect. All extragalactic radio sources are asymmetric and the intrinsic angular asymmetries are similar for the radio galaxies and the quasars, once they are deprojected. Notice, incidentally, that in the above analysis, one arm of the double source was kept fixed and the other oriented at a random angle within ϕ_{max} on the opposite side of the source. If the hot spots were ejected randomly within cones in opposite directions along the axis of the source, the total solid angle on either side of the source would be the same and so the cones would be somewhat smaller, with $\phi_{max} \sim 7$–10°.

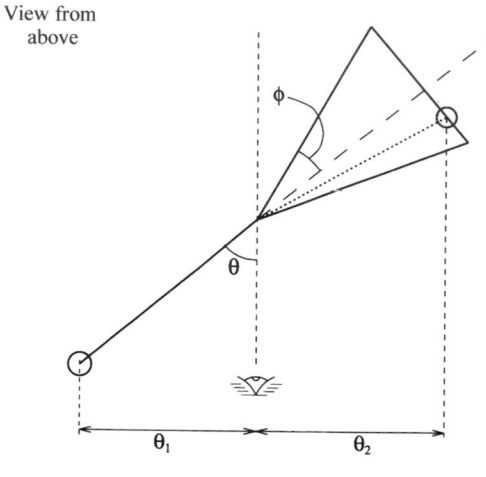

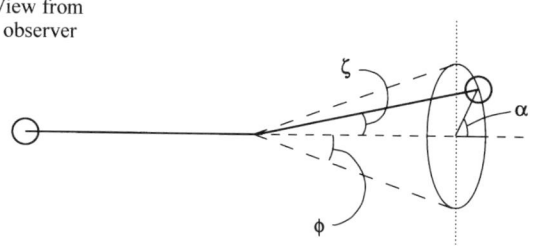

Fig 6.9. The simple source geometry used to model the distribution of asymmetry angles observed in the complete 3CR sample of sources.

In our third analysis, we have repeated the first analysis but have now assumed that one source component is ejected along the axis of the source and the other randomly within the cone of angle ϕ_{max}. We found that the asymmetries in x cannot be attributed to the distribution of intrinsic asymmetry angles ϕ but that the x distributions can be accounted for if the hot spots move at a significant relativistic velocity, with a velocity distribution similar to that shown in Fig. 6.6. We give more details of this analysis in our paper. We conclude that the environment of the source may make a significant contribution to the asymmetry, as noted by McCarthy, van Breugel, and Kapahi (1991), but that the differences in the distributions of the fractional separation quotients can be most simply attributed to the light travel time effect which implies that the hot spots move at significantly relativistic velocities.

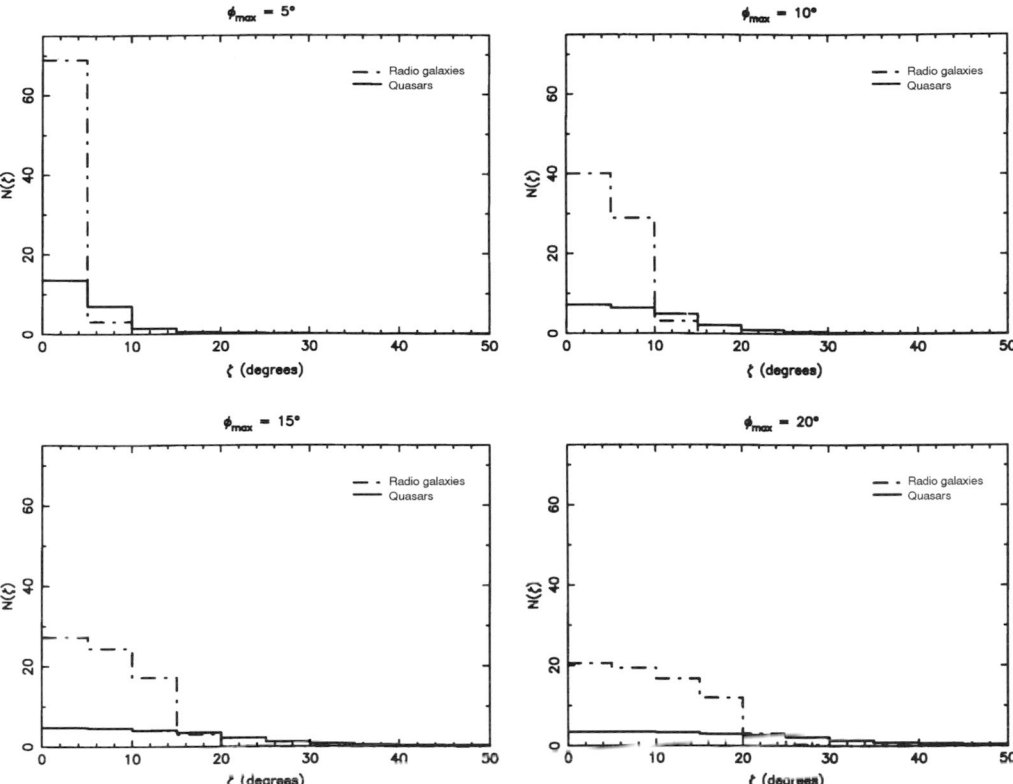

Fig 6.10. The predicted distributions of the asymmetry angle ζ for different assumed values of the intrinsic asymmetry angle ϕ_{max}.

Is there any other evidence that the hot spots are indeed moving at mildly relativistic velocities? Remarkably, a similar result has been found by an entirely independent argument. Liu, Pooley, and Riley (1992) determined the rate of advance of the hot spots from the parent galaxy in a sample of double radio sources using synchrotron ageing arguments in the equipartition magnetic fields of the source components. Laing (1993) has plotted these velocities as a function of radio luminosity, with the results shown in Fig. 6.11. The sources used in our sample all have radio luminosities greater than about $3 \times 10^{27}\,\mathrm{W\,Hz^{-1}}$, and it can be seen that the hot-spot velocities are similar to those found in the above analysis.

A consequence of this analysis is that I am now a convert to the unification hypothesis, illustrated in Fig. 6.4. Once this scheme is adopted, many other aspects of the story fall into place. For example, one-sided radio jets are frequently observed in radio quasars, but much more rarely in radio galaxies. There has been a debate as to whether this is an intrinsic effect, or

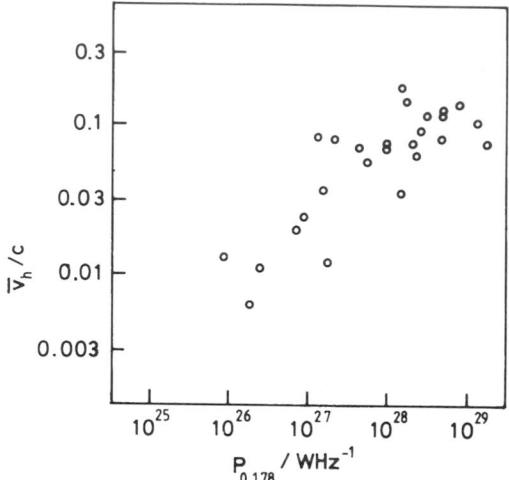

Fig 6.11. The inferred velocities of the hot spots in a sample of double radio sources derived by Liu *et al.* (1992). The diagram was presented by Laing (1993).

whether it is due to relativistic beaming of the radio emission of the jet. It is assumed that the jet ejected in the opposite direction is relativistically beamed away from the observer and so would be observed with much reduced intensity. An example of a one-sided jet in the radio quasar 3C 47 is illustrated in Fig. 6.12 (Fernini *et al.* 1991). There is growing evidence that the relativistic-beaming explanation is probably correct. Figure 6.12 also shows the depolarization of the extended source components between 5 and 1.4 GHz. It can be seen clearly that the northern component is significantly more depolarized at the lower frequency as compared with the southern component. This can be attributed to the fact that the source is embedded in a 'Faraday' medium, in which there are small-scale variations in the rotation measure. Consequently, the further the radio emission has to propagate through the medium, the more it is depolarized and the depolarization increases with decreasing frequency. The observations of 3C 47 indicate that the component on the jet side is nearer the observer than that on the (invisible) counterjet side. This is entirely consistent with the beaming picture in which the quasars are observed within 45° of the axis of the source and that the radiation of the jet is 'Doppler boosted'.

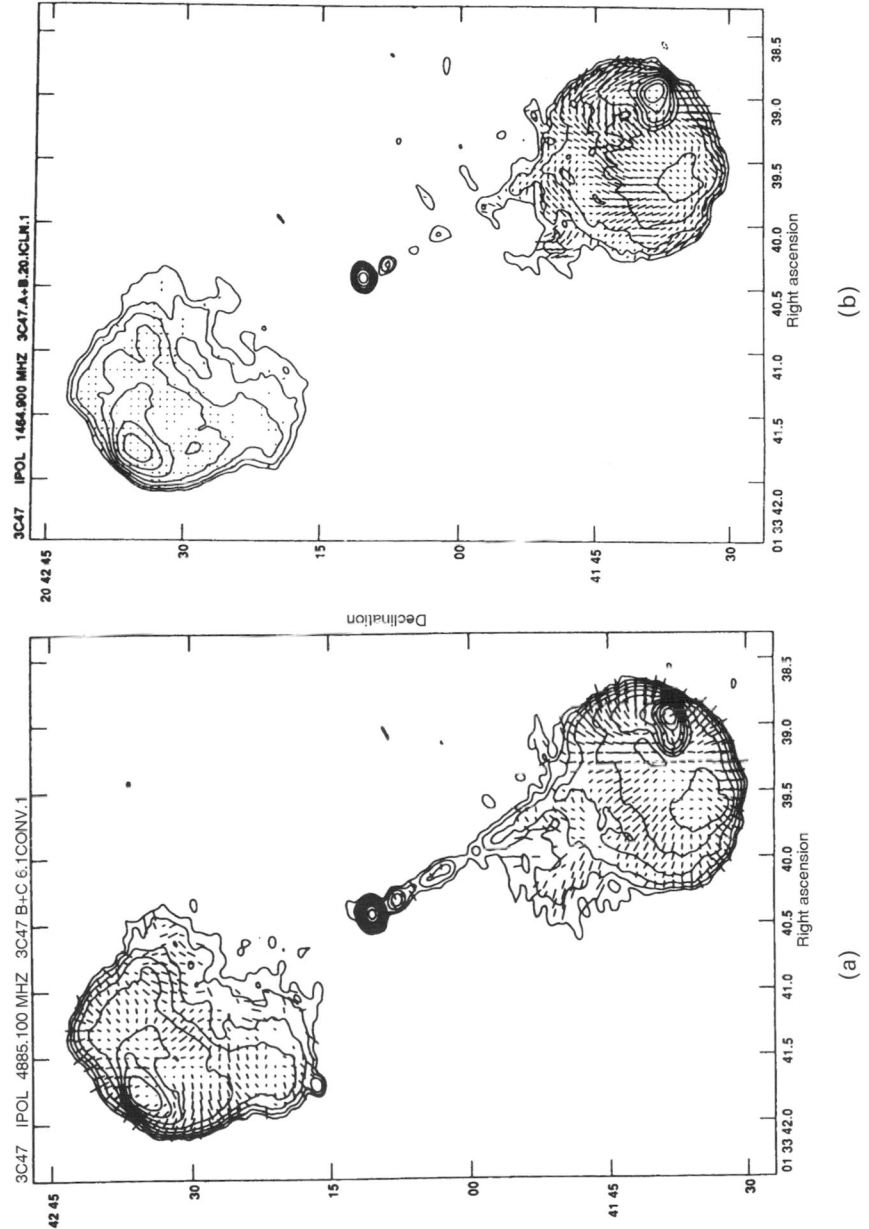

Fig 6.12. The radio structure of the radio quasar 3C 47 at (*a*) 5 GHz and (*b*) 1.4 GHz, showing the polarization vectors at the two frequencies (Fermini *et al.* 1991). It can be seen that the radio component on the counterjet side is more depolarized than that on the jet side.

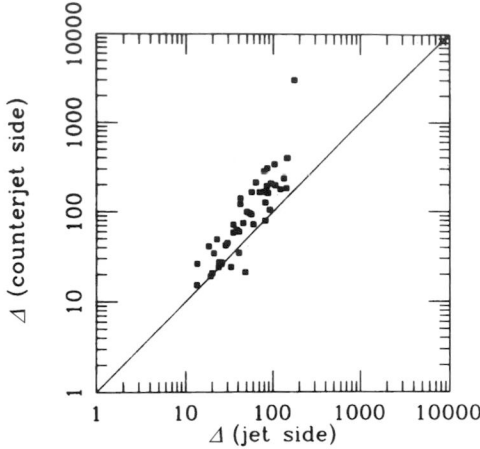

Fig 6.13. The Laing–Garrington effect, according to which the counterjet side is more strongly depolarized than the jet side (Laing 1993).

This correlation of counterjet side with greater depolarization was discovered by Laing (1988) and Garrington *et al.* (1988). A recent version of this correlation for a sample of 40 quasars and 7 radio galaxies with one-sided jets is shown in Fig. 6.13 (Laing 1993). There is a general trend for the counterjet side to be more strongly depolarized than the jet side, consistent with the ideas discussed above.

6.4 *Hubble Space Telescope* observations of 3CR radio galaxies

In my view, these arguments support the hypothesis that the radio quasars are the same types of object as the radio galaxies, but viewed at different angles with respect to the axis of the double radio source. If this is indeed the case, it has implications for observations of the parent bodies of quasars. Firstly, it implies that the obscuring tori must be optically very thick indeed in order to attenuate the intense optical emission of the nuclei of quasars. Secondly, we can understand what the parent bodies of the quasars look like by studying the radio galaxies.

We have recently been obtaining superb *Hubble Space Telescope* images of the 3CR radio galaxies at redshifts $z \sim 1$, which indicate what we would expect would be observed underlying the images of the 3CR quasars. Our *HST* programme consists of imaging a complete sample of 3CR radio galaxies in the redshift interval $0.6 < z < 1.8$, the majority of them having redshifts $z \sim 1$. We have completed a preliminary analysis of the *HST* images of three of these radio galaxies, as well as analysing new radio maps made with the VLA with a resolution of 0.15 arcsec, comparable to that of the *HST* observations (Longair, Best, and Röttgering 1995). We have also observed all these radio galaxies with the infrared camera IRCAM3 on the UK Infrared Telescope.

6.4.1 3C 368

The AGN 3C 368 has a redshift $z = 1.13$ and its optical spectrum has very strong, narrow emission lines. The *HST* image at a wavelength of about 800 nm, with the radio contours superimposed, is shown in Fig. 6.14.

The radio and optical structures are on similar physical scales but the optical image bears little resemblance to that of a giant elliptical galaxy. The appearance of the image is somewhat misleading because the bright

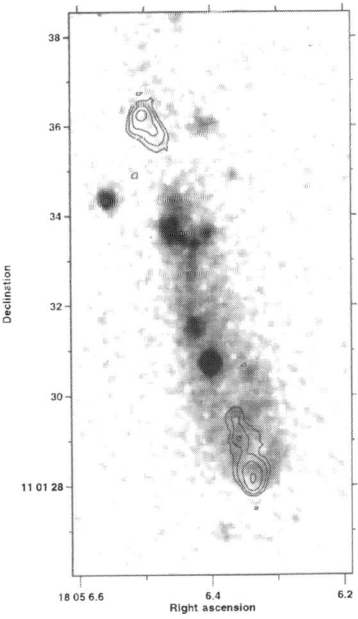

Fig 6.14. The *HST* image of the radio galaxy 3C 368 with the contours of its radio structure superimposed. The star-like object close to the centre of the image is an M star.

stellar object just south of the centre of the image has nothing to do with the radio galaxy at all – Hammer, Le Fevre, and Proust (1991) have shown that it is a nearby M star. Determining the location of the old stellar population of the galaxy and its nucleus is not a straightforward task since there is no central radio component in our VLA observations of its radio structure. We have obtained an excellent infrared image of the radio galaxy at 2.2 μm using the UK Infrared Telescope in Hawaii. After the emission from the M star is removed, we find that the centre of the residual image, which we assume to be the parent galaxy, lies about 1 arcsec to the north of the M star and coincides with a bright optical 'knot' in the *HST* image. An optical jet-like feature links the nucleus of the galaxy to the emission regions in the northern part of the image. The northern radio component lies beyond these emission regions, suggesting that the optical jet and the other complex structures have been caused by the passage of the jet of material responsible for the intense radio emission. To the south of the M star, the radio component lies towards the leading edge of a roughly elliptical optical emission region. The clear intensity minimum in the centre of this part of the image indicates that the optical emission originates from an ellipsoidal shell, not dissimilar to that found in bow shocks. Such a model for the emission in the southern part of the source has been proposed by Meisenheimer and Hippelein (1992).

The optical emission from the source is highly polarized and much of it is known to be line emission. At a wavelength of 690 nm, about 25% of the emission is line emission, whilst at 785 nm, about 40% is line emission. In the emission regions to the south of the image, there is very little continuum emission. Our interpretation of these observations is that these phenomena are associated with activity induced by the jets responsible for powering the radio-source components. It is an intriguing question whether or not the bright knots towards the north of the image are star-forming regions resulting from the passage of the radio jet. The strong optical polarization is inferred to be due to scattering of intense continuum emission from an obscured nucleus, the scatterers being either electrons liberated by the passage of the jet or dust associated with star-forming regions. No single theory of the alignment of the optical structures with the radio axis seems to be capable of explaining all the observations.

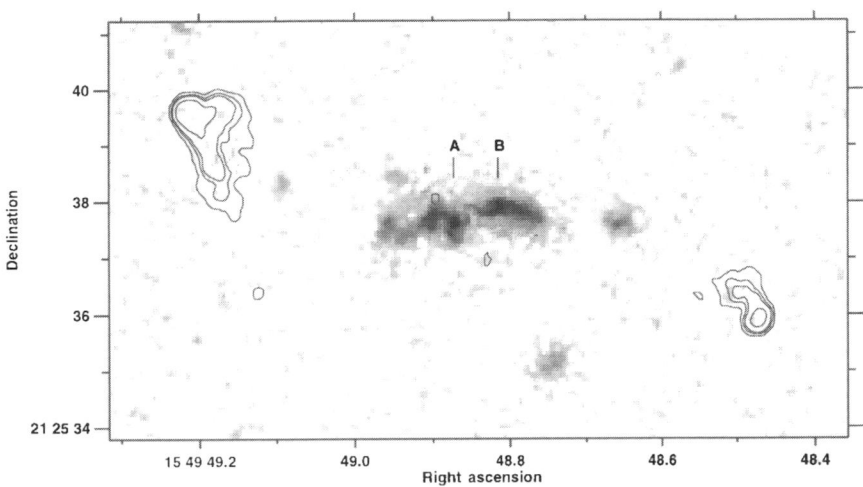

Fig 6.15. The *HST* image of the radio galaxy 3C 324 with the contours of the radio-source structure superimposed.

6.4.2 3C 324

The optical image of 3C 324 is much clumpier than that of 3C 368 and appears to consist of a number of 'interacting' components (Fig. 6.15).

In contrast, the 2.2-μm image appears to be a giant elliptical galaxy with maximum intensity centred upon the optical structure and coinciding with the 'dark lane' which cuts across the optical image between the components marked A and B. The axis of the optical image is misaligned by about 30° with respect to the radio axis defined by the outer radio hot spots, but it is aligned with an axis connecting the southern emission of the eastern lobe and the northern emission of the western lobe. This is the axis along which the transport of relativistic material is assumed to take place and it is clearly observed on the deep VLA images at 5 GHz presented by Fernini *et al.* (1993). Again there is a significant contribution of line emission to the total intensities observed in these wavebands, amounting to about 9% at 690 nm and about 26% at 783 nm.

Unlike 3C 368, there is no clear association between the radio hot spots and the optical structures in the case of 3C 324, the radio structure being on a somewhat larger scale than the optical image. We speculate that the structures observed in this case represent the aftermath of the passage of the jets powering the radio source.

6.4.3 3C 265

The radio galaxy associated with 3C 265 is surrounded by numerous companions or emission regions (Fig. 6.16). The radio structure is a factor of ten times more extended than the optical emission, and the direction of the hot spots in the extended components is indicated by the straight lines in Fig. 6.16.

If, however, the *HST* image is convolved to a resolution of 1.5 arcsec, the optical axis is aligned with the radio axis to within about 20°. A deep infrared image of the field at 2.2 μm shows only the parent galaxy of 3C 265 and the companion galaxy 5 arcsec north of the main body of the radio galaxy. It is assumed that the other structures are associated with large-scale gas clouds, possibly formed from massive cooling flows associated with the gas of a cluster about the giant elliptical galaxy (see, for

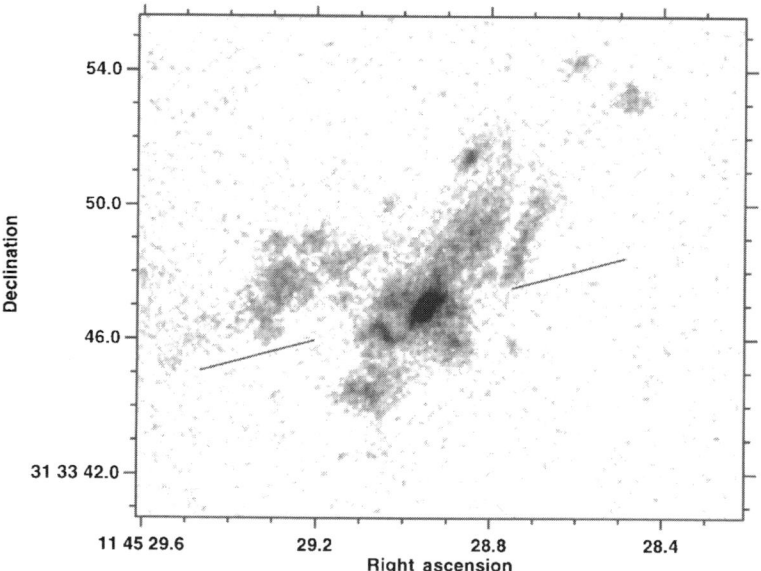

Fig 6.16. The *HST* image of the radio galaxy 3C 265. The radio structure is on a scale roughly ten times the size of the optical image, the direction of the hot spots in the outer radio structures being indicated by the straight lines.

example, Crawford and Fabian 1993). 3C 265 is among the most luminous radio sources known, and the fact that the structure extends to such large distances beyond the galaxy suggests that there must be confining gas which extends to a large distance about the galaxy. The emission regions may be associated with gas which has been compressed by the passage of the radio jet and the expansion of the radio lobes, and has cooled out of a diffuse gaseous halo about the galaxy. If this is indeed the case, these observations are of importance for the evolution of giant elliptical galaxies, because the cooling gas eventually falls onto the galaxy, a process which may be important in the formation of cD galaxies.

The upshot of these studies is that the alignments of the radio and optical structures are probably different manifestations of the interaction of the jets which power the radio sources with the surrounding medium. The three sources have been discussed in order of increasing size of the overall radio structure relative to the parent galaxy. In our interpretation of these observations, the interaction of the jets with the ambient medium results in different observational phenomena at different stages in the evolution of the radio source.

If the unification scheme described in Section 6.3 is correct, we now know what the galaxies underlying the radio quasars look like. They certainly do not look like giant elliptical galaxies, but rather have the appearance of projected versions of the types of structure seen in Figs. 6.14, 6.15, and 6.16. My expectation is that the *HST* observations of radio quasars at redshifts $z \sim 1$ will reveal structures which do not look like normal galaxies at all. In order to observe the parent galaxies of sources at redshifts of about 1, it is preferable to observe them in the near-infrared waveband, say, at 2 μm. This is a key set of observations for the Near Infrared Camera NICMOS which will be included in the next refurbishment mission of the *HST*. I continue to be impressed by the remarkable narrow dispersion found in the infrared apparent-magnitude–redshift relation for radio galaxies in the redshift interval $0.1 < z < 1.6$. From the present point of view, the important point is that the optical images of the galaxies at redshift 1 are strongly 'contaminated' by the line and continuum emission of star-forming regions, whereas the infrared observations show remarkably symmetrical structures which clearly represent the underlying old populations of the radio galaxies.

6.5 Superluminal radio sources

Analyses of the physics of the hot spots and lobes of double radio sources show that they must be continuously supplied with energy by jets from the nucleus. Evidence for these jets is found in the radio maps of quasars and radio galaxies, the case of Cygnus A being a good example of the type of jet necessary to power the hot spots which are observed towards the outer edges of the radio lobes (Fig. 6.17).

Evidently these jets originate in the nuclear regions, and the evidence that material is ejected relativistically from these regions is convincingly provided by the observation of superluminal motions in compact radio sources. This is the province of VLBI observations, which can probe to angular scales of 1 milliarcsecond or less.

The properties of superluminal sources have been surveyed by Marscher (1993), some of the key observations being as follows:

(i)   The components are observed to move out from the nucleus at transverse speeds up to ten times the speed of light.
(ii) The component identified with the nucleus has a flatter spectrum than the component which moves away from the nucleus.

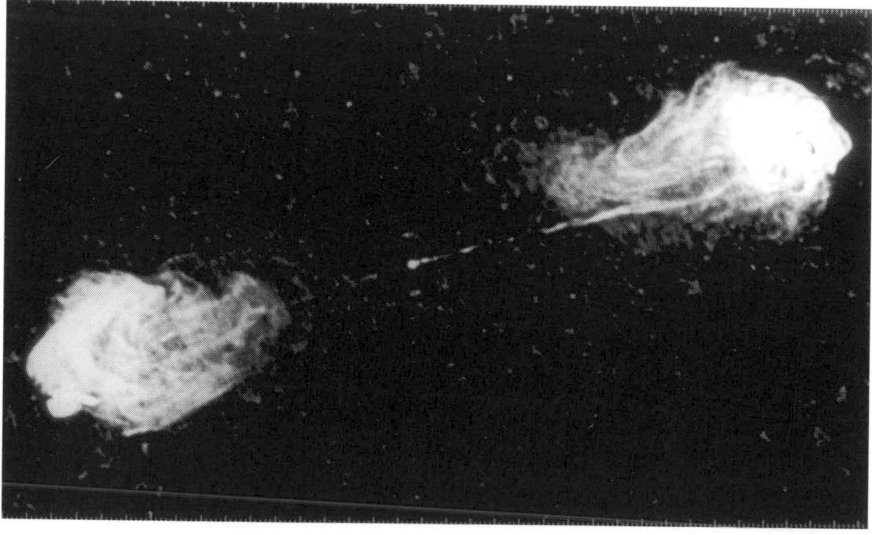

Fig 6.17. The structure of the radio source Cygnus A as observed by the VLA. (Perley, Dreher, and Cowan 1984).

(iii) The relativistically moving jets are almost all one sided.

(iv) The structures of the relativistically moving components often extend in the direction of the larger-scale radio jets, even if the jets and larger-scale structures are bent.

These properties can be naturally accounted for in terms of what has become the standard model for superluminal sources, first described in detail by Blandford and Rees (1978), although many of the ideas were foreshadowed in the prescient papers of Martin Rees in 1967 (Rees 1967). A typical representation of this model is shown in Fig. 6.18, which is taken from Marscher's review (Marscher 1993).

The relativistic jet is assumed to move out from the nucleus at a velocity v close to the speed of light with corresponding Lorentz factor $\gamma = (1 - v^2/c^2)^{-1/2}$. It is a simple calculation to show that, if a radio-source component moves at this velocity at an angle close to the line of sight, the maximum observed apparent velocity on the sky is $v_{\mathrm{obs}} = \gamma v$, if the jet is ejected at an angle $\theta \approx 1/\gamma$. Accompanying this motion, there is strong Doppler boosting of the intensity of radiation from the jet. The precise amount of boosting is model dependent but typically amounts to roughly a factor of $(1+z)^{-(3-4)}$, where I have written the result in terms of the redshift z, which, in the case of the component approaching the observer, is a large blueshift. It is a debatable point whether or not the Doppler shifts are sufficient to account for the one sidedness of the relativistic jets. My impression is that probably the asymmetries can be accounted for as a

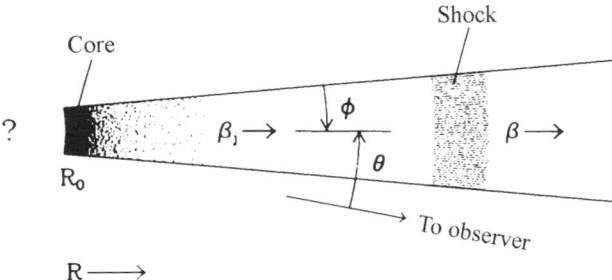

Fig 6.18. A typical model for the observation of superluminal motion in compact radio sources (Marscher 1993). The observer is located at an angle θ relative to the axis of ejection of the relativistic jet. Note that the base of the jet is identified with the compact, flat-spectrum source component and that this part of the jet also moves out relativistically from the nucleus.

result of Doppler boosting, but this is an issue which will have to be settled by the systematic observations which will be undertaken by the Very Long Baseline Array (VLBA).

There are some important points about the model illustrated in Fig. 6.18. It can be seen that the jet is assumed to be relativistic right down to its base, which, at compact enough scales, becomes synchrotron self-absorbed, and it then becomes impossible to study smaller-scale structures by this radio astronomical technique. Notice that, in the standard model, the compact core itself is also Doppler boosted in order to account for the frequency with which compact cores as well as relativistic jets are observed. It is not clear what the relativistic components observed in the jets are: are they relativistic shocks propagating along the jet, or are they physical objects being advected along the relativistic jet?

Since the jets are assumed to be pointing at a small angle to the line of sight, foreshortening effects are important in deriving their physical properties. A good example of this is the observed half-angle of the jet. If it is observed to have half-angle $\phi_{\rm obs}$, then the intrinsic opening half-angle must be $\phi = \phi_{\rm obs} \tan \theta$, where θ is the angle between the axis of the source and the line of sight. Marscher (1993) has shown that, in the cases of the sources NRAO 140 and 3C 345, the intrinsic opening angles of the jets must be only about $\phi \approx 1.5$–$2°$. Thus, the relativistic beams must already be very well collimated indeed by the time they have reached a distance of only 1–10 pc from the nucleus. These are very interesting numbers because the opening angles observed in extended sources such as Cygnus A (Fig. 6.17) have similar values. It therefore seems that the jets are able to remain highly collimated from scales of the order of 1 pc to scales of 100 kpc and greater in many cases.

Along the same line of reasoning, bends and kinks in the jets are all strongly foreshortened, and so, although the angles may look large on the radio maps, when they are deprojected by the $\tan \theta$ factor, the bends in the jets are quite small. I find it intriguing that the cone angle of about 7–10°, which we found in our analysis of the angular asymmetries in double radio sources, can probably encompass even the largest deviations observed in the superluminal sources.

In a few of the extreme blazars, which are assumed to be strongly Doppler-boosted sources, observed more or less precisely along the line

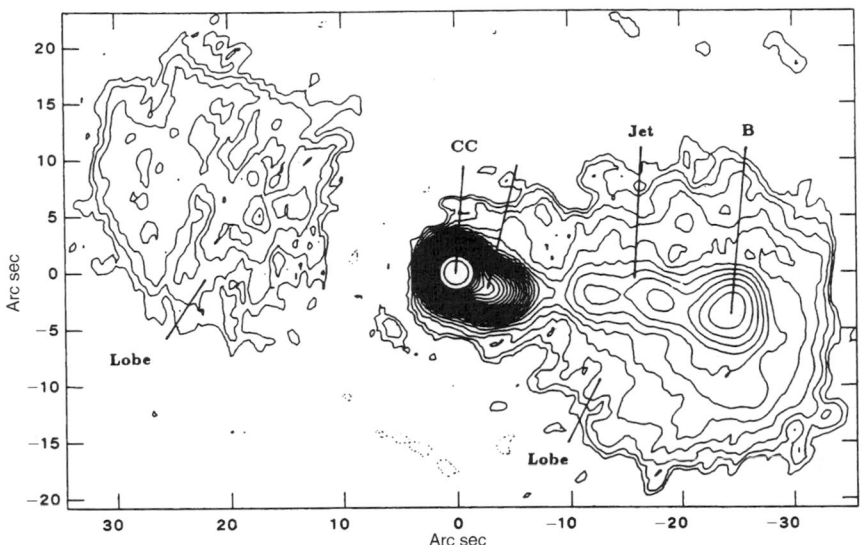

Fig 6.19. The highly variable radio source 3C 371 observed with high dynamic range (Wrobel and Lind 1990).

of sight, high dynamic range observations have enabled the underlying double radio sources to be observed. A good example of this is the blazar 3C 371, which looks to all appearances like a standard double radio source observed end on (Fig. 6.19).

Another unification scenario concerns the relation between the BL Lac objects and radio galaxies. A good case can be made for BL Lac objects being the relativistically beamed counterparts of the Fanaroff–Riley class-1 objects. The distinction between the two classes of source is clear morphologically, with the hot spots lying towards the edges of the source components in the class-2 sources and the maximum intensities of the components lying towards the nucleus in the class-1 sources. What Fanaroff and Riley (1974) found was that there is a very clear distinction in radio luminosity between the two classes of source (Fig. 6.20).

When radio observations with very high dynamic range are made of BL Lac objects, it is found that there are weak extended radio sources about the intense radio core and these have low radio luminosities and radio structures similar to the Fanaroff–Riley class-1 objects. It is then natural to interpret the radio emission of the BL Lac objects as the

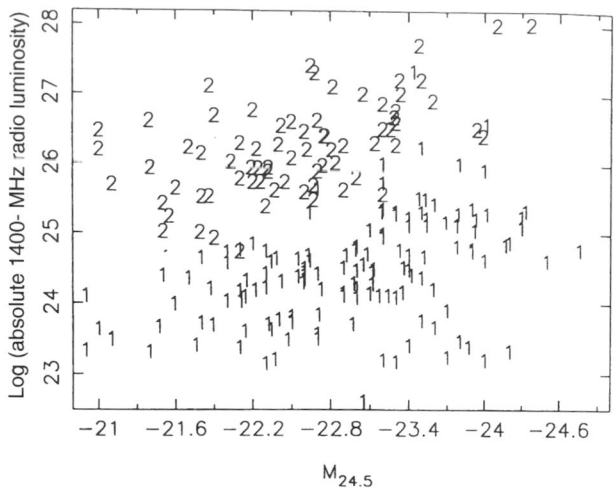

Fig 6.20. A plot of radio luminosity at 1.4 GHz against optical absolute magnitude, for Fanaroff–Riley classes 1 and 2 (Owen and Ledlow 1994).

relativistically beamed radio emission from jets emerging from the nucleus at angles very close to the line of sight. There is, however, a clear distinction between the FR1 and FR2 radio jets observed on large scales, in that normally the radio jets in the FR1 sources are double sided. Thus, unlike the radio jets in the radio quasars, the radio jets are not relativistically beamed once they leave the nuclear regions. It is assumed that the jets must therefore be rapidly decelerated in the case of the FR1 sources. The differences between the two types of strong radio source must be strongly correlated with the power of the beam powering them. In the standard model of double radio sources, the radio luminosity is directly related to the power of the jets powering them, and so it is likely that the different radio morphologies are associated with the jet power.

In the context of radio morphology, the radio–optical jet in M87 is an example of an FR1-type source, and it is intriguing that the motion of the knots in the jet have now been measured. Figure 6.21 shows a comparison of the radio and optical jets as presented by Biretta (1993). Aperture-synthesis observations with the VLA and VLBI observations have shown that the jet is moving relativistically from the nucleus. VLBI observations have shown that the radio jet moves out from the nucleus at about $(0.28 \pm 0.08)c$, while knot A, the brightest feature in the jet, moves at a

Fig 6.21. A comparison of (a) the optical emission of the jet of M87 with (b) a map of the radio emission showing the counterjet which coincides with a feature on the optical image. Knot A is the brightest feature in the optical jet (Biretta 1993).

projected velocity of $(0.52 \pm 0.03)c$. The absence of a prominent counterjet is probably partly due to relativistic beaming. That a counterjet is indeed present is inferred from the presence of optical and radio features at some distance from the nucleus, which can be interpreted as the interaction of the beam with the ambient medium.

Thus, it is quite plausible that, even in the Fanaroff–Riley class-1 sources, there are also relativistic beams which originate in the nucleus.

6.6 γ-ray sources in active galactic nuclei

One of the most remarkable discoveries of the *Compton Gamma-Ray Observatory* has been the observation of ultraluminous, extragalactic γ-ray sources at energies $\epsilon \geq 100\,\text{MeV}$. From the first two years of observation, 24 of these ultraluminous sources were detected, all of them associated with compact, radio-loud sources. The sources have flat or inverted radio spectra and many of them are found to be superluminal radio sources. Although γ-ray emission has been sought from their radio-quiet counterparts, none of them has been detected as a strong γ-ray source (Michelson 1994).

The spectra of these γ-ray sources indicate that, in terms of observed energy flux, most of the energy detected from these objects lies in the γ-ray region of the spectrum (Fig. 6.22). The γ-ray luminosities can be written in terms of a beaming factor f, which describes the fraction of the celestial sphere over which the source emits its observed intensity. Typically, the γ-ray luminosities amount to about $10^{41} f$ W, with a few examples up to ten times greater than this value. These are enormous luminosities if the sources were to radiate isotropically, that is, if $f \sim 1$.

The association of γ-ray sources with superluminal radio sources strongly suggests that the γ-ray emission is associated in some way with the relativistic jets inferred to be present in these sources. Two other pieces of evidence favour this picture. First, the γ-ray emission is highly variable. In the case of the γ-ray source associated with 3C 279, the γ-ray intensity changed by a factor of at least four over a period of days. Therefore, these enormous luminosities must originate within a very compact region.

The second argument concerns the enormous energy densities in high-energy γ-rays which must be present in these sources. All the γ-ray sources are also intense hard-X-ray emitters and so, in these very compact sources, the γ-rays lose all their energy in photon–photon collisions, which leads to the creation of electron–positron pairs. Blandford (1994) has shown that the result of this process is the formation of a 'γ-ray photosphere' about the nucleus: the γ-rays we observe cannot originate within this region.

The favoured model is one in which the source of the γ-ray emission is the same relativistic jet, which is responsible for the radio jet, but, because

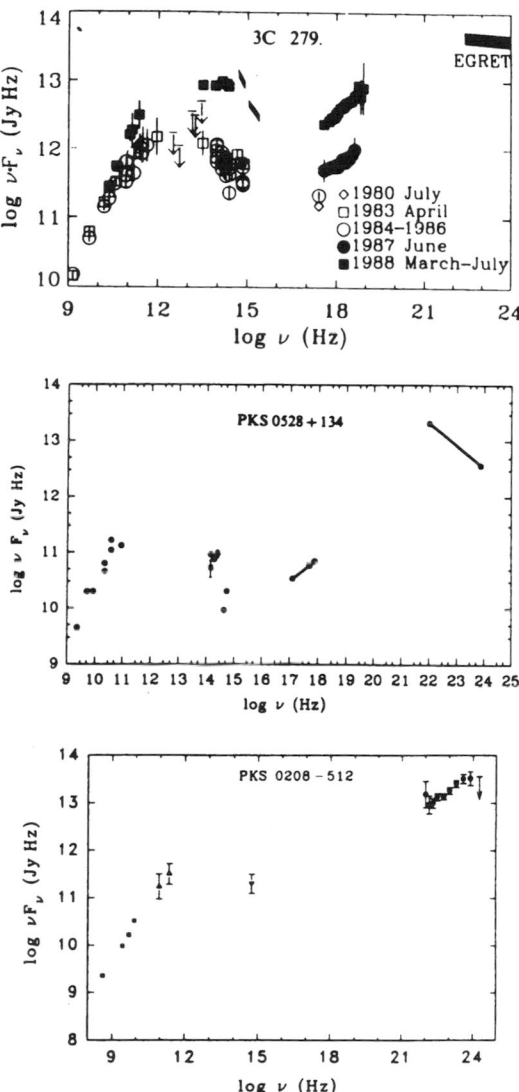

Fig 6.22. Three examples of the spectra of ultraluminous γ-ray sources. These spectra are plotted as νI_ν so that the ordinate represents the amount of energy radiated at each wavelength (Michelson 1994).

of the compactness of the source, the emission probably originates much closer to the nucleus, in the regions in which the radio core is synchrotron self-absorbed. Blandford (1994) argues that the most likely region in which the γ-rays are produced is at the point where the relativistic jet has just emerged from the γ-ray photosphere. The emission process is probably some version of inverse Compton scattering in which low-energy photons are boosted to enormous energies by the high-energy electrons present in the relativistic beam. The source of the photons which are scattered to high energies is a matter of speculation but various possibilities have been suggested. For example, the photons may be created in the inner regions of an accretion disk about a supermassive black hole. Alternatively, the photons from the accretion disk may be scattered into the beam by Thompson scattering by free electrons in the environment of the nucleus. A third possibility is that the soft photons are created within the relativistic jet itself. Blandford (1994) discusses the merits of each of these possibilities.

6.7 The innermost regions

Clues about the nature of the innermost regions come from X-ray observations of active galactic nuclei, X-ray spectroscopic studies being of particular importance. Pounds *et al.* (1990) have studied the spectra of those Seyfert galaxies which are strong X-ray sources and discovered that, although the spectra can be generally approximated by a power-law distribution through the 1 to 40 keV waveband, there are significant deviations from this relation, which are illustrated in Fig. 6.23.

It can be seen that there is a prominent peak in the vicinity of the 6.4-keV line of iron and there are depressions on either side of this maximum. These features have now been observed in the spectra of individual Seyfert galaxies and have been interpreted as evidence for a reflected X-ray component originating in the active galactic nucleus. The spectrum of X-ray radiation reflected from cold matter has been the subject of detailed study, an example of the results of calculations by Lightman and White (1988) being shown in Fig. 6.24.

The dashed line shows the energy spectrum of the incident X-ray radiation, which is assumed to be of power-law form, $I(\epsilon) \propto \epsilon^{-0.7}$, and the

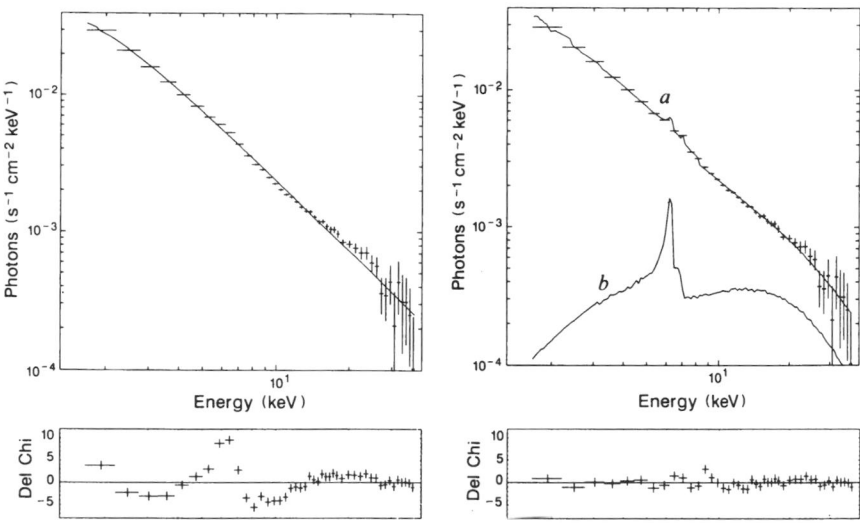

Fig 6.23. Left: The summed X-ray energy spectra of 12 Seyfert galaxies. Although generally following a power-law distribution, there are significant deviations from that relation as indicated by the residuals shown in the lower panel once the power law has been subtracted. Right: An improved fit to the observed X-ray spectra once a reflected X-ray component is included. (Pounds *et al.* 1990.)

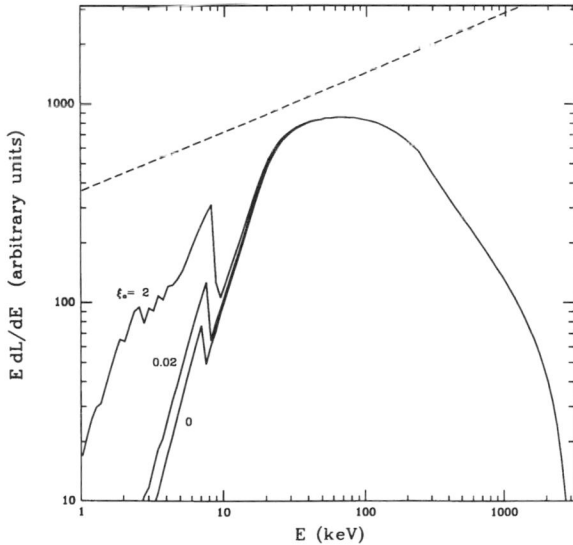

Fig 6.24. The spectrum of reflected X-rays from a semi-infinite slab of cold gas. The input spectrum is a power law of the form $I_\epsilon \propto \epsilon^{-0.7}$. The units on the ordinate are ϵI_ϵ. ξ is the differential ionization parameter which describes how the ionization state of the cold cloud is modified by the incident radiation. In the case $\xi = 0$, the matter remains cold (Lightman and White 1988).

reflected component is shown by the solid lines. At high energies, the principal energy-loss process is Compton scattering of the X-rays by cold electrons in the cloud. In each Compton scattering, a photon suffers an average fractional energy loss

$$\left\langle \frac{\Delta\epsilon}{\epsilon} \right\rangle = \frac{h\nu}{m_e c^2}. \qquad (6.2)$$

Thus, the highest-energy photons lose energy most rapidly, resulting in a pronounced steepening of the reflected spectrum at high energies. At low energies, the principal loss mechanism is photoelectric absorption, which increases dramatically at low energies. There are, however, pronounced absorption edges at which the absorption cross section changes abruptly, giving rise to pronounced features in the reflected spectrum, principally about the absorption edge of iron at about 6.4 keV. Figure 6.23 shows the effect of adding a reflected component to a power-law spectrum, and it can be seen that a much improved fit to the data is obtained.

Further evidence for the influence of dense gas upon the spectra of X-ray sources comes from the observation of some active galaxies which have very large column depths for absorption by neutral hydrogen. Pounds (1990) has reported the observation of column depths $\int N_H \, dl \geq 10^{23}$ hydrogen atoms cm^{-2} in the broad-line radio galaxy 3C 445. This dense gas is presumed to originate close to the nucleus.

An intriguing question concerns the location of the cold gas responsible for the absorption and the reflected component. Mushotzky, Done, and Pounds (1993) have attempted to synthesize all the data on the X-ray emission of Seyfert 1 and Seyfert 2 galaxies within the context of the unified model described in Section 6.2. In Figure 6.25, the continuum X-ray emission is shown as originating close to the central black hole. A small fraction of the continuum X-ray emission is reflected from cold clouds, resulting in the features observed in the X-ray spectrum.

The big problem is, 'Where is the cool gas located?' Is it related to the gas in the broad-line regions described by Don Osterbrock? There is a great deal yet to be understood about the distribution of gas in the vicinity of supermassive black holes.

There also remains the problem of understanding the origin of the form of the spectra of the X-ray emission from active galactic nuclei. While the

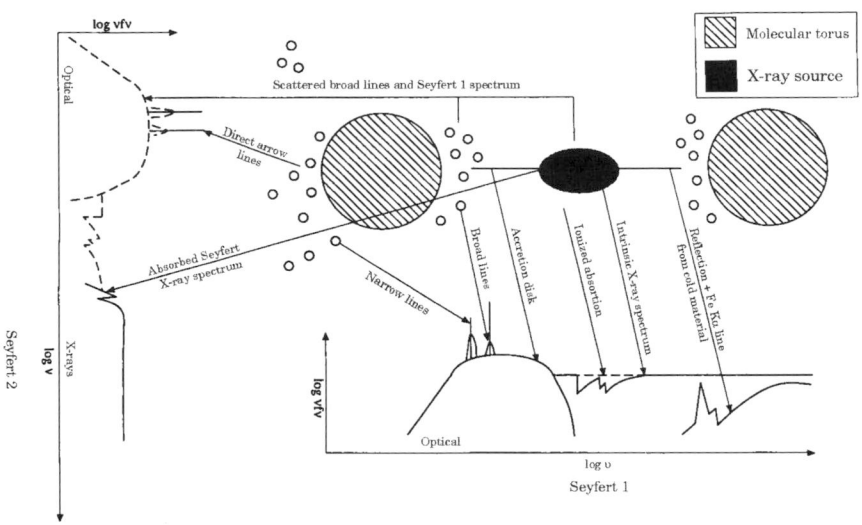

Fig 6.25. A schematic diagram illustrating the origin of the X-ray emission from Seyfert 1 and Seyfert 2 galaxies (Mushotzky, Done, and Pounds 1993).

reflection component adds some structure to the spectrum, it accounts for only about 10% of the X-ray emission, and there remains the problem of accounting for the spectrum of the hard-X-ray background, which shows a break at about 40 keV. The most likely origin for the background is that it is the integrated emission of active galaxies, but these must have the correct forms of spectra to match the X-ray background. Rashid Sunyaev reported an encouraging result from observations of the hard-X-ray spectra of Seyfert galaxies with the *GRANAT* X-ray telescope. Their spectra are similar in form to that of the X-ray background. According to Sunyaev, their spectra are exactly what would be expected from Comptonization of an input spectrum of soft ultraviolet and X-ray photons by very hot gas in the source region. The process of Comptonization in a very hot plasma with $kT_e \sim 25$ keV results in a power-law spectrum, with a spectral index of about 0.4–0.5 with a Wien cut-off at high energies (see, for example, Pozdnyakov, Sobol, and Sunyaev 1983). The existence of active galaxies with these forms of spectra is an important clue.

6.8 **How do we put it all together? Some key problems**

We have now assembled sufficient ingredients to account for essentially all the phenomena observed in active galactic nuclei. I enjoy very much the cartoon which Roger Blandford published in his Saas-Fee lecture notes on active galactic nuclei (Blandford 1991) and which is reproduced as Fig. 6.26. I like to think of this picture as a 'powers of ten' diagram in which we travel from the largest structures caused by the energy-release in the nuclei of active galaxies to the black hole in the centre.

It can be observed that, between us, Donald Osterbrock and I have covered most of the phenomena represented in these excellent cartoons. We have left the discussion of the black hole itself to Igor Novikov. It is left to the reader to select from this set of tools those needed to address any particular problem. There are a number of key issues which I will highlight in conclusion.

6.8.1 The collimation of jets

All models of extragalactic radio sources and γ-ray sources involve relativistic jets, but the means by which these are collimated is one of the major unsolved problems of high-energy astrophysics. Almost certainly, the mechanism will require the presence of magnetic fields, but they have to be tied onto something, or else they will simply fly apart. Important clues are provided by the discovery of superluminal radio sources within our own Galaxy, indicating that this phenomenon is not just associated with supermassive black holes (Mirabel and Rodríguez 1994).

6.8.2 What is the nature of the obscuring torus?

The case for obscuring tori in the nuclei of active galaxies seems to me to be increasingly persuasive but their nature and astrophysical role are not understood. Are they associated with the fuelling of accretion disks? Are they simply the outer regions of warped accretion disks? Are they objects similar to the molecular torus observed at the centre of our own Galaxy?

6.8.3 Are there really accretion disks in active galactic nuclei?

One of the worries about the accretion-disk models for active galactic nuclei is that there is no direct evidence for the disks themselves. The

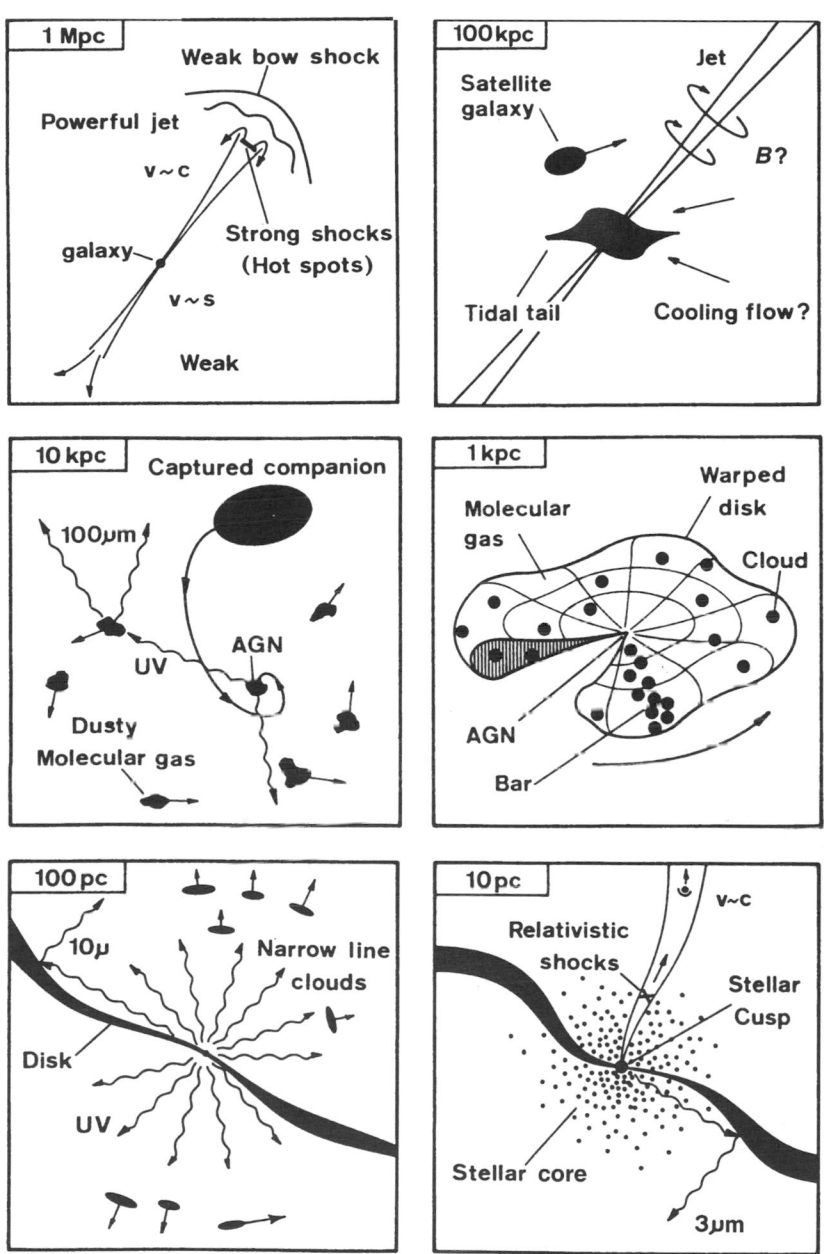

Fig 6.26. A 'powers of ten' diagram due to Blandford (1991), illustrating the different phenomena observed on different physical scales in active galactic nuclei.

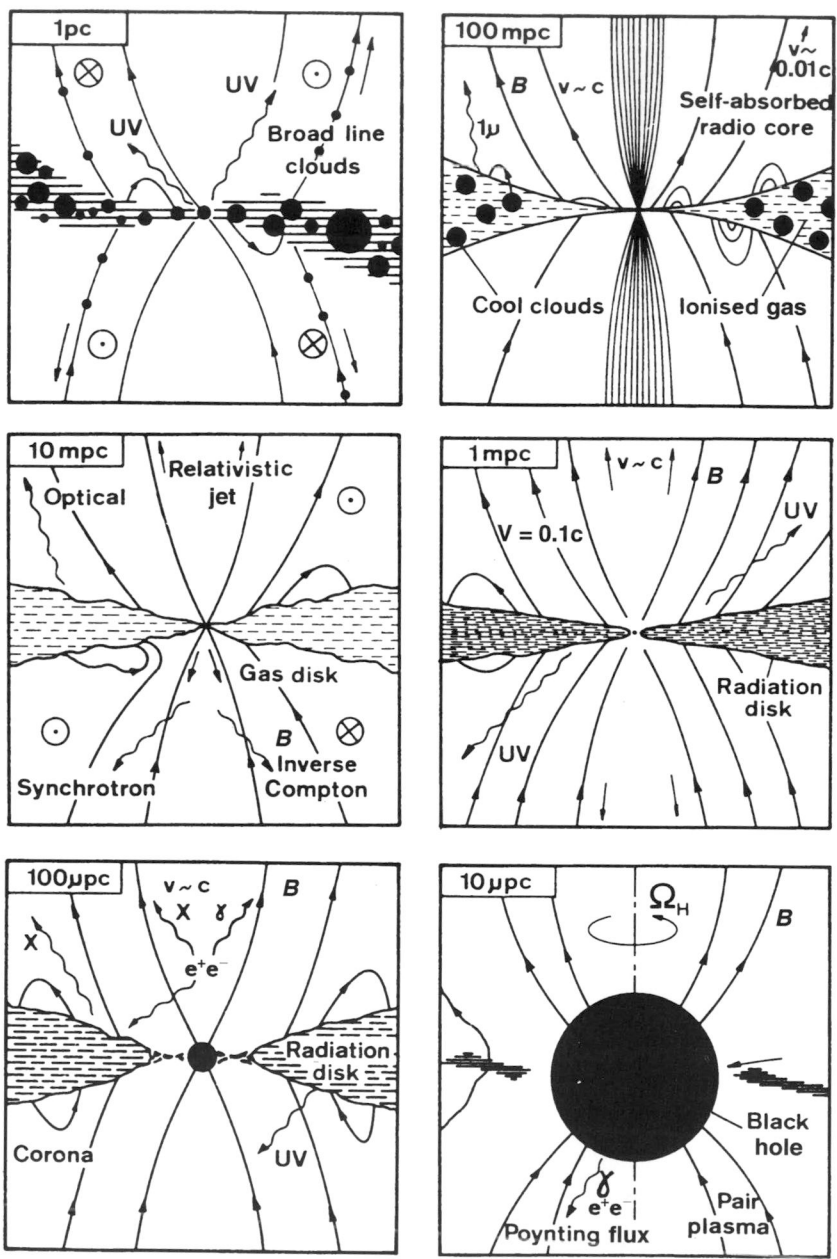

Fig 6.26. (*cont.*)

situation is in contrast to the accretion disks in sources within our Galaxy. Eclipse mapping of cataclysmic variables and dwarf novae have given convincing evidence that accretion disks are formed in these binary systems. There is also the remarkable evidence provided by X-ray observations of binary systems containing stellar-mass black holes. The Japanese astronomers have shown that, whilst the luminosity and temperature of the X-ray-emitting regions vary, they are consistent with matter accreting onto a black hole with a fixed last stable orbit. The variations in luminosity can be attributed to the increase in temperature of the matter close to the last stable orbit (Inoue 1992). Unfortunately, we do not have similar direct evidence for accretion disks about the supermassive black holes which must be present in the more extreme active galactic nuclei.

6.8.4 The blue bumps

Another intriguing problem concerns the existence and nature of the 'blue bump' observed in optical–ultraviolet spectra of active galaxies. Evidence for the bump is found by extrapolation from the readily observable ultraviolet waveband to wavelengths longer than 121.6 nm and from the soft-X-ray waveband at energies $\epsilon > 0.1$ keV to lower energies. This component may well be responsible for the ionizing flux needed to excite the emission-line regions in active galaxies. The theory of thin accretion disks suggests that the temperature of material at the last stable orbit about black holes scales as $M^{-0.25}$, and so, if solar-mass black holes radiate in the X-ray waveband, accretion disks about $10^8 \, M_\odot$ black holes should radiate at about 10^5 K, exactly the temperature needed to account for the blue bump. Is this just a coincidence? The big problem is that far-ultraviolet observations of extragalactic objects are very difficult even from space, because of photoelectric absorption in the interstellar gas.

6.8.5 What is the key distinction between radio-loud and radio-quiet objects?

It is striking that there seems to be a clear distinction between those objects which are radio-loud and those which are radio-quiet. It is most simply expressed in tabular form:

Radio-loud	Radio-quiet
(giant) elliptical galaxies	spiral galaxies
radio galaxies	Seyfert 1/2 galaxies
radio quasars	radio-quiet quasars
	BAL quasars

It seems that two separate populations exist, and within each of them a good case can be made for the unified schemes discussed in Section 6.2. It is important to test how precisely this dichotomy between radio-loud and radio-quiet objects can be sustained. It is remarkable that the outer structures of the parent galaxies seem to influence activity going on deep inside the nuclear regions, and yet this is what the data seem to be telling us.

6.8.6 What is the role of high-energy astrophysical activity in galactic evolution?

Ultimately, we have to understand the role which these forms of high-energy astrophysical activity play in the scheme of galactic evolution. In general terms, astrophysical systems tend towards their lowest-energy states, and it seems remarkable that, to do so, the nuclear regions have to generate beams of relativistic matter and enormous luminosities in so many different wavebands. Yet, that is how nature seems to have arranged things.

References

Antonucci, R.: 1993, Ann Rev A&A, **31**, 473

Antonucci, R., and Miller, J.: 1985, ApJ, **297**, 621

Banhatti, D.G.: 1980, A&A, **84**, 112

Barthel, P.D.: 1989, ApJ, **336**, 606

Barthel, P.D.: 1994, in *First Stromlo Symposium: Physics of Active Galactic Nuclei*, eds G.V. Bicknell, M.A. Dopita, and P.J. Quinn (San Francisco: Astron. Soc. Pacific Conference Series), p. 175

Best, P.N., Bailer, D.M., Longair, M.S., and Riley, J.M.: 1995, MNRAS, **275**, 1171

Biretta, J.: 1993, in *Astrophysical Jets*, eds D. Burgarella, M. Livio, and C. O'Dea (Cambridge: Cambridge University Press), p. 263

Blandford, R.D.: 1991, in *Active Galactic Nuclei*, by R.D. Blandford, H. Netzer, and L. Woltjer (Berlin: Springer-Verlag), p. 161

Blandford, R.D.: 1993, in *Astrophysical Jets*, eds. D. Burgarella, M. Livio, and C. O'Dea (Cambridge: Cambridge University Press), p. 15

Blandford, R.D.: 1994, in *First Stromlo Symposium: Physics of Active Galactic Nuclei*, eds. G.V. Bicknell, M.A. Dopita, and P.J. Quinn (San Francisco: Astron. Soc. Pacific Conference Series), p. 23

Blandford, R.D., and Rees, M.J.: 1978, in *BL Lac Objects*, ed. A.M. Wolfe (Pittsburg: University of Pittsburg Press), p. 328

Crawford, C.S., and Fabian, A.C.: 1993, MNRAS, **260**, L15

Fanaroff, B.L., and Riley, J.M.: 1974, MNRAS, **167**, 31P

Fernini, I., Leahy, J.P., Burns, J., and Basart, J.: 1991, ApJ, **381**, 63

Fernini, I., Burns, J.O., Bridle, A.H., and Perley, R.A.: 1993, AJ, **105**, 1690

Garrington, S.T., Leahy, J.P., Conway, R.G., and Laing, R.D.: 1988, Nature, **331**, 149

Hammer, F., Le Fevre, O., and Proust, D.: 1991, ApJ, **374**, 91

Inoue, H.: 1992, *Proc. Texas/ESO-CERN Symposium on Relativistic Astrophysics, Cosmology and Fundamental Particles*, eds. J.D. Barrow, L. Mestel, and P.A. Thomas (New York: New York Academy of Sciences), p. 86.

Kinney, A., Antonucci, R., Ward, M., Wilson, A., and Whittle, M.: 1991, ApJ, **377**, 100

Laing, R.A.: 1988, Nature, **331**, 149

Laing, R.A.: 1993, in *Astrophysical Jets*, eds. D. Burgarella, M. Livio, and C. O'Dea (Cambridge: Cambridge University Press), p. 95

Laing, R.A., Riley, J.M., and Longair, M.S.: 1983, MNRAS, **204**, 151

Lightman, A.P., and White, T.R.: 1988, ApJ, **335**, 57

Liu, R., Pooley, G.G., and Riley, J.M.: 1992, MNRAS, **257**, 545

Longair, M.S., and Riley, J.M.: 1979, MNRAS, **188**, 625

Longair, M.S., Best, P.N., and Röttgering, H.: 1995, MNRAS, **275**, L47

McCarthy, P.J., van Breugel, W.J.M., and Kapahi, V.K.: 1991, ApJ, **371**, 478

Marscher, A.P.: 1993, in *Astrophysical Jets*, eds. D. Burgarella, M. Livio, and C. O'Dea (Cambridge: Cambridge University Press), p. 73

Meisenheimer, K., and Hippelein, H.: 1992, A&A, **264**, 455.

Michelson, P.F.: 1994, in *First Stromlo Symposium: Physics of Active Galactic Nuclei*, eds. G.V. Bicknell, M.A. Dopita, and P.J. Quinn (San Francisco: Astron. Soc. Pacific Conference Series), p. 13

Miller, J.S.: 1994, in *First Stromlo Symposium: Physics of Active Galactic Nuclei*, eds. G.V. Bicknell, M.A. Dopita, and P.J. Quinn (San Francisco: Astron. Soc. Pacific Conference Series), p. 149

Mirabel, I.F., and Rodríguez, L.F.: 1994, Nature, **371**, 46

Mushotzky, R.F., Done, C., and Pounds, K.A.: 1993, Ann Rev A&A, **31**, 717

Owen, F.N., and Ledlow, M.J.: 1994, in *First Stromlo Symposium: Physics of Active Galactic Nuclei*, eds. G.V. Bicknell, M.A. Dopita, and P.J. Quinn (San Francisco: Astron. Soc. Pacific Conference Series), p. 319

Perley, R.A., Dreher, J.W., and Cowan, J.J.: 1984, ApJ, **285**, L35

Pounds, K.A.: 1990, MNRAS, **242**, 20P

Pounds, K.A., Nandra, K., Stewart, G.C., George, I.M., and Fabian, A.C.:
  1990, Nature, **344**, 132

Pozdnyakov, L.A., Sobol, I.M., and Sunyaev, R.A.: 1983, Astrophys and
  Space Sci Rev, **2**, 189

Rees, M.J.: 1967, MNRAS, **135**, 345

Ryle, M., and Longair, M.S.: 1967, MNRAS, **136**, 123

Stockton, H.S.: 1994, in *Frontiers of Space and Ground-Based Astronomy*,
  eds. W.Wamsteker, M.S.Longair, and Y.Kondo (Dordrecht: Kluwer
  Academic Publishers), p.87

Wall, J.V., and Peacock, J.A.: 1985, MNRAS, **216**, 173

Wrobel, J., and Lind, K.: 1990, ApJ, **348**, 135

6.9 **Discussion**

Question

(G.Burbidge): Malcolm, I understand this is an attempt at uni-fication. You can make it work for the radio flux. I still have not understood really what the relationship is between most of the QSOs, which are not radio-emitting QSOs, and this small class which you unify with radio galaxies. Maybe you show this at the end, but it is never made clear what the real difference is, because most of the power, most of the energetics, is still in the optical and the other bands, and not particularly in the radio bands.

Answer

(Longair): Yes, I accept what you are saying. I prefer the simple picture which Don Osterbrock described, in which the radio-quiet quasars are simply an extension of Seyfert 1 galaxies up to very high luminosities and are part of that population.

Question

(G.Burbidge): So, in other words, you put them in different types of galaxies?

Answer

(Longair): Yes.

Comment

(Sunyaev): I want to add to your description of the X-ray proper-ties of the active galactic nuclei. For the last five years at least, the observa-tions from the *GRANAT* spacecraft, which were confirmed by the *Compton Gamma Ray Observatory*, have shown that the spectrum, at least of NGC 4151, is just a copy of the galactic black-hole binary Cygnus X-1, that is, a Comptonized spectrum with a very strong cut-off at high ener-gies. It is very important, because here we see that we really are dealing with thermal plasma, hot plasma, in the vicinity of an accretion disk. These X-rays do not originate from the relativistic particles as inverse Compton scattering and do not originate from the electron–positron plasmasphere. At least in NGC 4151, we know there is no place for electron–positron emission: the spectrum is much stronger. I also know that the upper limits to the flux for at least 12 other Seyferts show practically the same: there is no strong flux, there is no γ-ray photosphere. I believe that it is very impor-tant. If you could mention something about this, it would be good. It is a lot of energy. The energy in the γ-ray and X-ray fluxes is comparable with the optical–ultraviolet region.

Answer

(Longair): I did mention this briefly. I do not know if Martin has a comment about these X-ray spectra?

Comment

(Rees): Two comments, Malcolm. First, about the UV bump and evidence for accretion disks. I think it is important to bear in mind that the UV bump and the $M^{-1/4}$ dependence is not peculiar to a particular accre-tion-disk model. That is just what you get by thermalizing the Eddington luminosity within a few Schwarzschild radii. One has to think of other discriminants for the actual dynamics: as you mentioned, the lines. I would like to make a comment about collimation. You are right in saying that magnetic fields have to be invoked, but I think there are two things that have happened in recent years. One, very recently, is the remarkable X-ray novae in the Galaxy, which are, in effect, speeded-up versions of these objects, where we see clearly the same sort of collimation happening, prob-ably around a black hole.

On the other hand, the fact that we see these remarkable narrow jets in protostars suggests that it is wrong to blame the collimation entirely on a relativistic effect, because there is certainly nothing relativistic in a protostar, but you get equally good collimation. I think this is suggesting that wound-up magnetic fields are probably important. In some of these objects they are related to the black hole, but I think it still is an open question to what extent relativistic effects are important in collimation. This also relates to the question of whether the alignment of the jets is with the spin of the black hole or with gas at larger radii, which is another open question.

Comment

(Sunyaev): I believe that it is great that we have finally observed two superluminous sources in the Galaxy which are connected, at the very least, with relativistic objects, neutron stars, or black holes, which are obviously in binaries. I wish, though, to mention a totally different thing. Today, we see jets everywhere where we see the real motion of matter, and it is great. It is also obvious that they must be artificial jets, connected with scattering, in practically any place where we see intergalactic gas, as, for example in M87. Today, for example, the extra luminosity of M87 is less than 10^{41} erg s^{-1}, which is negligible in comparison with the Eddington luminosity. According to just the scale of the dimensions of the intergalactic gas in M87, we can trace the activity of the nucleus for many hundreds of thousands of years, because it is 100 kpc in extent. If, at some moment, a few hundred thousand years ago, this gave, for example, a blazar, γ-rays, or X-rays, it is obvious that we must see the scattered component and the brightness. In reality, we are so close to M87 that we could see the scattered component, and it is not there. We measured spectra in the field of M87 and there is nothing at our level of sensitivity, which, for jets, is maybe 10^{38}–10^{39} erg s^{-1}, this sort of luminosity. There are no jets which are connected with scattering. I believe this is a problem. We must absorb in radio, in X-rays, and in the optical, both the artificial jets which are just connected with beamed radiation, which are just scattered by the gas inside, for example, M87. This would be also great because it shows us about the activity on much smaller scales than the real emission from the core, which is very strongly beamed. Or this beamed emission is coming from these clouds which are very far from the sources.

Question

(Rees): Just a question about the first part of your talk, where you refer to these velocities. Do you really believe those are velocities rather than upper limits? Obviously one cannot believe that the two lobes are intrinsically the same, and indeed, they cannot consistently be the same, so do you really believe those are more than upper limits?

Answer

(Longair): The story goes well beyond what I have had time to discuss in my talk, because we must look at intrinsic effects as well, for which there is some evidence. What we believe is that, for the powerful sources, the relativistic effect may be just as important as intrinsic effects. The big surprise which suggested that we should consider the high velocities seriously was the lack of symmetric double radio quasars. In the high-luminosity sources, I believe at least half the effect is relativistic. That would be my current position.

Question

(Sunyaev): There was an additional thing. Today, for the first time, I heard from you that many of your radio galaxies from the 3C catalogue have velocities of $0.2c$. For me, this is remarkable, because it is very close to the velocity of the jet in SS433, where we also have $0.24c$, and I believe this might be the link between Galactic objects and extragalactic objects. I am now very happy that we are discovering, case by case, a lot of similarities between the sources observed in the Galaxy and extragalactic sources.

Answer

(Longair): The opening angle for SS433 is about 28°, I think, and it is precessing: we found opening angles of about 10-15°! We almost believed (and were tempted to say) that the same sort of thing is happening on this vast scale in order to account for these symmetries. Clearly the hot spots are not coming out exactly collinear.

Question
(Lynden-Bell): Is that a precession?

Answer
(Longair): Quite possibly.

Comment
(Rees): Just on Rashid's point. The speeds you are measuring are the speeds of the hot spots. That does not mean that the jet is moving that slowly, so his data is consistent with the jets being relativistic in all his objects, rather than jets moving at a third of the speed of light, like in SS433. So it is not obvious that he is finding evidence for jets that are moving at the same speed as SS433. He is just finding the motion for hot spots at the end.

Comment
(E. Martín): R. Sunyaev has just pointed out what can be learnt from comparison with very young stellar objects which also show jets. I am very struck, because the key issues which you have mentioned are very similar to the key issues when dealing with jets in very young objects. Your last issue was the role of high-energy activity in the evolution of galaxies. Naively, it seemed to me that it could have a role in births: like the problem of getting rid of angular momentum.

Comment
(H. Zinnecker): [Shows image on viewgraph.] What you see here is a protostellar jet, for comparison. This is a totally embedded jet. It was observed at $2.12\,\mu$m in the line of molecular hydrogen...it is pure line emission. Here it is. This object has been given a name: it is HH212. This source is close to the Horsehead nebula in Orion. It is a totally obscured *IRAS* source, not detected at $10\,\mu$m, very cold, so that is why we observed this object in the first place. It is a submillimetre continuum source. It shows this very remarkable, highly symmetric jet. You see that there is a lot of coincidence on either side, between what you call hot spots. It is not clear what these hot spots are, but it could be that they are related to velocity fluctuations, just as has been discussed by M. Rees, in M87.

There is evidence for three ejection events: you see three bow shocks here. All this is to show, as E. Martín has just mentioned, that there are lots of similarities between galactic and extragalactic jets and, as M. Rees mentioned, these jets cannot be collimated by any relativistic effect. In fact, I should mention that the recent observations with the Owens Valley interferometer have shown that this, which looks like it is perpendicular here, is almost in the plane of the sky. So, within a few years, we can get proper motions of this system. I should tell you that what you see is not the jet itself, it is probably the direction of the jet, within the medium, or the shocks in the jet, so these proper motions may not be the velocity of the jet. Only if it is a ballistic jet, or an overdense jet, would we know. So, the questions here are, is it overdense? Is it underdense? What is it? How does it work? How does it evolve? My last remark is to ask what we might be able to learn from this in terms of extragalactic observations, since we think that, in star formation, these jets actually interact with the inflow. So the outflow and the inflow compete, and that is how you form the mass of the star. I wonder whether the masses of the galaxies, at least some galaxies, may be affected by the competition between accretion and outflows. You can show, for this jet, that CO is accelerated away from the star, owing to this very collimated object.

Question

(S. di Serego): Let me ask a very simple question. You have shown us these beautiful images of the 3CR radio galaxies. Can we call these objects galaxies any more? What amazes me is that the reason why we call these objects galaxies probably has nothing to do with stars, or very little to do with stars, but I am still very much interested in stars in these objects. You have talked about the infrared, and it is a pity you have not shown any infrared observations; certainly they are very relevant. However, let us keep to the optical. What I wanted to say was that, even if the structure we are seeing in the optical is just line emission, or scattered light, this line emission is a sign of heavy elements far away, and the scattering is also probably a sign of heavy elements because of the dust which does the scattering. This must have something to do with stars. Secondly, the other reason why I am interested in stars is that, as you know, these galaxies are the most distant stellar systems we know about and there is a recent one at $z = 4.2$. Still, I think we can learn something about galaxy formation using these objects. Obviously, they are so complicated that, before you can do that, you have to disentangle them very well from these different components, but I would like to know your opinion about studying these stars in such objects.

Answer

(Longair): I have to confess that, when I saw the *HST* data, I was amazed. I expected to see boring, faint elliptical galaxies with diffuse emission, and, instead of that, we discovered these amazing structures. So it was a big surprise, but if you look at the sensitivity of *HST* and the way the stellar spectra of galaxies are redshifted, it was always going to be very difficult in the optical waveband when the galaxies had redshifts of 1. Now, the images which are showing all of these distorted structures are detectable because of the excess surface brightness associated with them. It is still very difficult to see objects at a redshift of 1 unless you take very, very long integrations. When you try to carry out the same observations in the infrared wavebands it is easy. The reason why I am very confident that these are galaxies is that, when we study all of the $z = 1$ radio galaxies in the *JHK* wavebands, their colours are consistent with stellar continuum spectra of standard giant elliptical galaxies. I think the answer is that you have to shift all your thinking by a factor of at least two in wavelength, and then you can understand the properties of the stellar spectra. When the stellar spectrum shifts, it remains very flat through the 2-μm window, up to redshifts of the order of 2. Up to $z = 2$, I think you can do a good job in the infrared. As soon as you go to the really large redshifts, all the troubles that are now affecting observations in the optical waveband are going to reappear in the infrared waveband. So, I think it is going to be very difficult to disentangle precisely what is happening in the very large-redshift objects, simply because all the mess that we now observe in the optical waveband will reappear at 2 μm. The reason I am interested in working on the redshift 1 galaxies is that we can actually begin to do stellar astronomy. Another point is that, when the NICMOS near-infrared camera becomes available on the *HST* it will be possible to observe the stellar populations in the underlying galaxies.

Question

(G. Burbidge): It is the end of a long day, so I am going to ask what is probably a completely crazy question, or make a completely crazy remark, because I have got a lot of eminent pundits here. There has been a lot of excitement in the last half hour from Malcolm's talk and the discussion of jets and alignments and things of that kind. Yesterday, I showed several pictures of rather precise alignments: the jet in M87 with M84, another radio source, rather precisely, with not very different redshifts; an alignment between the jet of M87 and the jet in 3C 273 and the cloud next to it, with rather different redshifts; a line of QSOs emanating from Mrk 205, again, rather precisely aligned. Now, there is no comment about these. I would like the pundits here to go on record by saying that they believe that everything I showed in that connection was accidental, compared with all of these things which could be fitted together and everybody believes are quite fundamental!

Question

(E. Martín): The M dwarf star that we saw in the jet... perhaps we should search for a physical meaning behind it being there?

Answer

(G. Burbidge): That would be acceptable, of course, if you always found M dwarfs there: you might indeed wonder what is going on. There are other very interesting cases. To answer my own question, we think we know something about the objects which make up the alignment. We believe they are so totally different that there can be no connection other than that of accident, and I suppose that is what most of you will say under your breath. It is interesting that you will not go on record saying this.

Answer

(Longair): Trying to be as sympathetic as I can, I would go on record as saying that I believe that they are accidental coincidences. The reason I say that is that the statistics are difficult to do properly. An example that you quoted yesterday concerned the Peebles and Seldner result that some radio sources are associated with nearby galaxies. Mike Seldner and I demonstrated that indeed it is a correlation and that all the correlations were with existing radio galaxies which were known to be associated with groups and clusters of galaxies. That is an example where we could use statistics absolutely properly, because we knew the statistical properties of both the populations. In that particular example, there was no evidence of quasars being associated with nearby galaxies. My concern is that I have always found it difficult to determine the parent populations correctly.

Comment

(G. Burbidge): I do not claim that statistical arguments underlie the physical alignments. That is something rather different. It has been difficult to do this, but, for example, the association between 3C QSOs and the Shapley–Ames galaxies – that is a very good, as far as I know, a very reliable statistical calculation. All the others that I mentioned yesterday have never been questioned either. So, I was not getting to the general question of associating QSOs, but I am talking about specific cases and interesting pictures on the screen, where we see rather precise alignments, which, after all, however we do the statistics, are likely to be very rare. If you want to attribute everything that has been found by Arp, or anyone else, as accidents, then very well. But, as Hoyle always said, this makes us extremely accident prone. There comes a point in which my own personal view of this (because you may not believe this, but I sat on the fence a very long time about this) is that there simply have been too many examples of this kind. I also know the underlying problems that people have even in making the observations, so I do not believe that there have been thousands and thousands of cases which have been examined and only a few have been found, but I do not want to pursue this.

Question

(Sunyaev): Let me ask you a question on an entirely different subject. You were one of the persons in Ryle's group, which wrote a lot of papers about the cosmological evolution of radio sources and quasars; this is already about 20 years ago or more, maybe 30. My question is: where are all these dead quasars and radio galaxies? What is your point of view? Where are they today? Are they in the nuclei of the ordinary galaxies, or are they in between the galaxies, as that dark matter consisting of baryons which Lynden-Bell mentioned in his talk.

Answer

(Longair): If you simply regard the evolutionary effects as being changes in the mean luminosities of the objects with cosmic epoch, which is actually the simplest way of tying together the luminosity functions, then you can explain all the evolution. This is simply saying that the typical luminosities of these objects have changed with cosmic epoch and we do see the same types of objects in the present epoch, but they are just not as powerful as in the past.

Question

(Sunyaev): Then you are just telling us that, of the galaxies that we are observing around us, many of them were quasars, or enormously bright radio galaxies, at a redshift of 2 or 3?

Answer

(Longair): I think you have to remember the statistics which Osterbrock mentioned: the powerful radio galaxies and quasars are very rare objects. So we are not talking about all the galaxies doing this at all. It is only those that are powerful radio galaxies which exhibit the strong evolution. It is a very small population of objects.

Question

(Sunyaev): But if you recompute them in the time when…I believe that in your report, this was the number density around us today. If I take a model with a wholescale evolution to redshift 3, I obviously get a much larger number of objects.

Answer

(Longair): You can use the same comoving number of objects as today, but simply increase their luminosities.

Question

(J.-M. Rodríguez-Espinosa): I have a simple question, I guess. It is the following. Do you have any idea what is the nature of the obscuring tori. Or, more precisely, do you have the estimation of their optical depth? Because, eventually, if you go to long enough wavelengths, you should be able to see through them. Or if their optical depth is high enough that you do not see through them at any wavelength, they would get hot enough to dissociate all the molecules, and eventually disappear.

Answer

(Longair): I have been trying to get time on telescopes to make these observations. My impression is that there is not much evidence for being able to see through these tori at $3\,\mu$m. But, of course, it is a difficult experiment, because the sensitivities of array detectors at $3\,\mu$m, $5\,\mu$m, are too low to be able to detect these nuclei. It is a crucial experiment, but it will require a great deal of observing time.

Question

(Zinnecker): The question is, how bright is your M star? After all, maybe it is a nuisance, but it could also be an asset because you can use it to do adaptive optics. If it is bright enough to do adaptive optics in the infrared and you have an 8-m instead of a 2.4-m Space Telescope, then you can do diffraction-limited observations. Maybe that is when you are going to start to see your underlying galaxy! So, how bright is the M star?

Answer

(Longair): The star is quite faint: it is about 19th or 20th magnitude.

Comment

(Zinnecker): Oh! Then we have to forget it. Too bad!

7   The physics and astrophysics of black holes

Igor D. Novikov

7.1 **Introduction**

The astrophysics of black holes is a part of relativistic astrophysics. Relativistic astrophysics investigates celestial bodies and systems with extremely strong gravitational fields, such as neutron stars and black holes, including the problem of gravitational radiation. To start I will give a brief retrospective account of research in relativistic astrophysics over the last three decades, and outline current research and some of the future prospects. This will allow us to look to the next decade.

How far have we come in 30 years? What was the relativistic astrophysics of 30 years ago? First of all, there were only a few observational results, and theory dominated in this branch of science. There were only a few, separate investigations, by scientists who were experts in the field, with no coherent plan. Remember that 30 years ago there were no Nobel prizes in astronomy. Today, the situation has changed dramatically. There are numerous observational data and a close relationship between different directions in relativistic astrophysics. Now there is a large emphasis on unusual physics (the interior of neutron stars and black holes, in the early Universe). There have been six Nobel prizes in astronomy, four of them in relativistic astrophysics.

Let us discuss now the research devoted to relativistic compact objects and gravitational waves. Relativistic objects should arise when massive stars die, but 30 years ago it was tacitly assumed that neutron stars and black holes were too esoteric and most probably were the fruit of theorists' wishful thinking. Nobody tried to search for them. The death of stars was a sort of 'forbidden topic for discussion in polite society'. Theorists knew only some basic facts. Quasistellar objects (the most powerful sources of energy in the Universe) had just been discovered. There was only Weber's, rather simple, gravitational antenna (without any hope of detecting cosmic gravitational waves).

Today we know many hundreds of neutron stars – pulsars. Approximately 100 of them are in binary systems. Five black holes of stellar masses, in binary systems, have been discovered. There are about 20 other candidates for black holes of stellar masses. We believe that in eight galactic nuclei there are supermassive black holes with masses up to $10^9 M_\odot$ and more, and there are hundreds of candidates for supermassive black holes in active galactic nuclei. There is a detailed theory of physical processes in neutron stars and in the vicinity of black holes.

The general theory of relativity, including the effects of gravitational radiation, was tested to an accuracy of 0.5% using observations of a binary pulsar. The relativistic gravitational lensing effect has become a critical probe of the Universe.

Today, in the USA, the Laser Interferometer Gravitational Wave Observatory is being built. There are also VIRGO and other important projects for gravitational wave antennae. Among important plans for the future, I would like to mention the observational and theoretical investigation of the physical processes in active galactic nuclei. Another important problem for theory is the question: what happens inside a black hole? In the near future cosmic gravitational radiation will probably be detected. Astrophysicists need a detailed theory of the sources of gravitational radiation.

Modern astrophysics considers three types of black hole in the Universe:

(i) Black holes of stellar masses, which were born when massive stars died.

(ii) Supermassive black holes with masses up to $10^9 M_\odot$ and more ($M_\odot = 2 \times 10^{33}$ g is the mass of the Sun) at the centres of galaxies.

(iii) Primordial black holes which might appear from the large-scale inhomogencities at the very beginning of the expansion of the Universe. Their masses can be arbitrary, but primordial black holes with $M \leq 10^{15}$ g will have radiated away their mass by the Hawking quantum process in a time $t \leq 10^{10}$ years (the age of the

Universe). Only primordial black holes with masses $M > 10^{15}$ g could exist in the contemporary Universe.

The history of the idea of black holes and their astrophysics is given in Israel (1987). General problems of the astrophysics of black holes are discussed in Zel'dovich and Novikov (1971), Novikov and Thorne (1973), Shapiro and Teukolsky (1983), Blandford (1987), Novikov and Frolov (1989), Lamb (1991), and Frolov and Novikov (1997). In this chapter I use material from this last book.

7.2 The problem of the origin of the stellar black holes

The following question is the key one. How heavy must a star be in order to turn into a black hole? The answer is not simple. A light enough star ends up as a white dwarf or a neutron star. Both these types of celestial body have upper limits to their masses. For white dwarfs, it is the Chandrasekhar limit, which is about 1.2–$1.4 M_\odot$ (see Shapiro and Teukolsky 1983; Kippenhahn and Weigert 1990). For neutron stars, it is the Oppenheimer–Volkoff limit (Oppenheimer and Volkoff 1939). The exact value of this limit depends on the equation of state at a matter density larger than the density of nuclear matter, $\rho_0 = 2.8 \times 10^{14}$ g cm^{-3}. The modern theory gives $M_{\mathrm{max,OV}} \approx (2$–$3) M_\odot$ (see Baym and Pethick 1979; Lamb 1991) for this maximum mass. Some authors have discussed the possibility of the existence of so-called 'quark stars' and 'hadronic stars' (see Alcook, Farhi, and Olinto 1986; Bahcall, Lynn, and Selipsky 1990; Madsen 1993). At present, there is no evidence for such stars.

Rotation can increase $M_{\mathrm{max,OV}}$ only slightly – up to 25% (Friedman and Ipser 1987; Haswell et al. 1993). Thus one can believe that the upper mass limit for neutron stars should not be greater than $M_0 \approx 3 M_\odot$ (Lamb 1991; Cowley 1992; McClintock 1992). If a star at the very end of its evolution has a mass greater than M_0 it must turn into a black hole. This does not mean that black-hole progenitors are all normal stars (on the main sequence of the Herzsprung–Russell diagram; see, for example, Bisnovaty-Kogan 1989; Kippenhahn and Weigert 1990) with masses $M > M_0$. The point is that the final stages of evolution of massive stars are poorly understood. Steady mass loss, catastrophic mass ejections, and even disruption in supernova explosions may also be possible (see, for example, Kippenhahn and Weigert 1990). These processes can essentially reduce the initial stellar mass at the end of its evolution. Thus, the initial mass of black-hole progenitors could be much greater than M_0.

There are different estimates for the minimum masses of progenitor stars, M_*, that form black holes. For example, $M_* \approx 10 M_\odot$ (Shapiro and Teukolsky 1983) and $M_* \approx 30 M_\odot$ (Lipunov 1987) or even $\geq 40 M_\odot$ (van den Heuvel and Habets 1984; Schild and Maeder 1985). Note that the evolution of stars in close binary systems differs from the evolution of single stars because of mass transfer from one star to another (see Novikov 1974; Masevich and Tutukov 1988), and the conclusions about masses of black-hole progenitors in this case could be essentially different (see the end of Section 7.5). In further discussion in this section we shall focus our attention on the fate of single stars.

One can estimate how many black holes were created by stellar collapse in our Galaxy during its existence. The spectrum of stellar masses at present for stars on the main sequence, in the solar neighbourhood, is approximately known from the observational data together with the theory of stellar structure.

The lifetime of massive stars is less than T_0 - the age of the Galaxy ($T_0 \approx 10^{10}$ years). We assume a constant birth rate and the same constant death rate for massive stars during the lifetime of the Galaxy. Now, if we suppose that all stars with $M > M_*$ on the main sequence (progenitors) must turn into black holes, we can calculate the birthrate of black holes in the solar neighbourhood. If this rate is the same everywhere, we can estimate the total number of black holes in the Galaxy and the total mass of all stellar black holes in it.

Much work has been devoted to such estimates but, because of the great uncertainties, progress here, starting from the pioneering work by Zwicky (1958), Schwarzschild (1958), Hoyle and Fowler (1963), Novikov and Ozernoy (1964), and Hoyle, Fowler, Burbidge, and Burbidge (1964) is very slow. A review of more recent estimates is given by Shapiro and Teukolsky (1983). Because of the large uncertainty in the lower mass limit M_* for the stars collapsing to black holes, we give the following expression for the rate of black-hole formation in the Galaxy (see Novikov 1974):

$$\frac{\mathrm{d}N}{\mathrm{d}t} \approx 0.1 \left(\frac{M_*}{3M_\odot}\right)^{-1.4} \mathrm{yr}^{-1}. \tag{7.1}$$

At present, we can probably repeat the conclusion of Novikov and Thorne (1973):

> For stars with masses greater than ~12 to $30M_\odot$ a supernova explosion may produce a black hole. If this tentative conclusion is correct, then no more than ~1 per cent of the [visible] mass of the Galaxy should be in the form of black holes today; and new black holes should be created at a rate not greater than ~0.01 per year.

Thus, the total number of stellar black holes in the Galaxy can be of the order of $N \approx 10^8$ or less. What is their observational appearance? The most important physical process which leads to observable manifestations of a black hole's presence is gas accretion (Zel'dovich 1964; Salpeter 1964).

7.3 Disk accretion onto black holes

For the purpose of finding and investigating black holes, two specific cases of accretion are of particular importance: accretion in binary systems and accretion onto the supermassive black holes which probably reside at the centres of galaxies.

In both cases the accreting gas has angular momentum $L \gg r_g c$, where r_g is the size of the black hole, and c is the velocity of light. As a result, the gas elements go into Keplerian orbits around the hole, forming a disk or a torus around it. The inner edge of the disk is in the region of the last stable circular orbit. Close to a rotating black hole, Lense–Thirring precession drags the gas around it in the equatorial plane (see Bardeen and Petterson 1975). Viscosity plays a crucial role in the accretion. It removes angular momentum from each gas element, permitting it to spiral gradually inwards towards the black hole. As the gas reaches the inner edge of the disk it then spirals down the hole practically in a state of free fall. At the same time the viscosity heats the gas, causing it to radiate. The sources of viscosity are probably turbulence in the gas disk and random magnetic fields. Unfortunately, we are no nearer to having a good physical understanding of the effective viscosity.

Large-scale magnetic fields can also play an important role in the physics of accretion.

The properties of the accreting disk are determined by the rate of gas accretion. In the case of disk accretion, the rate is an independent (external) parameter and is determined by the evolution of a binary system, or by the conditions in a galactic nucleus. As a result, the rate of accretion can be much higher than in the spherical case, and observational manifestations are much more prominent.

An important measure of any accretion luminosity of a black hole is provided by the Eddington critical luminosity

$$L_{\rm E} = 4\pi GM_{\rm h}\mu m_{\rm p}c/\sigma_{\rm T} = (1.3 \times 10^{38}{\rm erg\ s}^{-1})\mu\left(\frac{M_{\rm h}}{M_\odot}\right). \tag{7.2}$$

Here μ is the molecular weight per electron, $m_{\rm p}$ is the rest mass of the proton, and $\sigma_{\rm T}$ is the Thomson cross section. It is the luminosity at which the radiation pressure just balances the gravitational force of the mass $M_{\rm h}$, for a fully ionized plasma. Eddington (1926) derived this value for a discussion of stellar equilibrium; Zel'dovich and Novikov (1964) introduced it into the discussion of black holes in quasars and accretion.

A useful measure of the accretion rate is the so-called 'critical accretion rate',

$$\dot{M}_{\rm E} = L_{\rm E}c^{-2}, \tag{7.3}$$

where $L_{\rm E}$ is given by eq. (7.2). We shall use the dimensionless expression $\dot{m} \equiv \dot{M}/\dot{M}_{\rm E}$. Excluding the innermost parts of the disk, relativistic effects are not important in the physical processes of accretion.

Lynden-Bell (1969) was the first to propose the model of gaseous disk accretion onto a black hole. Shakura (1972), Pringle and Rees (1972), and Shakura and Sunyaev (1973) built Newtonian models of the accretion disk. Finally, Novikov and Thorne (1973), and Page and Thorne (1974) gave the theory of disk accretion in the framework of general relativity.

The first models were rather simple. They focused on the case of a moderate rate of accretion, $\dot{m} < 1$. Subsequently, the theory for $\dot{m} \sim 1$ and $\dot{m} \gg 1$ was developed. It takes into account various complex processes in radiative plasma and various types of instability; a review is given by Lamb (1991). However, these processes have no direct relation (or, at least, almost no relation) to specific properties of spacetime in the vicinity of a black hole. Thus, we will not describe them here in detail. We will only give some theoretical estimates of observational properties of accreting disks around black holes in binary systems and in galactic nuclei.

The source of luminosity in disk accretion is gravitational energy released when gas elements spiral down in the disk. Most of the gravitational energy is released and most of the luminosity is emitted from the inner parts of the disk. The total energy radiated by the gas element must be equal to the gravitational binding energy of the element when it is in the last stable circular orbit. This energy, for a mass m_*, is

$$E_{\text{bind}} \approx 0.057 m_* c^2 \tag{7.4}$$

for a non-rotating hole, and

$$E_{\text{bind}} \approx 0.42 m_* c^2 \tag{7.5}$$

for a maximally rotating hole.

The total luminosity of the disk is

$$L \approx 0.057 \dot{M} c^2 \approx \left(3 \times 10^{36} \text{erg s}^{-1}\right) \left(\frac{\dot{M}}{10^{-9} M_\odot} \text{yr}^{-1}\right) \tag{7.6}$$

for a non-rotating hole, and

$$L \approx 0.42 \dot{M} c^2 \approx \left(3 \times 10^{37} \text{erg s}^{-1}\right) \left(\frac{\dot{M}}{10^{-9} M_\odot} \text{yr}^{-1}\right) \tag{7.7}$$

for a maximally rotating hole.

Here, $\dot{M}$ is an accretion rate. This parameter is an arbitrary external parameter which is determined by the source of gas (for example, by the flux of the gas from the upper atmosphere of the companion star in the binary system). We normalize $\dot{M}$ to a value of $\dot{M}_0 = 10^{-9} M_\odot \text{yr}^{-1}$ because it is probably the typical rate at which the normal star is dumping gas onto its companion black hole (see Section 7.4). The ratio $\dot{M}_0/\dot{M}_{\text{crit}} \equiv \dot{m}_0 \approx 0.1 (M_h/M_\odot)^{-1} \leq 1$ for cases of interest (see Section 7.5). As estimates showed (see Lamb 1991), under this condition a geometrically thin disk (with thickness $h \ll r$) will probably be formed. This is the so-called standard disk model (Shakura and Sunyaev 1973; Novikov and Thorne 1973). In this model, the electron and ion temperatures are equal and the disk is effectively optically thick. Temperatures of the gas in the inner parts of the disk reach $T \approx 10^7$–10^8 K. In this region, electron-scattering opacity modifies the emitted spectrum so that it is no longer a blackbody. The total spectrum of the disk radiation has a power law $F \sim \nu^{1/3}$ with an exponential cut-off at high frequencies (Shakura and Sunyaev 1973). The innermost regions of such 'standard' disks are probably unstable.

In the model proposed by Thorne and Price (1975), Shapiro, Lightman and Eardly (1976) the ions in the inner region are hot: $T_i \approx 10^{11}$ K, but the electrons are considerably cooler: $T_e \approx 10^9$ K. This inner disk is thicker than in the 'standard' model and produces most of the X-ray emission.

Further development of the theory of disk accretion led to more complex models. Abramowitz *et al.* (1988, 1989) have demonstrated that, when the luminosity reaches the critical value (corresponding to $\dot{m}$ of the order of unity), radiation pressure in the inner parts of the disk dominates the gas pressure and the disk is thermally and viscously unstable. For especially large $\dot{m}$, the greater part of the energy of the plasma is lost by advection into the black-hole horizon. This process stabilizes the gas flow against perturbation.

For high-mass accretion rates when the thickness of the accretion disk becomes comparable to the radius. In this case the inner edge of the disk can be closer to the black hole than the marginally stable circular orbit because of the essential pressure gradient.

In conclusion, we note that, in some models of disk accretion, pair production can be important; a review of these aspects is given by Lamb (1991).

For new developments see, for example, Bjornsson and Svensson (1991, 1992), Narayan and Yi (1994), Abramowicz *et al.* (1995), and Artemova *et al.* (1995). Many aspects of the physics of accretion disks, including the development of instabilities have been discussed by Mineshige and Wheeler (1989), Wheeler, Soon-Wook Kim, and Moscoso (1993), and Moscoso and Wheeler (1994). We believe that this new development using modern achievements of plasma physics is a key one in the modern astrophysics of black holes.

7.4 Evidence for black holes in stellar binary systems

Probably the best evidence for the existence of black holes comes from studies of X-ray binaries. Galactic X-ray sources were first discovered by

Giacconi *et al.* (1962). Hayakawa and Matsuoko (1964) pointed out that X-rays might be produced by the accretion of gas in close binary systems. However, they discussed accretion into the atmosphere of a normal companion star, rather than onto a compact companion. Novikov and Zel'dovich (1966) were the first to point out that the accretion onto compact relativistic objects (neutron stars and black holes) in binary systems should produce X-rays. They also inferred that Sco X-1, which had just been discovered, might be a neutron star in a state of accretion. After that the observational data were analysed by Shklovsky (1967). Models for X-ray sources have been discussed in some detail by Prendergast and Burbidge (1968). They argued that the gas flow forms a disk around the compact object, with approximately a Keplerian velocity distribution.

A new era started in December 1970 when the X-ray satellite *UHURU* was launched. This satellite has provided much new data about the sources (see Giacconi *et al.* 1972); see Novikov (1974) for the history of the early period of the investigations of the observational evidence for black holes in stellar binary systems. In this section we summarize the observational evidence for black holes in stellar binary systems which at present appear to be the best-documented cases (for an overview see McClintock 1992; Cowley 1992).

The argument is as follows:

(i) The X-ray-emitting object in a binary system is very compact, and therefore cannot be an ordinary star; thus it is either a neutron star or a black hole. This argument comes mainly from the analysis of the features of X-ray emission.

(ii) Analysis of the observational data to determine the orbital motion in the binary system makes it possible to obtain the mass of the compact object. Here the most important are data on the observed velocity of the optical companion star. Note that the Newtonian gravitational theory is always enough for the analysis. The technique of weighing stars in binaries is well known in astronomy.

If the mass of the compact component is greater than the maximum possible mass of neutron stars, $M_0 \approx 3M_\odot$ (see Section 7.2), then it is a black hole.

It is worth noting that the evidence is somewhat indirect because it does not confront us with the specific relativistic effects that occur near black holes and which are peculiar to black holes alone. However, it is the best that modern astronomy can propose so far. In spite of these circumstances we believe that this logic of the arguments is reliable enough.

According to the generally accepted interpretation, only three presently known systems have the necessary observational confirmation, and it is believed strongly that the compact X-ray-emitting companions of the systems are black holes. Some characteristics of the three leading black-hole candidates are summarized in Table 7.1.

The plausible masses for the compact objects in these systems are essentially larger than $M_0 \approx 3M_\odot$. In Table 7.2 there are estimates of the minimum masses of these three candidates. These estimates were obtained by various methods and have different reliability. In most cases, even the minimum masses are greater than M_0. Thus, these three objects are strong black-hole candidates.

Using data from Table 7.1 and expressions (7.2), (7.3), and (7.5–7), one can estimate $\dot{m} \equiv \dot{M}/\dot{M}_E$. In all cases $\dot{m} < 1$ or $\dot{m} \approx 1$.

Several other systems have been suggested as black-hole candidates (a review is given by Cowley 1992; Tutukov and Cherepashchuk 1993). Three systems, LMC X-1 (mass of compact companion $M_c \geqslant 2.91 \pm 0.08$), Nova Muscae 1991 ($M_c \geqslant 3.1 \pm 0.4$), and V404 Cyg ($M_c \geqslant 6.26 \pm 0.31$), could be considered good candidates. The total number of systems that are frequently mentioned as possible candidates for black holes is about 20. All seriously discussed candidates are X-ray sources in binary systems. Some of them are persistent; others are transient. For example, Cyg X-1 and LMC X-1 are persistent, and A0620−00 is transient.

In the more than 20 years since the discovery of the first black-hole candidate, Cyg X-1, only a few new candidates have been added. This is in contrast to the rapid increase of the number of identified neutron stars. At present, many hundreds of neutron stars have been identified in the Galaxy. About 100 of them are in binary systems (Lamb 1991). One might conclude that black holes in binary systems are exceedingly rare objects. Probably that is not true and the small number of identified black-hole candidates is related to the specific conditions which are necessary for their observable manifestation (for a summary, see Cowley 1992). Inoue (1991) estimates the total number of just the soft-X-ray transient black-hole candidates in the Galaxy to be $(1–3) \times 10^3$. Thus, such systems may

Table 7.1. *Black-hole candidates in binary systems (mainly from McClintock 1992).*

Property	Cyg X-1	LMC X-3	A0620−00
X-ray luminosity (erg s^{-1})	2×10^{37}	3×10^{38}	10^{38}
Spectral type of the optical companion	O9.7Iab	B3II–III	K5V
Distance (kpc)	2.5(?)	55	1(?)
Orbital eccentricity	0.00 ± 0.01	0.13 ± 0.05	0.01 ± 0.01
Orbital period (days)	5.6	1.7	0.32
Plausible masses ($M_\odot$)			
(a) Optical companion	33	6	0.5
(b) Compact companion	16	9	5

Table 7.2. *Estimates of the minimum mass (in $M_\odot$) for the compact objects.*

Cyg X-1	Ref.[a]	LMC X-3	Ref.	A0620−00	Ref.
3	(1, 2)	6	(5)	3.2	(11)
7	(3)	2.5	(6)	2.90 ± 0.8	(12)
$3.4(d/2 \text{ kpc})^2$	(4)[b]	2.3	(7)	4.5	(12)
		6	(7)	3.30 ± 0.95	(13)
		3	(8, 9)	6.6	(13)
		4	(10)	3.1 ± 0.2	(14)
				3.82 ± 0.24	(15)
				6	(16)

[a] (1) Webster and Murdin (1972) (9) Bochkarev *et al.* (1988)
(2) Bolton (1972a, b; 1975) (10) Treves *et al.* (1990)
(3) Gies and Bolton (1986) (11) McClintock (1988)
(4) Paczynski (1974) (12) McClintock and Remillard (1990)
(5) Paczynski (1983) (13) Johnston *et al.* (1989)
(6) Mazeh *et al.* (1986) (14) Johnston and Kulkarni (1990)
(7) Cowley *et al.* (1983) (15) Haswell and Shafter (1990)
(8) Kuiper *et al.* (1988) (16) Cowley *et al.* (1983)
[b] d = distance from the Solar System.

be as frequent as neutron-star binaries. The estimate of the formation rate for black-hole binaries is given in van den Heuvel and Habets (1984).

At present we know only one persistent X-ray source in the Galaxy and two in another galaxy, the LMC, which are strong candidates for black holes (Cyg X-1, and LMC X-1 and LMC X-3 respectively). The probable explanation was provided by van den Heuvel (1983). He estimates that the evolutionary stage when a black-hole binary continuously radiates X-rays may last only 10^4 years. We can detect it during this short period only.

Thus, the population of black-hole binaries may be much larger than we can presently see.

7.5 Supermassive black holes in galactic centres

Since the middle of this century, astronomers have increasingly often come across violent or even catastrophic processes associated with galaxies. These processes are accompanied by powerful releases of energy and are fast, not only by astronomical, but also by earthly standards. They may last a few days, or even just a few minutes. Most such processes occur in the central area of galaxies: galactic nuclei.

About 1% of all galactic nuclei eject radio-emitting plasma and gas clouds, and are themselves powerful sources of radiation in the radio, infrared, gamma, and especially the 'hard' (short-wavelength) X-ray regions of the spectrum. The full luminosity of the nucleus is, in some cases, $L \approx 10^{47}$ erg s^{-1} and millions of times the luminosity of the nuclei of more stable galaxies, such as ours. These objects have been termed active galactic nuclei (AGNs). Practically all the energy of activity and of the giant jets released by galaxies originates from the centre of their nuclei.

Quasars are a class in themselves among AGNs. Their total energy release is hundreds of times greater than the combined radiation of all the stars in a big galaxy. At the same time, the average linear dimensions of the radiating areas proper are small: a mere one-hundred-millionth part of the linear size of a galaxy.

Quasars are the most powerful energy sources registered in the Universe to date. What processes are responsible for the extraordinary outbursts of energy from AGNs and quasars? For an overview of the problem, see Blandford (1987) and Wallinder (1993).

Learning about the nature of these objects involves measuring their sizes and masses. This is not at all easy. The centre of the emitting areas of AGNs and quasars is so small that a telescope view reveals them just as point sources of light. The job has been made much simpler by the fact that the brightness of the quasar 3C 273 is not constant, and sometimes it changes very fast: in just a week. After that discovery, even faster changes (as fast as a few hours or even less) were detected in other galactic nuclei. From those changes one could get the dimensions of the central parts of the nuclei. The conclusion was that they were no more than a few light-hours in diameter, that is, comparable to the Solar System in size.

In spite of the rather small linear dimensions of quasars and many galactic nuclei, their masses turned out to be enormous. They were estimated for the first time by Zel'dovich and Novikov (1964), using formula (7.3.2). For quasistatic objects the luminosity cannot be greater than L_E. The comparison of the observed luminosity with expression (7.3.2) gives an estimate of the lower limit of the central mass. In some quasars this limit is $M \approx (10\text{-}10^3) \times 10^6 M_\odot$. These estimates are supported by data on the velocities of stars and gas clouds within the galactic nuclei accelerated in the gravitational fields of the centre of the nuclei. We will discuss this in the next section.

Large mass but small linear dimensions prompted the guess that it could be a black hole. Zel'dovich and Novikov (1964) and Salpeter (1964) suggested that the centres of quasars and AGNs could harbour supermassive black holes accounting for all the extraordinary qualities of these objects.

It is interesting to note that the estimates of masses of black holes in AGNs ($M \approx 10^7 M_\odot$ and more) are close to the estimates of masses of super-massive invisible 'stars' in the Universe (black holes, in our terminology) which were done at the end of the 18th century by Mitchell and Laplace. They speculated on supermassive stars, which would generate a gravitational field strong enough to trap light rays so that they would be invisible.

In the modern history of the idea of supermassive black holes, the crucial step was taken by Lynden-Bell (1969). He derived and applied a theory of thin accretion disks in orbit around massive black holes. It is now generally accepted that in AGNs there are supermassive black holes with gas (and maybe also dust) accretion disks. One of the most important facts derived from observations, and especially radio observations, is the existence of directed jets from the nuclei of some active galaxies. Sometimes there is evidence that radio components move away from the nucleus with ultrarelativistic velocity. The observation of an axis of ejection strongly suggests that there is some stable compact gyroscope, probably a rotating black hole. In some cases one can observe evidence that there is also precession of such a gyroscope. Probably the essential role in the physics of processes in the centres of AGNs is played by black-hole electrodynamics. A review of the physics of AGNs is given by Lamb (1991) and by Svensson (1994).

In the model of supermassive black holes with accretion disks for AGNs one requires sources of fuel – gas or dust. The following sources of fuelling were discussed: gas from a nearby galactic companion, a result of the interaction of the host galaxy and the companion, interstellar gas of the host galaxy, disruption of stars by high-velocity collisions in the vicinity of a black hole, disruption of stars by the tidal field of a black hole, and some others (see Shlosman, Begelman, and Frank 1990). An excellent review of all the problems of supermassive black holes in galactic centres, including the problem of their origin and evolution, is given by Rees (1990a) and Haehnelt and Rees (1993).

Clearly, the processes taking place in quasars and other nuclei are still a mystery in many respects, but the suggestion that we are witnessing the work of a supermassive black hole with an accretion disk seems rather plausible.

7.6 Dynamical evidence for black holes in galactic nuclei

So far we have considered supermassive black holes as the most probable explanation of the nuclei of some galaxies. Is there more conclusive evidence of the presence of black holes?

First of all, massive black holes should not only be in the active galactic nuclei but also in the centres of 'normal' galaxies, including nearby galaxies and our Milky Way (Rees 1990a). They are quiescent because they are now starved of fuel (gas). Observations show that the activity of galactic nuclei was more common in the past. Thus, now 'dead quasars' (massive black holes without fuel) should be common.

How could they be detected? It was pointed out that black holes produce cuspy potentials and hence they should produce cuspy density distributions of the stars in the central regions of the galaxies. Some authors argued that the brightness profiles of the central regions of the particular galaxies imply that they contain black holes. Kormendy (1993) emphasized that arguments based only on surface-brightness profiles are inconclusive. The point is that a high central number density of stars in the small-radius core could be a consequence of the dissipation, and the cuspy profile could be a result of the anisotropy of the velocity dispersion of stars. Thus, these properties are not sufficient evidence for the presence of a black hole.

The reliable way to detect black holes in galactic nuclei is analogous to the case of black holes in binaries. That is, one needs proof that there is a big dark mass in a small volume, and that it can be nothing other than a black hole. In order to give such a proof we need stellar kinematics, as well as surface photometry of the galactic nuclei.

The mass M inside a radius r is given by the formula (Kormendy 1993)

$$M(r) = \frac{v^2 r}{G} + \frac{\sigma_{\mathrm{r}}^2 r}{G}\left[-\frac{\mathrm{d}\ln I}{\mathrm{d}\ln r} - \frac{\mathrm{d}\ln \sigma_{\mathrm{r}}^2}{\mathrm{d}\ln r} - \left(1 - \frac{\sigma^2}{\sigma_r^2}\right) - \left(1 - \frac{\sigma_\phi^2}{\sigma_{\mathrm{r}}^2}\right)\right], \qquad (7.8)$$

where I is the brightness, v is the rotation velocity, and σ_r, σ, σ_ϕ are the radial and two tangential components of the velocity dispersion. These values must be obtained from observations after a few special corrections. More complex formulae are used for more sophisticated models.

Now we can consider the mass-to-light ratio M/L (in solar units) as a function of the radius. This ratio is well known for different types of stellar populations. As a rule, this ratio is between 1 and 10 for elliptical galaxies and globular clusters (old stellar populations dominate there). If, for some galaxy, the ratio M/L is almost constant at a rather large radius r (and has a 'normal' value between 1 and 10) but, starting from some small r, rises rapidly towards the centre to values much larger than 10, then there is evidence for a central dark object, probably a black hole.

As an example, consider the galaxy NGC 3115, which is at a distance of 9.2 Mpc (Kormendy and Richstone 1992). For this galaxy $M/L \approx 4$ and is almost constant over a large radius range at radii $r > 4''$ (in angular units).

This value is exactly normal for the bulge of this type of galaxy. At a radius $r < 2''$ the ratio M/L rises rapidly up to $M/L \approx 40$. If this is due to a central dark mass, added to a stellar distribution with constant M/L, then $M_{\text{b.h.}} = 10^{9.2 \pm 0.5} M_\odot$.

Is it possible to give another explanation of the large mass-to-light ratio in the central region of a galaxy? We cannot exclude the possibility that a galaxy contains a central compact cluster of dim stars. But it is unlikely. The central density of stars in the galaxy NGC 3115 is not peculiar; it is the same as in the centres of globular clusters. The direct observational data (spectra and colours) of this galaxy do not give any evidence of a dramatic population gradient near the centre. Thus, the most plausible conclusion is that there is a massive central black hole.

Unfortunately, it is most difficult to detect massive black holes in giant elliptical galaxies with active nuclei, where we are almost sure black holes must exist because we observe their active manifestation (Kormendy 1993). The reason is that there are fundamental differences between giant elliptical galaxies (the nuclei of some of them are among the most extreme examples of AGNs) and dwarf elliptical galaxies and spiral galaxies. Dwarf ellipticals rotate rapidly and their star velocity dispersions are nearly isotropic. Giant elliptical galaxies do not rotate significantly and they have anisotropic velocities. It is not so easy to model these dispersions. In eq. (7.8) they lead to uncertainties.

Giant elliptical galaxies have large cores and shallow brightness profiles, and because of this, projected spectra are dominated by light from large radii, where a black hole has no effect.

The type of technique described above was used for a search for black holes in galactic nuclei. Another possibility is the observation of the rotational velocities of gas in the near vicinity of a galactic centre. To date, black-hole detections have been reported for the following galaxies (for an overview see Kormendy 1993; see also van der Marel *et al.* 1994; Miyoshi *et al.* 1995): M32, M31, NGC 3377, NGC 4594, the Milky Way, NGC 3115, M87, and NGC 4258. Special investigations were performed in the case of the galaxy M87 (see Dressler 1989 for an overview of earlier works, and Lauer *et al.* 1992, Ford *et al.* 1994, and Harms *et al.* 1994 for the *Hubble Space Telescope* observations). This is a giant elliptical galaxy with an active nucleus and a jet from the centre. At present, there is firm

Table 7.3. *Estimates of the mass of the central black hole of a small sample of galaxies.*

Galaxy	Mass of black hole ($M_\odot$)
M31	2×10^7
M32	$(2-5) \times 10^6$
Milky Way	3×10^6
NGC 4594	10^9
NGC 3115	10^9
NGC 3377	10^8
M87	2.4×10^9
NGC 4258(M106)	3.6×10^7

stellar-dynamical evidence for a black hole with mass $M \approx 3 \times 10^9 M_\odot$ in the galaxy. The presence of a black hole in M87 is especially important for our understanding of the nature of the central regions of the galaxies because, in this case, we also observe the activity of the 'central engine'.

In conclusion, we give a table (Table 7.3) of estimates of the masses of black holes in the nuclei of some galaxies (Kormendy 1993; van der Marel 1995; Miyoshi *et al.* 1995).

Progress in this field is very fast, and in the near future, our knowledge of evidence for supermassive black holes in galactic nuclei will be more profound. We want also to mention the possibility of the formation of binary supermassive black-hole systems in the process of merger of galaxies (see, for example, Rees 1990b and Polnarev and Rees 1994). The radiation of gravitational waves in such a system leads to the decay of the orbits of black holes. Eventually they coalesce. The final asymmetric blast of gravitational radiation may eject the merged hole from the merged galaxy.

7.7 Primordial black holes

Let us consider now the possibility of the existence of primordial black holes. The smaller the mass of matter, the greater the density to which it must be compressed in order to create a black hole. Powerful pressure develops at high densities, counteracting the compression. As a result, black holes of mass $M \ll M_\odot$ cannot form in the contemporary Universe. However, the density of matter at the beginning of the expansion of the Universe was enormously high. Zel'dovich and Novikov (1967a, b) and then Hawking (1971) hypothesized that black holes could have been produced at the early stages of the cosmological expansion of the Universe. Such black holes are known as primordial. Very special conditions are needed for primordial black holes to form. Lifshitz (1946) proved that small perturbations in a homogeneous isotropic hot Universe (with the equation of state of the matter $p = 8/3$) cannot produce appreciable inhomogeneities. A hot Universe is stable under small perturbations (see Bisnovaty-Kogan *et al.* 1980). Large deviations from homogeneity must exist from the very beginning in the metric describing the Universe (i.e., the gravitational field had to be strongly inhomogeneous) even though the spatial distribution of matter density close to the beginning of the cosmological expansion was very uniform. When the quantity $l = ct$, where t is the time elapsed since the Big Bang, grows in the course of expansion to a value of the order of the linear size of an inhomogeneity of the metric, the possibility appears of the formation of a black hole with the mass contained by the time t in the volume l^3. The formation of black holes with masses substantially smaller than stellar masses was thus possible, provided that such holes were created at a sufficiently early stage (see below).

Primordial black holes are of special interest because Hawking's quantum evaporation is important for small-mass black holes, while only primordial black holes can have such masses. (Note that quantum evaporation of massive and even supermassive black holes may be essential for the distant future of the Universe.)

First of all, the following two questions arise:

    (i) How large must the deviations from the metric of a homogeneous isotropic model of the Universe be for black holes to be born?

    (ii) What is the behaviour of the accretion of the surrounding hot matter to the created hole and how does the accretion change the mass of the hole?

The second question arises because of a remark made in the pioneering paper of Zel'dovich and Novikov: if a stationary flux of gas into the black hole builds up, the black-hole mass grows at a catastrophically fast rate. But if such stationary accretion does not build up immediately after the black hole is formed, accretion is quite negligible at later stages because the density of the surrounding gas in the expanding Universe falls off very rapidly. Both questions can be answered via numerical modelling. The required computations for the case of spherical symmetry were carried out by Nadezhin *et al.* (1977, 1978) and Novikov and Polnarev (1980).

The main results of these computations are as follows. The dimensionless amplitude of metric perturbations, g, necessary for the formation of a black hole is about 0.75–0.9. The uncertainty of the result reflects the dependence on the perturbation profile. Recall that the amplitude of the metric perturbation is independent of time as long as $l = c\,t$ remains much less than the linear size of the perturbed region. If g is less than 0.75–0.9, the density perturbations created transform into acoustic waves after $l = c\,t$ increases to about the size of the perturbation. This answers the first of the questions formulated above.

As for the second question, the computations show that the black-hole mass at the moment of formation incorporates 10–15% of the mass within a scale of $l = c\,t$. This means that the accretion of the gas to the newborn black hole cannot become catastrophic. Computations confirm that the gas falling into the black hole from the surrounding space only slightly increases its mass.

If, in the early history of the Universe, there were periods when the pressure was reduced for a while, then all pressure effects were not important. The formation of primordial black holes (PBHs) under these conditions was discussed in the papers of Khlopov and Polnarev (1980) and Polnarev and Khlopov (1981).

Other exotic formation mechanisms include cosmic phase transitions (Kodama, Sosaki, and Sato 1982; Hawking, Moss, and Stewart 1982; Kardashev and Novikov 1983; Naselsky and Polnarev 1985; Hsu 1990) or the collapse of loops of cosmic strings (Hawking 1989; Polnarev and Zembowicz 1991; Polnarev 1994). A review of the problem is given by Carr (1992).

The discovery, by Hawking, that black holes can evaporate by thermal emission made the study of PBH formation and evaporation of considerable astrophysical interest, because the evaporation effects of small black holes

[287]

are potentially observable. Their absence from observational searches therefore enables powerful limits to be placed upon the structure of the very early Universe (Zel'dovich and Novikov 1967a, b; Carr and Hawking 1974; Novikov *et al.* 1979; Carr 1983; Carr, Gilbert, and Lidsey 1994).

When we consider black holes of solar-mass size or supermassive black holes, Hawking radiation is quite negligible, but for small enough PBHs, it becomes the controlling influence in the black hole's evolution and very important for their possible observational manifestations. It is easy to conclude that PBHs of $M \approx 5 \times 10^{14}$ g or less will indeed have evaporated entirely in the 10^{10} years or so in the Universe's history. PBHs a little more massive than this initially will still be evaporating in the modern Universe. Their rate of evaporation is large enough that the stream of energetic particles and radiation they emit can be turned into a strong observational limit on their presence.

Searches for PBHs attempt to detect a diffuse photon (or another particle) background from a distribution of PBHs, or to search directly for the final emission stage of individual black holes. Using the theoretical spectra of particles and radiation emitted by evaporating black holes of different masses, one can calculate the theoretical backgrounds of photons and other particles produced by a distribution of PBHs emitted over the lifetime of the Universe. The level of this background depends on the integrated density of PBHs with initial masses in the considered range.

A comparison of the theoretical estimates with the observational cosmic ray and γ-ray backgrounds places an upper limit on the integrated density of PBHs with initial masses in this range. According to estimates by MacGibbon and Carr (1991), this limit corresponds to $\sim 10^{-6}$ of the integrated mass density of the visible matter in the Universe (matter in the visible galaxies). The comparison of the theory with other observational data gives weaker limits (for an overview see Halzen *et al.* 1991; Carr 1992; Coyne 1993).

The search for high-energy γ-ray bursts as a direct manifestation of the final emission of the evaporating (exploding) individual PBHs has continued for more than 20 years. No positive evidence for the existence of PBHs has been reported (see Cline and Hong 1992, 1994).

A population of PBHs whose influence today is small may have been more important in the earlier epochs of the evolution of the Universe.

Radiation from PBHs could perturb the usual picture of cosmological nucleosynthesis, distort the microwave background, and produce too much entropy in relation to the matter density of the Universe. As we mentioned above, limits on the density of PBHs, now or at earlier times, can be used to provide information on the homogeneity and isotropy of the very early Universe, when they were formed. For an overview see Novikov *et al.* (1979) and Carr *et al.* (1994).

The final state of the black-hole evaporation is still unclear. There is a possibility that the end point of the black-hole evaporation is a stable relic. The possible role of such relics in cosmology was first discussed by MacGibbon (1987), for a recent review, see Barrow, Copeland, and Liddle (1992).

In this section we have reviewed the current search for evidence for black holes in the Universe. Our conclusion is the following: right now (at the start of 1997), we are almost 100% sure that black holes of stellar masses exist in the binary systems. Probably we should say the same about the supermassive black holes in the centres of many galaxies. However, so far there is no evidence for the existence of PBHs in the Universe.

7.8 The physics of black holes

The branch of physics that is now referred to as black-hole physics was born and actually took shape as a full-blooded scientific discipline during the past three decades, at the junction of the theory of gravitation, astrophysics, and classical and quantum field theory. Profound links were found between black-hole theory and such seemingly very distant fields as thermodynamics, information theory, and quantum theory. By now, a fairly detailed understanding has been achieved of the properties of black holes, their possible astrophysical manifestations, and the specifics of the various physical processes involved.

For a review of the physics of black holes see the book by Novikov and Frolov (1989). In this short section I will outline three key problems which are of the greatest interest, and which are the focus of attention. A review of all these problems is given by Frolov and Novikov (1995) and references may be found in this book. The first is the problem of the entropy of black holes. A black hole, considered as a part of a thermodynamical system, possesses the Bekenstein–Hawking entropy (in natural units) $S = A/4l_p^2$, where A is the area of a black-hole surface and l_p is the Planck length.

Two questions arise. The first is: why is the entropy so big? And the second is: why is it universal and does it not depend on the number and characteristics of the fields? One proposal to explain the dynamical origin of the entropy of a black hole was by identifying its dynamical degrees of freedom with the physical modes propagating in the black-hole interior. The universality of the entropy is connected with the fact that, in a state of thermal equilibrium, the parameters of the internal dynamical degrees of freedom of a black hole depend on the temperature of the system in the universal way.

The second key problem is connected with the question: can one see what happens inside a black hole? The standard answer to this question is 'no'. Recently, it was demonstrated that such a possibility does exist and, in principle, one could search a black hole's interior practically without disturbing the metric describing the black hole. A *gedanken experiment* was discussed in which the black hole's interior could be studied by means of a traversable worm hole, and it was shown that many habitual concepts concerning the properties of black holes (including the impossibility of extracting energy and information from black holes) must be treated with some caution.

The last key problem is connected with the possibility of creating, in principle, time machines (closed time-like curves), allowing one to travel into the past. Two crucial questions are discussed. The first is: do the laws of physics prevent time machines from ever being made? The second is: do the laws of physics deal with closed time-like curves, and if they do, how?

7.9 Conclusions

I am an optimist and believe that many of the problems of the physics and astrophysics of black holes which look hopeless or fantastic now will be solved in the next century.

Acknowledgements

This work was supported in part by the Danish Natural Science Research Council through grant N9401635 and in part by Danmarks Grundforskningsfond through its support for the establishment of the

Theoretical Astrophysics Centre. The author is grateful to the Instituto de Astrofísica de Canarias and the Fundación del Banco Bilbao-Vizcaya for the organization of the excellent conference on which this book is based.

References

Abramowicz, M.A., Szuskiewicz, E., and Wallinder, F.H.: 1989, in *Theory of Accretion Discs*, eds. F. Meyer, W.J. Duschl, J. Frank, and E. Meyer-Hofmeister, NATO ASI Series, Vol. **290** (Dordrecht: Kluwer), p. 141

Abramowicz, M.A., Czerny, B., Lasota, J.P., and Szuzkievicz, E.: 1988, ApJ, **332**, 646

Abramowicz, M.A., Chen, X., Kato, S., Lasota, J.F., and Regev, O.: 1995, ApJ, **438**, L37

Alcook, C., Farhi, E., and Olinto, A.V.: 1986, ApJ, **310**, 261

Artemova, J., Bisnovaty-Kogan, G., Björnsson, G., and Novikov, I.: 1995, ApJ, **456**, 119

Bahcall, S., Lynn, B.W., and Selipsky, S.B.: 1990, ApJ, **362**, 251

Bardeen, J.M., and Petterson, J.A.: 1975, ApJL, **195**, L65

Barrow J.D., Copeland, E.J., and Liddle, A.R.: 1992, Phys Rev D, **46**, 645

Baym, G., and Pethick, C.J.: 1979, Ann Rev A&A, **17**, 415

Bisnovaty-Kogan G.S.: 1989, *Physical Problems of the Theory of Stellar Evolution* (Moscow: Nauka)

Bisnovaty-Kogan G.S., Lukash, V.N., and Novikov, I.D.: 1980, in *Proceedings of the Fifth European Regional Meeting in Astronomy, Liege, Belgium*, G. 1.1

Bjornsson G., and Svensson, R.: 1991, ApJ, **371**, L69

Bjornsson G., and Svensson, R.: 1992, ApJ, **394**, 500

Blandford, R.D.: 1987, in *300 Years of Gravitation*, eds. S.W. Hawking and W. Israel (Cambridge: Cambridge University Press), p. 199

Bochkarev, N.G., Sunyaev, R.A., Kruzina, T.S., Cherepashchuk, A.M., and Shakura, N.I.: 1988, Sov Astron J, **32**, 405

Bolton, C.T.: 1972a, Nature, **235**, 271

Bolton, C.T.: 1972b, Nature Phys Sci, **240**, 124

Bolton, C.T.: 1975, ApJ, **200**, 269

Carr, B.J.: 1983, in *Quantum Gravity*, eds. M.A. Markov and P.C. West (New York: Plenum Press), p. 337

Carr, B.J.: 1992, *Report on Zel'dovich's meeting*, Moscow

Carr, B.J., and Hawking, S.W.: 1974, MNRAS, **168**, 399

Carr, B.J., Gilbert, J.H., and Lidsey, J.E.: 1994, Phys Rev D, **50**

Cline, D.B., and Hong, W.: 1992, ApJL, **401**, L57

Cline, D.B., and Hong, W.: 1994, BAAS, **185**, 116.08

Cowley, A.P.: 1992, Ann Rev A&A, **30**, 287

Cowley, A.P., Schmidtke, P.C., Crampton, D., and Hutchings, J.B.: 1990, ApJ, **350**, 288

Cowley, A.P., Crampton, D., Hutchings, J.B., Remillard, R., Penfold, J.E.: 1983, ApJ, **272**, 118

Coyne, D.G.: 1993, in *International Symposium on Black Holes, Membranes, Wormholes and Superstrings* (Singapore: World Scientific Publishing Co.), p. 159

Dressler, A.: 1989, in *Active Galactic Nuclei, IAU Symposium 134*, eds. D.E. Osterbrock and J.S. Miller (Dordrecht: Kluwer), p. 217

Eddington, A.S.: 1926, *The Internal Constitution of the Stars* (Cambridge: Cambridge University Press)

Ford, H.C. *et al.*: 1994, ApJ, **435**, L27

Friedman, J.L., and Ipser, J.R.: 1987, ApJ **314**, 594

Frolov. V., and Novikov, I.D.: 1997, *Physics of Black Holes; New Methods and Perspectives*, in preparation

Giacconi, R., Gursky, H., Paolini, F.R., and Rossi, B.B.: 1962, Phys Rev Lett, **9**, 439

Giacconi, R., Murray, S., Gursky, H., Kellogg, E., Schreier, E., and Tananbaum, H.: 1972, ApJ, **178**, 281

Gies, D.R., and Bolton, C.T.: 1986, ApJ **304**, 371

Haehnelt, M., and Rees, M.: 1993, MNRAS, **263**, 168

Halzen, F., Zas, E., MacGibbon, J.H., and Weekes, T.C.: 1991, Nature, **353**, 807

Harms R.J. *et al.*: 1994, ApJL, **435**, L35

Haswell, C.A., and Shafter, A.W.: 1990, : ApJL, **359**, L47

Haswell, C.A., Robinson, E.L., Horne, K., Stiening, R., and Abbott, T.M.C.: 1993, ApJ, **411**, 802

Hawking, S.W.: 1971, MNRAS, **152**, 75

Hawking, S.W.: 1989, Phys Lett B, **231**, 237

Hawking, S.W., Moss, I., and Stewart, J.: 1982, Phys Rev D, **26**, 2681

Hayakawa, S., and Matsuoka, M.: 1964, Prog Theor Phys Suppl, **30**, 204

Hoyle, F., and Fowler, W.A.: 1963, Nature, **197**, 533

Hoyle, F., Fowler, W.A., Burbidge, G., and Burbidge, E.M.: 1964, ApJ, **139**, 909

Hsu, S.D.U.: 1990, Phys Lett B, **251**, 343

Inoue, H.: 1991, *28th Yamula Conference: The Frontiers of X-ray Astronomy,* ed. K. Koyama.

Israel, W.: 1987, in *'300 Years of Gravitation,'* eds. S.W. Hawking and W. Israel (Cambridge: Cambridge University Press), p. 199

Johnston, H.M., and Kulkarni, S.R.: 1990, in *Accretion-Powered Compact Binaries* (Cambridge: Cambridge University Press)

Johnston, H.M., Kulkarni, S.R., and Oke, J.B.: 1989, ApJ, **345**, 492

Kardashev, N.S., and Novikov, I.D.: 1983, in *Early Evolution of the Universe and Its Present Structure,* IAU Symp. 104, eds. G.O. Abell and G. Chincarini, (Crete: Reidel Pub.) p. 463

Khlopov, M.Yu., and Polnarev, A.G.: 1980, Phys Lett B, **97**, 383

Kippenhahn, R., and Weigert, A.: 1990, *Stellar Structure and Evolution* (Berlin: Springer-Verlag)

Kodama, H., Sasaki, M., and Sato, K.: 1982, Prog Theor Phys, **68**, 1979

Kormendy, J.: 1993, in *The Nearest Active Galaxies*, eds. J.E. Beckman, H. Netzer, and L. Colina (Madrid: Consejo Superior de Investigaciones Científicas)

Kormendy, J., and Richstone, D.: 1992, ApJ, **393**, 559

Kuiper, L., van Paradijs, J., and van der Klis, M.: 1988, A&A, **203**, 79

Lamb, F.K.: 1991, in *Frontiers of Stellar Evolution*, ed. D.L. Lambert (Astronomical Society of the Pacific), p. 299

Lauer, T.R. *et al.*: 1992, AJ, **103**, 703

Lifshitz, E.M.: 1946, Zh Eksp Teor Fiz, **16**, 587

Lipunov V.M.: 1987, *Astrophysics of Neutron Stars* (Moscow: Nauka)

Lynden-Bell, D.: 1969, Nature, **233**, 690

MacGibbon, J.H.: 1987, Nature, **329**, 308

MacGibbon, J.H., and Carr, B.J.: 1991, ApJ, **371**, 447

Madsen, K.: 1993, *Proceedings of 2nd International Conference on Physics and Astrophysics of Quark–Gluon Plasma* (Singapore: World Scientific)

Masevich, A.G., and A.V. Tutukov: 1988, *Stellar Evolution: Theory and Observations* (Moscow: Nauka)

Mazeh, T., van Paradijs, J., van den Heuvel, E.P.J., and Savonije, G.J.: 1986, A&A, **157**, 113

McClintock, J.F.: 1988, in *Supermassive Black Holes*, ed. Minas Kafatos (Cambridge: Cambridge University Press), p. 1

McClintock, J.E.: 1992, in *Frontiers of X-ray Astronomy*, eds. Y. Tanaka, and K. Koyama (Tokyo: Universal Acad. Press), p. 333

McClintock, J.E., and Remillard, R.A.: 1990, BAAS, **21**, 1206

Mineshige, S., and Wheeler, J.C.: 1989, ApJ, **343**, 241

Miyoshi, M., Moran, J., Herrnstein, J., Greenhill, L., Nakai, N., Diamond, P., and Inoue, M.: 1995, Nature **373**, 127

Moscoso, M.D., and Wheeler, J.C.: 1994, in *Interacting Binary Stars*, ed. A.W. Shafter (Astron Soc of the Pacific, Conf. Series, Vol. 56), p. 100

Nadezhin, D.K., Novikov, I.D., and Polnarev, A.G.: 1977, in *Proc. of the 8th Int. Conference of the GRG Soc.* (Waterloo, Canada), p. 32

Nadezhin, D.K., Novikov, I.D., and Polnarev, A.G.: 1978, Astr Zh, **55**, 216

Narayan, R., and Yi, I.: 1994, ApJ, **428**, L13

Naselsky, P.D., and Polnarev, A.G.: 1985, Astr Zh, **29**, 487

Novikov, I.D.: 1974, in *Astrophysics and Gravitation, Proceedings of the 16th Solvay Conference on Physics (1973)* (Brussels: Edition de l'Université de Bruxelles), p. 317

Novikov, I., and Frolov, V.: 1989, *Physics of Black Holes* (Dordrect: Kluwer)

Novikov, I.D., and Ozernoy, L.M.: 1964, Pub. Lebedev Phys. In-t, A-17

Novikov, I.D., and Polnarev, A.G.: 1980, Astr Zh, **57**, 250

Novikov, I.D., and Thorne, K.S.: 1973, in *Black Holes, Les Astres Occuls*, eds. C. Dewitt and B.S. DeWitt (London: Gordon and Breach), p. 343

Novikov, I.D., and Zel'dovich, Ya.B.: 1966, Nuovo Cim Suppl, **4**, 810, addendum 2

Novikov, I.D., Polnarev, A.G., Starobinsky, A.A., and Zel'dovich, Ya.B.: 1979, A&A, **80**, 104

Oppenheimer, J.R., and Volkoff, G.: 1939, Phys Rev, **55**, 374

Paczynski, B: 1974, A&A, **34**, 161

Paczynski, B.: 1983, ApJL, **273**, L81

Page, D., and Thorne, K.: 1974, ApJ, **191**, 499

Polnarev A.G.: 1994, A&A Transactions, **5**, 35

Polnarev, A.G., and Khlopov, M.Yu.: 1981, Astr Zh, **58**, 706

Polnarev, A.G., and Rees, M.: 1994, A&A, **283**, 301

Polnarev, A.G., and Zembowicz, R.: 1991, Phys Rev D, **43**, 1106

Prendergast, K.H., and Burbidge, G.R.: 1968, ApJ, **151**, L83

Pringle, J.E., and Rees, M.J.: 1972, A&A, **21**, 1

Rees, M.: 1990a, Science, **247**, N4944, 16 February, p. 817

Rees, M.: 1990b, Scientific American, November, p. 26

Salpeter, E.: 1964, ApJ, **140**, 796

Schild, H., and Maeder, A.: 1985, A&A, **143**, L7

Schwarzchild, M.: 1958, *Structure and Evolution of the Stars* (Princeton: Princeton University Press)

Shakura, N.I.: 1972, Astr Zh, **16**, 756

Shakura, N.I., and Sunyaev, R.A.: 1973, A&A, **24**, 337

Shapiro, S.L., Lightman, A.P., and Eardly, D.M.: 1976, ApJ, **204**, 187

Shapiro, S.L., and Teukolsky, A.A.: 1983, *Black Holes, White Dwarfs, and Neutron Stars. The Physics of Compact Objects* (New York: Wiley)

Shklovsky, I.S.: 1967, ApJL, **148**, L1

Shlosman, I., Begelman, M., and Frank, J.: 1990, Nature **345**, 679

Svensson, R.: 1994, ApJSS, **92**, 585

Thorne, K.S., and Price, R.H.: 1975, ApJ, **195**, L101

Treves, A., Belloni, T., Corbet, R.H.D., Ebisawa, K., Falomo, R. *et al.*: 1990, ApJ, **364**, 266

Tutukov, A.V., and Cherepashchuk, A.M.: 1993, Astr Zh, **70**, 307

van den Heuvel, E.P.J.: 1983, in *Accretion-Driven Stellar X-ray Sources*, eds. W.H.G. Lewin and E.P.J. van den Heuvel (Cambridge: Cambridge University Press), p. 303

van den Heuvel, E.P.J., and Habets, G.M.H.J.: 1984, Nature, **309**, 598

van der Marel, R.P.: 1995, in *Highlights of Astronomy* 10, *Proceedings of the XXII General Assembly of the IAU* (The Hague, August 1994), ed. J. Bergeron (Dordrecht: Kluwer)

Wallinder, F.H.: 1993, Comments on Astrophys, **16**, 331

Webster, B.L., and Murdin, P.: 1972, Nature, **235**, 37

Wheeler, J.C., Soon-Wook Kim, and Moscoso, M.: 1993, in *Cataclysmic Variables and Related Physics*, 2nd Technion Haifa Conference, Eilat, Israel (Annals of the Israel Physical Society **10**), p. 180

Zel'dovich, Ya.B.: 1964, Dokl Akad Nauk USSR, **155**, 67

Zel'dovich, Ya.B., and Novikov, I.D.: 1964, Sov Phys Dokl, **158**, 811

Zel'dovich, Ya.B., and Novikov, I.D.: 1967a, Astr Zh, **10**, 602

Zel'dovich, Ya.B., and Novikov, I.D.: 1967b, *Relativistic Astrophysics* (Moscow: Nauka)

Zel'dovich, Ya.B., and Novikov, I.D.: 1971, *Relativistic Astrophysics Vol. 1, Stars and Relativity* (Chicago: University of Chicago Press)

Zwicky, F.: 1958, Handbuch der Physik, **51**, 766

7.10 Discussion

Question

(Osterbrock): In your talk about black holes in the nuclei of galaxies, you concluded that they probably do exist. What is the main reason that led you to that conclusion? It seemed to contradict the conclusion that you presented of Kormendy.

Answer

(Novikov): Of course, it will depend on personal taste. Different astronomers, different physicists can give different conclusions from the same observational data, as you know. I believe the situation is the following. We know that, at least in some galaxies, we can observe a very rapid increase of the mass–luminosity ratio in the very central region. We can conclude that here there is a very compact, very heavy body. Small and heavy. After that we have to include in our discussions a theory about what this body could be. From the theory we know that it can only be a black hole. Thus, if we go along with this logic, it is the proof. Of course, you are right: if we want to give a strict proof we should observe the relativistic velocities of stars and gas in the very vicinity of a black hole. That would be strict proof. Only in this case can we say that it is a relativistic object. Probably, in the future, you could give this strict proof, but I still believe that we do not have any other possibilities: only black holes on the basis of these observations. The important point is that after this type of observation, the conclusion about the black hole is not a mere speculation about the possible reason for the activity, as it was in the very beginning when we tried to explain the activity of quasars and other objects, using the hypothesis about the nature of the central engine. Now we really see something very, very compact and heavy. The only explanation of that, on the basis of contemporary physics, could be black holes.

Question

(Osterbrock): I would agree, but what do you think is the best case? – Which of these objects that have been mentioned which people have tried to observe?

Answer

(Novikov): In most cases, we observe the most compact objects in ordinary galaxies without any activity, but at least one example, which I gave at the end, corresponds to an active galactic nucleus, so probably it is proof of it.

Question

(M. Burbidge): Can you imagine the importance of considering magnetic fields and magnetohydrodynamics. This is usually ignored in these discussions.

Answer

(Novikov): Yes, relativistic magnetohydrodynamics is very important for discussion of the processes in the neighbourhood of the black hole. Theoreticians, though, start their discussions with simple things, for simplicity. Only recently were the electrodynamics and magnetodynamics of black holes developed.

Question

(M. Burbidge): I have been puzzled for some time about pulsars, where there is a rotation axis and the magnetic axis always seems to make an angle with the rotation axis. Is there any physical explanation, or attempt at an explanation of that? Why do they not coincide? Is there some reason?

Answer

(Novikov): They can coincide but probably they do not, because there is no physical reason for them to coincide. You know, for example, in the case of our Earth, the orientations of the magnetic axis and the rotational axis are different. So, the same can be true in the case of pulsars. But, of course, it depends on the physics of pulsars and very complex processes in them. That is a separate subject. Today I am discussing the astrophysics of black holes, so allow me to concentrate mainly on that, but I agree that physicists prefer mainly to discuss theories without magnetic fields, unfortunately. For example, the theory of accretion disks without magnetic fields. Now it is clear that that is not enough.

Question

(Longair): Could you give us an example of what we would have to do in order to be able to see inside a massive black hole? I would like to know about this device. How can you do it?

Question

(Sandage): Is there a machine?

Question

(Longair): Yes! Tell me what to do. Tell me, in principle, even if it is very difficult.

Answer

(Novikov): Allow me to describe it. It is possible to do the following. We know that if we contract some mass, we can observe the formation of a well in space. Let us consider a two-dimensional analogy: a plane represents a flat space. If one contracts a massive body, the gravitational field increases, the geometry of this two-dimensional space changes, and one can observe the formation of a well with matter at the bottom. Now one can imagine the contraction of two bodies and formation of two wells. If now, just before the formation of two black holes, the two bottoms connect, and if we make a hole in the matter, we obtain a worm hole. Unfortunately, this worm hole is not traversable; because of the huge gravitational field in it, it tends to shrink and prevent any possibility of propagation, even at the velocity of light. To stabilize the worm hole, we have to have some kind of unusual, exotic matter here, with special, peculiar properties: matter with huge and negative pressure. As a result, we can have negative gravity there. This negative gravity would stabilize the worm hole. In theory it is possible and, more than that, we know that the laws of physics do not forbid it, but how can we realize it? What kind of technology can we use for it? We do not know yet.

Question

(Longair): The key words for me were when you said 'if the black holes move in the correct way, then you can get the worm hole to form'. What did you mean by 'it moves in a special way'?

Answer

(Novikov): Unfortunately, it is not so simple, because this picture is in superspace. In reality, there is not this additional dimension in the Universe. There is some kind of probability for these two bottoms to be connected. We do not know if we can increase this probability, but we can calculate it. It is small, but not zero.

Question

(Rees): As you said earlier, Igor, we have good evidence for compact bodies in the nuclei of galaxies and we could probably say that they are compact enough that they cannot have star clusters and therefore have to be, in some sense, black holes. Then there is a question I think other people were getting at which was: how can we show that black holes have the properties implied by the Kerr metric. As I see it, there are a number of possibilities and I would like you to comment on these. If we could show that some of the energy, maybe in the radio jets, came from the Blandford-Znajek process (that is, the relativistic effect, where the energy is being extracted from the spin of the hole, and not from the accretion) - that would be one test. Another would be if we could show that there was Lense-Thirring drag in the inertial frame in the inner edge in the accretion disk. That would be another test. Another would be if we could find evidence for a star in a very close relativistic orbit around the black hole. The star might not be visible, but it might modulate the observed radiation. I wonder if you think any of these is more likely than any other, or if you have any other ways in which you could really show that these objects have the Kerr metric. It does seem to me that the best hope is probably in these massive black holes, rather than the stellar-mass black holes, because stellar-mass black holes would be probed by gas dynamics, which is messy, whereas, in the case of these massive black holes, at least there is the prospect of having a star which is like a point mass orbiting around them. So, it is slightly cleaner in the case of super-massive black holes, because we do not have the problems of high-density physics and we do not have to rely entirely on gas dynamics. So, I would just like to know if you have any other ideas.

Answer

(Novikov): Thank you very much for this detailed list. In my talk I mentioned only very briefly the general ideas. I agree completely that all these effects could be important. I would like to add one more possibility. If there is a close binary black hole, for example, we could observe, in the future, gravitational radiation, if they are close enough. This would also be some direct proof. The gravitational radiation can be observed also in the case of the motion of a compact star in the vicinity of a black hole.

Comment

(Lynden-Bell): I would just like to make a general comment. We have been told that, in order to get through the worm hole, we need some very exotic material which has never been seen by anybody, and we know of no material that is capable of producing negative pressures, or tensions much bigger than ρc^2. We know no material remotely as strong as that. However, everybody who believes in inflation has assumed this stuff exists. It is now so general to believe in inflation that, I think, Igor does not think very much of having to invoke this strange material, in order to get through a worm hole. I just want to emphasize that, when you believe in inflation, you are believing that there are substances which have such a large negative pressure, that there is gravitational repulsion and that this is a most unreasonable and unusual substance. It is not against the laws of physics, but it is against all experience. That is my reason for not believing in inflation; it is also my reason for not believing in Igor's holes, but I must warn you, I must give you a terrible health warning: I was one of those people who wondered why certain people were wasting their time studying neutron stars, which clearly did not exist.

Comment

(Novikov): I believe that we know some matter that has such peculiar properties. Quantum fields, in certain quantum states, can have such properties. Of course, there is not a huge amount of this matter, but, still, physicists can observe it in their laboratories.

Question

(Lynden-Bell): What matter?

Answer

(Novikov): For example, if we take two metal plates and put them very close to each other. Between them, there would be vacuum polarization which has exactly these properties. We can observe this: it is an observable fact. We can propose some other ideas. As I said, we do not know exactly how we can create a huge amount of this matter and cause the motion of this matter in the appropriate way. That is impossible for us now. But first, we know about the existence of these effects, we know exactly why. It is not fantasy, it is just the facts of modern physics. Not just theory, but also experiment.

Question

(Lynden-Bell): You are telling me a very interesting thing, which I did not know, or understand. This is that you have the Casimir effect, and that the Casimir effect actually has a situation like this and that you can measure the negative gravity. Can one measure the negative gravity?

Answer

(Novikov): No, not negative gravity, but physicists measure this peculiar property when we have either negative energy density or pressure, negative first and larger than ρc^2, for this quantum matter. Of course, we cannot measure the gravity of this state, but we know from general relativity what it should be.

Comment

(Lynden-Bell): You must talk to me afterwards. All I understood about the Casimir effect was that it was an attraction between two plates and I feel I understand that.

Answer

(Novikov): This is a non-gravitational attraction.

Question

(Lynden-Bell): ...a non-gravitational attraction – an electrical attraction?

Answer

(Novikov): Yes, it is the manifestation of the Casimir effect.

Comment

(Reeves): On the same subject, there are two indirect pieces of evidence in favour of existing matter with an equation of state $\rho = -P$. One is the success of the quark–hadron phase transition (I have said a few words about that in Chapter 10). There you need something of that order. Also the success of electroweak symmetry, which also implies the existence of Higgs fields, scalar fields, which can in some cases develop the transition of this equation of state. This is all explained in your book with Zel'dovich, *The Structure and Evolution of the Universe*, which I can recommend for those of you who are interested in this subject. It is very well explained and the foundations of the existence of the equation of state of that type are, I think, indirect, but quite real. I think there is more to it than you seem to say.

Comment

(Lynden-Bell): Fine!

Answer

(Novikov): Thank you. When I talked about the physics of unusual matter, I thought about these effects which you have described.

Question

(Sandage): May I ask two questions? The famous question of time travel into the past: is it possible to kill your mother before you are born and, if it is, does this deny causality?

Answer

(Novikov): The short answer is 'no'. I believe that, in the case of the existence of a time machine, we can travel from the future into the past, and it means we can influence the past. But we cannot do many things, for example: kill ourselves, the younger version of ourselves, or kill our mothers, grandmothers, and so on.

Question

(Sandage): Why not?

Question

(Sunyaev): Why not? If you are a very angry person...

Answer

(Novikov): Even if you are a very angry person it cannot happen. Man has free will, right? The concept of free will, 'I want!' So, if I come into the past, I can take a knife from my pocket and kill the younger version of myself. I prefer to speak about myself...right? because I want to do that. But we know that there are some laws of physics which restrict our free will. Of course, it is very difficult, or even impossible, for physicists to calculate such complex processes as free will and the travel of human beings. But we can perform calculations of simple processes and demonstrate what kind of restrictions exist. The processes which exist in the case of the time machine must be self-consistent. This is our key point. It means that some events are forbidden because of the laws of physics. Allow me to give a very simple example. We know that all laws of physics forbid something. I can wish to do the most fantastic things. I want to walk along this ceiling and along this wall without any special equipment. It is my free will but I know it is impossible!

Comment

(Sandage): Great example! Yes!

Answer

(Novikov): ...because of the laws of physics. We know that, in the case of processes with a time machine, the restrictions exist; we know that not from fantasy but from our calculations. These restrictions are different, not so trivial as in the mentioned example. Thus, the restrictions in the case of a time machine are not something absolutely new: all laws of physics forbid something.

Question

(Sandage): OK the second question concerns a bit of history. In the Caltech campus there was this marvellous, controversial man, Fritz Zwicky. It is said in Caltech that he had visions and predictions of the neutron stars in the 1930s. What is your opinion, having read the history, if that is the case?

Answer

(Novikov): Yes, I believe so.

Question

(M. Burbidge): It has just occurred to me that your negative gravity has an analogy with the Hoyle, Burbidge, and Narlikar C field, which is supposed to come into play when there is a very strong gravitational field. We know nothing experimentally on what happens in a very strong gravitational field. I wonder if you would comment on that?

Answer

(Novikov): I believe that, for a very strong gravitational field, gravity has the property of attraction, not repulsion. So, for real repulsive gravity, for negative gravity, we would have special properties of matter. Not a special gravitational field, but special matter. This is the point of view of modern physics.

Comment

(Lynden-Bell): But the C field helped. It was a form of matter.

Answer

(Novikov): I know. It was a hypothesis about some kind of C field. Yes, I agree completely. It was the beginning of the development of the idea which was, in some sense, analogous to the idea of inflation of the Universe. Probably the properties of the repulsive gravitational action of the inflation now are different from the properties of the C field, but some analogy does exist. I agree completely.

Question

(G. Burbidge): But it is very closely related to the inflationary concept, which you also invoked here?

Answer

(Novikov): Yes, I agree with this.

Question

(Sunyaev): Yes, I would like to come to the present time. I was a little shocked when Igor said that the existence of supermassive black holes is a matter of tests. I myself believe that maybe it is a question of tests when we are discussing any particular object. We do not have real facts about any of those objects which we have discussed today as supermassive black-hole candidates but the sum of our knowledge about the variability of these sources, about their enormous energy release. For me, the fast time variability is extremely important and, when we compare this with the laws of physics, which are accepted today, this shows, at least to me, that the question of the black holes is not a matter of tests. It is just astrophysics giving us, practically, at our level of understanding, only one choice. It is that all these events connected with quasars, with active galactic nuclei, these silent giant bodies in the nuclei of many galaxies – all this together shows us that these are black holes. I agree with Igor: in any particular case, this might be the subject of tests but, all together, the general picture today, if you believe present physics, just present relativistic physics, shows that there is only one way to understand the whole variety of phenomena which we are observing in the nuclei of galaxies.

Answer

(Novikov): I agree completely, and have tried to express the same point of view. So excuse me if my explanation was not so clear. Still, each time, we can be asked, why only this possibility, but not the other? But I agree completely.

Question

(E. Gaztañaga): I think that, in order to construct those worm holes and time machines, besides the problem of the negative gravity, it is important to realize that it is not classical physics but quantum gravity during the construction. You told us that there are many uncertainties in the theory of quantum gravity. My question is: do you think it would be possible to have observational information to construct the theory of quantum gravity that you later need to produce these worm holes?

Answer

(Novikov): My answer is the following. First of all, I agree with your statement. Secondly, it involves the possibility of looking for any observational evidence which can give us additional ideas about the way to construct the theory of quantum gravity. I believe that the one way to get this observational information today is the attempt to investigate the consequences of inflation at the very beginning of our Universe. Only this way. I hope this can give really good additional ideas about the processes during the first moments of existence of the Universe and about quantum gravity, which was very important then. In the future, we can probably get some ideas from the investigation of black holes, if we can put into practice the idea which I proposed. If we can look inside a black hole, near the central singularity, it can give us direct information about quantum gravitational processes, but that is only for the future; today – only the beginning of the Universe.

Question

(A. Mampaso): Can you say a little more about this 10^4 years for the lifetime of observable black holes? Is this calculated considering the number of candidates and the number of binary stars, or is it a theoretical calculation?

Answer

(Novikov): Of course, these calculations have a basis in observations. We can use the theory of the evolution of very close binaries for the case when one component of these binary systems is a supergiant, or a giant, or another type of star which could be good for the observational effects, and another component is a black hole. We can combine, after that, our knowledge about different types of binary systems and our knowledge of the theory of evolution and get these conclusions. But, as I said, it is only a very speculative conclusion because we have no really detailed theory. It should be done, and probably now we have enough information to do it. Somebody should do this.

Question

(H. Castañeda): Is the theory compatible with logical principles (such as the relation of cause and effect) to avoid paradoxes in the theoretical possibility of constructing a time machine?

Answer

(Novikov): The answer is the following. Using a time machine we can travel into the past and we can meet ourselves. No problem. We can talk with ourselves and so on, but some restrictions exist. Allow me to give a simple example from mechanics, in the case of the existence of a time machine. Let's imagine the following simple mechanical process. At the beginning, without any time machine, let us suppose that we have one billiard ball, which moves on the surface of a table. There is a hole in this table and we push the ball in such a way that this ball moves and falls down into this hole. OK? After that, we can take another ball and push it in such a way that the motion of the second will make it collide with the first one and change the direction of its motion. We can organize the collision between them in any way, for example, such that the direction of the first one changes drastically and it moves after that in a different direction and never reaches the hole. OK, so far there is no problem. It is a simple mechanical problem with initial conditions. Now, let's imagine an analogous experiment, but with a time machine. Let's suppose that we have the time machine. The time machine consists of two holes. If you go into one of them, you will appear from another one, in the past. That is enough for our discussion. Now, we can choose the following initial conditions: let's suppose that we have only one ball. Let us push it in such a way that it moves in the direction of the entrance of this time machine. If it comes to this entrance it must appear from another hole in the past before it reaches the entrance. Now we can organize the motion in such a way that this older version of the same ball from the future comes and moves in a trajectory which crosses the first one. Because of our free will, we can choose the initial conditions freely and we can try to organize a strong collision between them. As a result, the trajectory of the first changes drastically and it never reaches the entrance. Paradox! If this ball never reaches the entrance it cannot appear from another hole. That is a paradox, but a real calculation demonstrates that we made a mistake in our simple discussion. From the very beginning we have to take into account the collision between the two versions of the ball (the younger and the older one). Our discussion had a logical mistake. At the beginning we tried to discuss things the following way. Let us consider the motion of the first ball without any collision. After that, let's take into account this collision, but that's wrong. From the very beginning we should take into account their collision and that means that, after the collision, the first ball should move with a trajectory which is not the same as the initial trajectory. If we take into account this fact and if we push the younger version of the ball in the direction of the entrance of the time machine, we never obtain the strong collision between them: the 'older' ball strikes its 'younger' self gently, and the trajectory does not change drastically. It changes a little bit. This ball will reach the hole and, because of this changed trajectory, we have no strong collision, but only a gentle glancing blow: now we have a self-consistent picture. This is a very simple example, but something like that must happen in more complex situations. If you are interested, I can give you the references about the discussion of many facts related to that.

Question

(R. Watson): The ball has not got consciousness. What if some-body with a remote control started guiding it to actually hit the ball?

Answer

(Novikov): I assume we are very far from astrophysics now! These problems have been discussed, and we can discuss this after my talk, probably. I can give the references for a discussion of this type. It is possible to imagine some engine inside the ball and a remote control, and more complex situations. These were discussed. The answer is, in any case, that there is at least one self-consistent solution to this problem. In some cases, many of them. In the case of billiard balls, there are even an infinite number of self-consistent solutions of the problem. If we repeat the experiment absolutely identically many times, the results would be different, according to quantum mechanics, by the way. But that is an absolutely different point and we cannot discuss it now.

Comment

(Sandage): We talked about prestidigitators earlier on. I think that you are the original magician! ...You wanted to follow up?

Question

(Watson): It was to do with the making of a time machine. You are relying on a quantum fluctuation to make the actual connection between the two. Would it not be more logical that the quantum fluctuation were forbidden, so that you could not actually make the worm hole in the first place.

Answer

(Novikov): What fluctuations could be forbidden?

Question

(Watson): Your argument about self-consistency, that the quantum tunnelling to connect the worm hole, if that were forbidden...

Answer

(Novikov): As far as I can understand, you are trying to find physical reasons to prevent the construction of a time machine.

Comment

(Watson): Exactly!

Answer

(Novikov): Because you dislike it, right?

Question

(Watson): Maybe the Universe dislikes it.

Answer

(Novikov): Maybe, but I believe we have to do the following: we have to discuss everything which does not contradict modern physics and our knowledge. If we can find something which would prevent the construction of a time machine, it would be very, very important, but so far we do not know any effects about which we can say that they definitely prevent the creation of a time machine. Not all experts agree with me, but that is my considered opinion.

Question

(R. Rebolo): Coming back to the extraction of energy from black holes. Could you put on the table an example, an astronomical example, of a possible situation in which, using black-hole binaries, or a supermassive black hole, or whatever, you could extract energy. Have you thought in which energy band it would appear?

Answer

(Novikov): You mean the extraction of energy using a worm hole? Well, unfortunately I cannot give any example because I cannot imagine the existence of a worm hole in the real Universe today. Probably they exist because there are hypotheses about the possibility that they exist from the very beginning of the expansion of the Universe. Unfortunately,

I do not know any concrete proposals or examples of that. What kind of observations could be arranged for that? – I do not know.

Question

(H. Zinnecker): I have two questions. When you were talking about exotic material and so forth, I just wondered: is there any chance that this material which you are discussing has anything to do with the other huge problem we have in astronomy, namely dark matter.

Answer

(Novikov): I believe not. I believe that they are two separate problems. Definitely, the main part of dark matter must be exotic, but exotic from another point of view. The dark matter could consist of heavy neutrinos, or photinos, or gravitinos, or something like that, but not of matter with negative gravity, because we know that the dark matter must create gravitational attraction.

Question

(Zinnecker): Is Sagittarius A* a black hole?

Answer

(Novikov): I do not know! As a conclusion, allow me to say one thing. I want to emphasize that not all specialists in general relativity believe in the possibility of the existence, or the construction and so on, of a time machine. For example, Stephen Hawking is against the idea. So, you should remember that. I am one of the enthusiasts, but there are many pessimists.

8 Galaxy formation and quasars – progress and prospects

Martin J. Rees

8.1 The 'standard' hot Big Bang – is it a plausible model?

The clearest evidence that the Universe has evolved from a hot dense beginning is, of course, the microwave background radiation. Its spectrum is now known, from the FIRAS experiment on *COBE*, to be a very precise blackbody, the deviations being no more than one part in ten thousand – indeed, any spectral distortions due to high-z activity, hot intergalactic gas, etc. are smaller than many people expected. Also, the light-element abundances have remained concordant with the predictions of Big Bang nucleosynthesis, thereby giving us confidence in extrapolating back to when the Universe was a few seconds old (see Chapter 10). These developments give us greater confidence in the 'hot Big Bang' – or, at least, in an extrapolation back to times of the order of one second – than would have been warranted ten years ago. Several things could have happened which would have refuted the picture, but they have not happened.

For instance:

(i) Objects could have been found where the helium abundance was far below the $\gtrsim 23\%$ predicted to have been synthesized in the early Universe.

(ii) The background spectrum at millimetre wavelengths could have been weaker than a blackbody with temperature chosen to fit the Rayleigh–Jeans part of the spectrum. While it is easy to think of processes that could augment the submillimetre spectrum, it would be harder to account for a deficit in this band if the radiation was indeed a relic of an opaque early phase.

(iii) A stable neutrino might have been discovered in the mass range 100 eV – 1 MeV. Such particles would then contribute much more than the critical density, if the Universe had once been hot enough (above 10 MeV) for the number of neutrinos to equilibrate with the number of photons.

The evolution of our Universe from the stage when it was a millisecond old has, by now, the same scientific status as several other branches of astrophysics – late stages of stellar evolution, for instance. The broad outlines and the basic physical ideas seem to be understood; most astronomers have moved on to consider more detailed issues, many of which, of course, remain confusing.

There are still, however, advocates of drastically different views. Geoffrey Burbidge has described the alternative picture that he has developed in collaboration with Hoyle and Narlikar; so it is perhaps worth mentioning how measurements of anisotropies in the microwave background might distinguish this model from the 'standard' hot Big Bang. It is a generic feature of all models which attribute the entropy in the microwave background to a dense Big Bang that the dominant opacity on the last scattering surface would be electron scattering. This means that the surface lies at the same redshift, and has the same thickness, whatever microwave observing frequency is used. Any angular fluctuations attributed to a 'last scattering surface' at high redshift should be the same at each frequency: if the same strip of sky were scanned at two frequencies, the temperature fluctuations would track each other closely (on the Rayleigh–Jeans part of the spectrum this is still true even when there is a Sunyaev–Zel'dovich contribution). But in the model of Burbidge *et al.*, the relevant opacity (due to carbon 'whiskers' etc.) depends strongly on frequency. Scans at different frequencies are therefore probing 'surfaces' at different distances. One would therefore not expect the same fluctuations, except maybe on the very largest angular scales.

Progress in understanding the 'Big Bang' brings into focus a new set of issues that are certainly problems for the future: the physical processes, occurring during the first millisecond, which determine such key features of the present Universe as the baryon/photon ratio, the fluctuations, etc. Any inferences about the ultra-early eras remain tentative because the basic physics is itself uncertain. It is only when the Universe has cooled down below 100 MeV that 'conventional' physics becomes adequate and we can have confidence in quantitative models.

8.2 What is the dark matter, and how much of it is there?

8.2.1 Some history

It would be hard to imagine a bigger question than the nature of 90%, and maybe even 99%, of what is in the Universe. Intimations of dark matter date back, of course, to studies of motions in clusters of galaxies in the 1930s. These were later supplemented by analyses of motions within the local group, particularly the classic paper of Kahn and Woltjer (1959), and by radio and optical study of rotation curves in the outer parts of disk galaxies. This is not a historical review – in this book we are supposed to look to the future. I would nevertheless like to go back 20 years, to 1974, because that was when a consensus about the dynamical dominance of dark matter was crystallized, particularly in two important papers.

One of the classic 1974 papers, by three Estonian astronomers, Einasto, Kaasik, and Saar, stated that

> the mass of galactic coronae exceeds the mass of populations of known stars by 1 order of magnitude, as do the effective dimensions. The mass–luminosity ratio rises to 100 for spiral and 120 for elliptical galaxies. With $H = 50\,\mathrm{km\,s^{-1}\,Mpc^{-1}}$ this ratio for the Coma cluster is 170.

In the second paper, by Ostriker, Peebles, and Yahil at Princeton, it was stated that

> currently available observations strongly indicate that the mass of spiral galaxies increases almost linearly with radius to nearly 1 Mpc, and that the ratio of this mass to the light within the Holmberg radius is $200(M/L)$.

These particular inferences have been buttressed enormously by progress in the last 20 years. But it is remarkable that the net conclusions have not drastically changed.

Another indirect constraint on the amount of dark matter in baryonic form comes from the abundances of light elements predicted by cosmic nucleosynthesis. As has been well known since the 1960s, these abundances depend on the baryon density during the few minutes over which the Universe cools through the temperature range from 1 MeV to 100 keV, and therefore (since the present background temperature is known) can be related directly to the present baryon density. The predicted helium abundance increases only slowly with density, but the measurements of helium are now precise enough to provide a significant upper limit. Deuterium, however, is a more sensitive diagnostic of the primordial baryon density. Since it is an intermediate product in the synthesis of helium, more deuterium survives in a Universe of low baryon density. Moreover, it is now much clearer than it was in the 1960s that deuterium is best explained as a relic of the early Universe – we owe much of this insight to Hubert Reeves and his colleagues, and Hubert discusses it in his contribution.

Estimates of deuterium as a measure of baryon density have improved, particularly through a better understanding of the relationship of deuterium and ^{3}He. There has recently been a flurry of interest in cosmic deuterium, stimulated by the claim of a high relative abundance of deuterium to hydrogen, of the order of 3×10^{-4} in a high-redshift damped Ly α absorption system along the line of sight to a quasar (Songalia *et al.* 1994). If this result were to stand up, it would push down the permitted baryon density, completely ruling out the possibility that most halo dark matter could be baryonic unless one abandons other standard assumptions. However, it would be wise to suspend judgement on this issue. The alleged deuterium line is a weak satellite of a very strong feature attributed to high-column-density HI. It is indeed displaced by 80 km s^{-1}, equivalent to the expected isotropic shift, from the centre of a strong hydrogen feature, and there is only a chance of a few per cent of finding a random weak line in the Lyman forest in this position. But there may very well be an excess of weak 'satellite' lines close to any damped Ly α system (caused by gas associated with the same 'protogalaxy'). Until we are sure that there are more systems displaced by 80 km s^{-1} than by, say, 60 or 100 km s^{-1} the significance of this claim for a high abundance of deuterium must remain in doubt. Further data, particularly from the Keck Telescope, ought to settle this question within the next couple of years.

The year 1974, plainly a vintage one for this subject, also saw the publication of a review by Gott, Gunn, Schramm, and Tinsley (1974). These authors adopted a synoptic approach, and tried to seek consistent ranges for the density parameter Ω and the Hubble constant. They considered three constraints. The first was the requirement that the Universe (the age of which depends on the Hubble constant and, in Friedmann models, on Ω) should be older than the oldest stars. The second was an upper limit on the baryon density from deuterium. And the third was a lower limit to Ω, of the order of 0.1, set by the amount of dark matter that was reliably established by dynamical arguments. Gott *et al.* claimed that there was a very small window, with Ω of the order of 0.1 and a low Hubble constant, such that the age constraints could be satisfied and all the reliably established dark matter could be baryonic.

What has changed regarding Gott *et al.*'s arguments in the last 20 years? The uncertainties about the Hubble constant and stellar ages are unfortunately still with us – Allan Sandage is still trying to convince the community that the uncertainties are small. There is firmer evidence on the dark-matter density in clusters of galaxies, though its net import does not substantially change the old estimates. However, the issue of extra dark matter between clusters, maybe even sufficient to provide the critical density, is now a more lively one, and I shall return to it later. There is also a much greater willingness to invoke non-baryonic matter.

8.2.2 Baryonic dark matter

The possibility of baryonic dark matter in stars or stellar remnants was addressed with particular thoroughness by Carr, Bond, and Arnett (1984). These authors showed, through a variety of arguments that are now well known, that there were two possible mass ranges. Dark matter could exist in black holes in the mass range between a few hundred and 10^6 solar masses, which could be a remnant of a population of early massive stars that ended their lives collapsing via the pair-production instability. Some constraints on high-mass objects in our Galaxy are set by the lack of evidence for accretion onto those passing through the disk, and so forth. The other possibility is brown-dwarf or planetary-mass objects, similar to stars except they are below the threshold of around 0.07 solar masses needed to trigger hydrogen fusion. There has been important progress involving gravitational microlensing and the search for evidence of lensing in our own Galaxy. Here again, we can be optimistic that the total amount of mass in the lensing objects (and perhaps their typical masses) will be clarified within the next couple of years.

8.2.3 Clusters of galaxies

Traditionally, the strongest evidence for dark matter in clusters of galaxies comes from application of the virial theorem to galaxy motions – this line of argument goes back to Zwicky and others in the 1930s. But there are now two other lines of attack. Maps of the X-ray brightness profile and temperature are now good enough to allow estimates of the depth (and even the profile) of the gravitational well confining the hot X-ray-emitting gas. And very faint background galaxies whose shapes are distorted, often into conspicuous arcs, by the effects of light bending due to the cluster's gravitational field, will soon offer very direct information about the total mass distribution, whatever the matter contributing that mass may be. One of the early highlights of the data from the post-refurbishment *HST* is a superb picture of the cluster Abell 2218, with a redshift of 0.18, by Kneib *et al.* (1996), which shows very large numbers of obvious background arcs. (This latter result would have especially pleased Zwicky, who was also among the first astronomers to emphasize that gravitational lensing might be observable.)

It now seems more logical to discuss the structure and dynamics of cluster masses in an order different from the traditional one. We can infer the shape of the potential well more directly than before by reconstructing it from the observed distortion by gravitational lensing of background galaxies. It is then possible to address whether the observed spatial distribution of galaxies, and the spread in their velocities, is consistent with an isotropic equilibrium in that particular potential; if it is not, the angular distribution of velocities must be more complex or the system must be out of equilibrium. X-ray maps will reveal whether the gas has a temperature and density profile consistent with that potential. Any discrepancy would motivate us to consider whether the gas is partially supported by rotation, macroscopic bulk motion, magnetic pressure, relativistic particles, etc. (the gas can, of course, be somewhat inhomogeneous, but the clumping factor is constrained because gas confined in the potential well cannot be on a very much higher adiabat than the gas that dominates the X-ray emission).

Clustering must be viewed in the more general context of overall cosmic structure formation. Numerical simulations of this are now a heavy industry, and an increasingly sophisticated one. Most of these simulations are based on the assumption that the dominant gravitating stuff is non-baryonic. So let us briefly consider this option.

8.2.4 Non-baryonic dark matter

One of the main changes since Gott *et al.* wrote their 1974 paper has been the much greater willingness to invoke non-baryonic matter. Non-zero neutrino masses are no longer thought theoretically unacceptable, and there is a willingness to invoke new kinds of particles, particularly those predicted by supersymmetric theories. One topic which does not figure on the programme for this meeting, but which is undoubtedly one of the most exciting areas of particle astrophysics, is direct searches for non-baryonic matter. Primordial neutrinos seem impossible to detect by feasible current techniques, and axions present a very severe experimental challenge. But there is substantial interest in detecting heavy neutral particles, such as the lightest stable supersymmetric particles. These techniques involve detecting the recoil in the rare event when one of these particles, which would pervade the entire halo moving with speeds of about $10^{-3}c$, interacts with a nucleus in an experimental detector.

We should certainly give every possible encouragement to those of our colleagues who have accepted the challenge to detect dark matter. Even the optimist cannot predict success with great confidence, but the attainable upper limits are themselves becoming significant, and detection of such particles would tell us what 90% of the Universe is made of, as well as perhaps discovering an entirely new class of particles that cannot be produced terrestrially.

Progress in fundamental physics (by theorists or in terrestrial laboratories) may sometime tell us whether supersymmetric particles actually exist, and what their masses and cross sections are likely to be. Calculating how many should survive from the first millisecond, and what they contribute to the dark-matter density, would then be as straightforward in principle as calculating helium and deuterium surviving from the first few minutes. Experiments may also tell us that neutrinos have a mass –

in which case they would definitely contribute to the dark matter, unless we abandon our extrapolation even back to times of the order of 1 second.

8.2.5 Simulations of structure formation

The lack of firm guidance from high-energy physics need not inhibit us from exploring what consequences the various non-baryonic candidates would have for structure formation (or cosmogony). The essential distinction here is between those, such as light neutrinos, which move fast enough to homogenize on small scales, and those that are sufficiently slow moving to be treated as 'cold'. The other essential input into the calculations is some assumption about the initial fluctuations.

The outcome of simulations of structure formation can be confronted with the data by comparing the predicted relative amplitude of clustering on different scales with what is actually observed at the current epoch, and (on large scales) by fitting the amplitude of the microwave background fluctuations. It is the dark-matter distribution which is most directly predicted by theory: this is controlled only by gravity; it is therefore easier to simulate than the behaviour of the baryonic component. Observationally, the dark matter is probed by the peculiar velocities of galaxies and (at large redshifts) by gravitational lensing.

Although the purely gravitational interactions are the easiest to simulate, it is now feasible to incorporate the dynamics of gas (including shock waves, radiative cooling, etc.), and thereby model, for instance, the temperature and density profiles of gas in clusters. These can then be compared with X-ray data.

Most of our information on structures in the real Universe concerns the spatial distribution of galaxies. Unfortunately, the relation between the galaxies and the dark matter is too complicated to be adequately simulated. Any scheme which relates the galaxies to the dark matter via a single 'biasing parameter' is plainly oversimplified: the galaxy formation efficiency may depend on environment, etc., in many ways. One cannot be optimistic about the prospects of modelling, *ab initio*, how individual galaxies form. Some parameters (e.g. from the efficiency of star formation, the initial-mass function, and the feedback effects from stellar winds and supernovae) will have to be fitted empirically. But one can make progress by testing the consistency of the assumptions with data on galactic evolution: having chosen parameters to match the luminosity function and morphologies of present-day galaxies, one can then check whether this choice accounts for the way galaxy populations depend on redshift.

We know that bound systems on galactic scales exist back to redshifts of 5 from quasars, neutral-hydrogen clouds, etc. The mere existence of structures at such early epochs is in itself an important constraint on models for structure formation.

A benchmark for comparison of observations has been the so-called 'standard' cold-dark-matter (CDM) model (see Ostriker 1993 for a review). This involves a package of five assumptions:

(i) The primordial spectrum has the Harrison–Zel'dovich form (index $n = 1$ in the conventional notation), and the fluctuations are Gaussian.

(ii) The Universe is dynamically dominated by cold non-baryonic matter which interacts only gravitationally with everything else.

(iii) Galaxies are related to dark matter by a simple biasing prescription.

(iv) Neutrino masses are taken to be zero.

(v) The density is taken to be equal to the critical value, in other words, Ω is unity.

It is now well known that this five-item package, the 'standard' CDM model, has difficulty reconciling small and large-scale structure and the microwave background fluctuation amplitude. However, there are several modifications which are physically motivated and by no means simply *ad hoc*. First, the primordial-fluctuation spectrum could be tilted, so that the amplitude increases slowly with scale (so that $n < 1$). Indeed most inflationary models predict that this should occur. Also one could consider models where Ω is different from unity: these are either open or else flat, with the extra curvature made up by a non-zero cosmological constant.

It may turn out that neutrino masses are not exactly zero. The so-called 'hybrid' or 'mixed' dark-matter models (MDM), in which a neutrino has a mass of a few eV, surmount some of the difficulties of standard CDM. If experimentalists find such evidence for neutrino masses, believers in CDM would delightedly incorporate it in their existing models, ending up with a better fit.

8.3 Is $\Omega = 1$?

Another change since Gott *et al.*'s classic paper is that there is now a strong theoretical prejudice in favour of Ω being unity, stemming from the attractiveness of the general concept of an inflationary Universe. Many people seem happy to accept whatever non-baryonic matter and biasing would be needed to meet this requirement. Inflationary models naturally predict that the Universe expands enough to stretch itself flat, in the sense that the Robertson–Walker curvature radius would become enormously larger than the present Hubble scale. Anything different from a flat Universe would, as is well known, involve fine tuning in the expansion factor. This tuning is implausible at the level of a few per cent, even in the more optimistically contrived scenarios. However, most variants of inflation provide an even stronger argument in favour of flatness: if the Universe had inflated only enough to make the present Robertson–Walker curvature of the order of the Hubble radius, there would be quadrupole or dipole effects in the microwave background of the order of unity. (There may, however, be ways to evade this latter constraint.)

What, then, is the observational case for or against a critical density? Until a few years ago, there was essentially no evidence for more than 20% of the critical density. I think everyone would agree that any such evidence is still tentative. However, there are (as Donald Lynden-Bell discusses) many attempts being made to measure and interpret large-scale streaming motions. Extra dark matter between clusters, if it is inhomogeneous on supercluster scales, may be revealed by searches for correlated distortions in the images of distant galaxies induced by 'weak' gravitational lensing.

Some of the classical 'geometrical' methods should soon become more helpful. The Hubble diagram for supernovae may be extended to high enough redshifts to reveal the deceleration parameter; further studies may firm up the earlier tentative evidence from the angular diameters of high-redshift sources in favour of a high density. On the other hand, if the Hubble constant error bars are reduced, and the Hubble time 'stabilizes' at less than 15 billion years, this will obviously argue against a critical density in which the time since the Big Bang is only two-thirds of the Hubble time.

Some other, rather less direct lines of evidence on Ω will attract increasing attention. Two of these involve clusters of galaxies. The first is an inference from the irregular shapes of most clusters, indicating that they have undergone recent mergers of subcomponents each comprising a substantial fraction of the total mass. In a low-density Universe, the growth of structure would almost have ceased, because gravity cannot compete with the kinetic energy of expansion. On the other hand, growth continues right up to the present if Ω is high. The prevalence of conspicuous substructure points, therefore, towards a high Ω, though this evidence is not yet quantitative enough to tell us whether the full critical density is required.

A quite distinct argument, again based on clusters, suggests a low Ω. X-ray data show that the baryon fraction in a cluster, mainly in hot gas, is typically between 10 and 20%. (The exact fraction depends, of course, on the Hubble constant.) This has been inferred from detailed study of the Coma cluster (S. White *et al.* 1993), and also from X-ray data on a sample of 19 clusters (White and Fabian 1995). When the baryonic fraction of the mass in the cluster is compared with the baryonic fraction in the Universe allowed by standard Big Bang nucleosynthesis, there is a contradiction if Ω is more than about 0.3. A high Ω requires either abandonment of standard nucleosynthesis (despite the evidence that this is in good shape) or a mechanism that can concentrate baryons (by a factor of about 3) by pushing them relative to dark matter through distances as large as the turnaround radius of a cluster. Resolution of this dilemma may come from a combination of small effects and uncertainties, but at the moment it seems a serious argument against a standard cosmological model with the full critical density.

Another quite different estimate of Ω will soon come from studies of microwave background fluctuations. The *COBE* data refer to angular scales of $10°$. However, several other groups are now reporting fluctuations on angular scales of the order of $1°$. These latter scales seem to display a larger amplitude than found by *COBE* (see Scott, Silk, and White 1994 for a recent review). This is precisely what is expected on scales smaller than the horizon at recombination, because of the contribution from Doppler motions, etc. This horizon subtends about $2°$ in an Einstein–de Sitter model, but the angle scales as $\Omega^{\frac{1}{2}}$ in low-density hyperbolic models. Firm evidence for an upturn in the background fluctuation amplitude on angular scales of $1°$ or $2°$ would be hard to reconcile with a low Ω, where any Doppler contribution would be restricted to angular scales below $1°$. When the next generation (post-*COBE*) of space experiments have flown, the power spectrum of the microwave background fluctuations will probably have been so well measured that we will be able to infer most of the cosmological parameters from such data alone. However, we will not have to wait so long before getting firm constraints on Ω from the general shape and angular scale of the 'Doppler peak'.

The inflationary requirement of flatness can be reconciled with a low matter density if there is a non-zero cosmological constant. In addition to accounting for the observations that suggest low Ω, it has further advantages. Because the expansion would now be accelerating, the age of the Universe can actually be higher than the inverse Hubble time. An acceptable fit to large-scale structure can be obtained in the CDM model, without dropping any of the other assumptions (i)–(iv) above (though 'fine tuning' is still required, for the cosmological constant to be of appropriate strength). It will probably be a 'classical' test – the measurement of the deceleration parameter – that eventually pins down the cosmological constant. Supernovae are good standard candles – better than galaxies. They can be used to determine a Hubble diagram out to $z = 0.5$, and I think there is a high chance that, within two or three years, such measurements will be precise enough to decide among a Λ-dominated model (with $q \simeq -0.6$, at low redshifts, in the best-fitting case), a low-density open model ($q \simeq 0.1$), and a flat (CDM or MDM) model ($q = 0.5$).

The pace of the subject is fast. Most of the current emphasis is on topics that barely existed a few years ago. Within the next few years we may get definitive answers to the basic questions – What is the dark matter? Does it provide the critical density? Is the cosmological constant actually zero?

8.4 The relation of AGNs to the central bulges of galaxies

4.1 When did the dark age end?

It took about half a million years for the Universe to cool down to 3000 K. Thereafter, further expansion shifted the primordial radiation into the infrared, and a dark age began, which persisted until the first non-linear perturbations developed into bound systems and released enough nuclear or gravitational energy to light up the Universe again.

The most remote quasar so far detected has $z = 4.89$ (Schneider, Schmidt, and Gunn 1991). The population genuinely seem to be 'thinning out' at redshifts above 2.5–3: the comoving density of quasars falls by at least 3 for each unit increase in z (Shaver 1995). The high-z quasars tell us that, after about one billion years, the dark age had certainly ended, and black holes of as much as $10^8 M_\odot$ had accumulated. But what about redshifts larger than five? What happened during the timespan from a million to around a billion years? The answer depends on the relation between quasars and galaxies, and on when galaxy formation began. There are three options:

(i) In CDM the first structures are loosely bound systems of subgalactic scale, which start to form at z as high as 20 or 30. However, the galaxies themselves only form more recently. The z distribution of quasars can be fitted if the hole mass depends in a plausible way on the depth of the potential well and the density. In a CDM model, we would expect the quasar density to fall off steeply beyond $z = 5$, though the intergalactic medium would have been originally ionized, perhaps as early as $z = 20$, by stars in shallow potential wells of subgalactic scale.

(ii) In the MDM model, there would have been less structure at early times. It is then problematic even to account for the high redshifts already observed; quasars certainly could not extend to much higher redshifts. Indeed, even the ionization of the IGM at higher redshifts would be a problem, since the fluctuation spectrum in MDM suppresses the amplitude on subgalactic scales, so even these do not form early.

(iii) The so-called primordial isocurvature baryon (PIB) model leads to non-linear structures soon after recombination. Bound systems that condense out early could evolve directly into black holes, even before the virialization of galactic-scale potential wells which seem a prerequisite for black-hole formation in CDM and MDM.

Recent progress in the study of active galactic nuclei (AGNs) brings into sharper focus the question of how and when supermassive black holes formed, and how this process relates to galaxy formation. There is hope of learning more, not only by better quasar statistics but also by observing ordinary galaxies in the same redshift range (the quasar absorption-line spectra, discussed in Chapter 4, offer another probe).

The next step is to push back further towards the 'dark age'. The primordial medium could have been re-ionized by the quasars themselves, or perhaps, even before quasars formed, by the first stars (the quasar spectrum has a flatter UV slope, and therefore is likely to be dominant for ionizing He even if not for H). But can we probe the pre-ionization era, when the hydrogen was still in atomic form? The best hopes here are the 21-cm line. This makes a much smaller contribution to the background continuum than either to the relic radiation or to the synchrotron emission from non-thermal sources, but would display distinctive angular and spatial structure, owing to incipient large-scale clusters, and perhaps also to patchiness in the original heating. Lyα from smaller scale pre-galactic structure may also be detectable by next-generation telescopes.

Another probe of the dark age would be post-reionization scattering of the microwave background, which could attenuate the fluctuations imprinted at recombination and imprint secondary fluctuations with distinctive angular structure and polarization.

8.4.2 Quasars and young galaxies

Whether there are quasars with $z \gg 5$ is therefore an important unsolved problem. Discovery of higher redshifts would push back our estimates of when galaxies formed, unless we adopt the radical view (which could be maintained if the PIB model were right) that these quasars are not closely connected with galaxies.

I have discussed elsewhere (Rees 1993; Haehnelt and Rees 1993) the processes whereby, when the stellar bulge of a galaxy forms, part of the gas may collapse into a black hole. The formation and growth of the hole then manifests itself as a quasar. It is hard to predict how long the quasar-level activity would last. However, a natural time scale is set by the 'Salpeter time' – the time it takes a hole to double its mass if it accretes at the rate necessary to supply the Eddington luminosity. This time is of course proportional to the efficiency with which mass is converted into radiation; but if the latter were around 10%, the Salpeter time would be 4×10^7 years.

Two things could prevent a hole from forming:

   (i) the material may condense, with nearly 100% efficiency, into stars that are all of low mass (so that none is expelled out again); or

  (ii) gas may remain in a self-gravitating disk for hundreds of orbital periods, without any instability redistributing angular momentum and allowing the inner fraction to collapse.

Neither of these options seems at all likely. So I think it is natural to expect that, when big bulges form, a fraction of the baryons, some already processed through stars, condense into a massive central black hole. However, the issue will not be settled without detailed computer modelling of stars and gas in the central 100 pc of a newly formed bulge. We need to know at what stage the gas stops being able to form stars (because of radiation pressure, magnetic fields, or whatever) and evolves instead into a supermassive object. The central hole's actual mass would depend on the angular momentum of the protogalaxy, the depth of the potential well, and no doubt other parameters as well.

If typical high-z quasars are indeed associated with this process (or its immediate aftermath), their lifetimes would be a few tens of millions of years. There would then have been 50 generations of quasars during the period over which the population density rises and falls. At redshifts of 2–3, where quasars were most common, their comoving space density was only a few per cent of the present density of bright galaxies. These 'demographic' arguments therefore tell us that every galaxy could have passed through a quasar phase, and that by $z = 2$ (2–3 billion years) most bulges have formed with central holes of 10^6–10^9 solar masses. An important project for the future would be to refine this very rough argument to take account of the broad luminosity function for AGNs.

8.5 Quasars and their remnants: probes of general relativity?

5.1 Dead quasars in nearby galaxies

Photometry and spectroscopy of several nearby galaxies reveal that stars near the centre are under the gravitational influence of a massive dark object (for an up-to-date assessment, see the review by Kormendy and Richstone 1995).

Much more compelling evidence for central black holes, however, has recently come from a quite different technique: probing gas motions by measuring the emission in the 1.3-cm line of the water molecule in the nearby peculiar spiral galaxy NGC 4258 (Watson and Wallin 1994; Miyoshi *et al.* 1995). The spectral resolution in this microwave line is high enough to pin down the velocities with an accuracy of $1 \, \mathrm{km \, s^{-1}}$. Such observations, combined with VLBI mapping with a resolution of 0.5 milliarcseconds (100 times sharper angular resolution than the *HST*), have revealed, right in the galaxy's core, a disk with rotational speeds following an exact Keplerian law around a compact dark mass, which is hard to interpret as anything other than a black hole. The circumstantial evidence for black holes has been gradually growing for 30 years, but this remarkable discovery clinches the case completely.

Should we be surprised that these putative holes are so quiescent? Their environment could be almost free of gas, so that very little gets accreted; moreover, when the accretion rate is low, it is also inefficient, in that the cooling is so slow (because of the low densities) that only a small fraction of the binding energy can be radiated before the gas is swallowed. However, there is one unavoidable gas supply. Around these holes is a high concentration of stars – the stars whose motions are analysed by Kormendy, Richstone, and others. The stars interact with each other gravitationally; every few thousand years one of them would get deflected onto an almost radial orbit that approaches the hole closely enough to be ripped apart by tidal forces. Some of the debris would then swirl inwards, providing a high transient fuel supply.

The predicted flares offer a robust diagnostic of the massive holes in quiescent galaxies. They would attain high luminosity – the total photon energy radiated could be a thousand times more than the photon output of a supernova. They would, however, not be standardized – what is observed depends on the hole's mass and spin, the type of star, the impact parameter, and the orbital orientation relative to the hole's spin axis and the line of sight. What is observed may also depend on absorption in the galaxy. To compute what happens involves relativistic gas dynamics and radiative transfer in an unsteady flow with large dynamic range which possesses no special symmetry and therefore requires full 3-D calculations – a challenge to those who have many gigaflops at their disposal.

The tidal-disruption phenomenon presents a worthwhile challenge to observers as well as to theorists. Any survey that accumulates several thousand 'galaxy years' of exposure should detect such a flare. There are two possible strategies. Searches of relatively nearby galaxies can be carried out using CCD detectors, which do not discriminate against the central regions that would be burnt out if photographic techniques were used. Alternatively, thousands of more distant galaxies can be monitored in just a small selected patch of sky. (The latter has the disadvantage that it would be harder to check that the galaxy was otherwise completely quiescent.)

The case for a black hole in our own Galactic Centre was until recently ambiguous, but is now strengthened by the discovery of stars within the central 0.1 pc (Eckart and Genzel 1996). The most recent stellar-disruption event may have left traces that could still be detectable in our own Galaxy. Up to 10^{53} ergs of ionizing radiation could be released by accretion of the captured debris – more photons than would be emitted by steadier UV sources in the entire 10^5-year interval between successive disruptions. Moreover, the half of the star that is ejected may have left traces in some of the strange patterns of the gas within the central 2 pc. The mean energy input may be dominated by these rare events.

8.5.2 Do these holes have a Kerr metric?

There is growing evidence that supermassive black holes exist. But do they have the exact properties predicted by general relativity (and described by Igor Novikov in Chapter 7)? In 1975 Chandrasekhar wrote

> In my entire scientific life the most shattering experience has been the realization that an exact solution of Einstein's equations, discovered by the New Zealand mathematician Roy Kerr, provides the absolutely exact representation of untold numbers of massive black holes that populate the Universe.

Chandra refers, of course, to the 'no-hair' theorems which prove, on the assumption that Einstein's equations are correct, that any stationary black hole is characterized just by the two defining parameters of the Kerr solution: mass and spin.

But one cannot yet claim that any observed features of AGNs offer a clear diagnostic of a Kerr metric. All we can really infer is that 'gravitational pits' exist, deep enough to allow several per cent of the rest mass of infalling material to be converted into kinetic energy, and then radiated away from a region compact enough to vary on time scales as short as an hour. General relativity has been resoundingly vindicated in the weak-field limit (by high-precision observations in the Solar System, and of the binary pulsar) but we still lack quantitative probes of the strong-field domain.

The fate of the debris from a tidally disrupted star (mentioned above) depends crucially on relativistic precession. Many detailed attempts to model AGNs use specific features of the Kerr metric. For instance, the process suggested by Blandford and Znajek (1977), which may create the energetic plasma jets in radio sources, involves direct extraction of a Kerr hole's spin energy. Corroborating these ideas – finding evidence as strong (albeit circumstantial) as the evidence that nuclear fusion powers ordinary stars – is a challenge for the future.

For straightforward thermodynamic reasons, the well-studied optical emission lines from AGNs come from a volume much larger than the hole itself. In contrast, the much hotter gas that emits thermal X-rays can be concentrated in the innermost parts of an accretion flow. X-ray lines should therefore display substantial gravitational redshifts, as well as large Doppler shifts. X-ray telescopes, foremost among them the Japanese *ASCA* satellite, are now achieving good enough spectral resolution to reveal lines (e.g. Tanaka *et al.* 1995). In the next few years, such observations should offer a probe for gas flow in the region where relativistic effects are large.

But gas dynamics is always messy and intractable. A much cleaner probe of the metric would be a star orbiting close to the hole whose orbit would precess (Karas and Vokrouhlicky 1993). Could a star attain such an orbit? Ordinary solar-type stars would undergo physical collisions, rather than large-angle 'Coulomb' deflections, if their relative speed were more than $1000 \, \mathrm{km \, s^{-1}}$; their orbits cannot therefore, by stellar-dynamical processes, achieve very high binding energies, corresponding to orbital speeds a substantial fraction of c. Neutron stars or white dwarfs, on the other hand, could exchange orbital energy by close encounters with each other until some got close enough that they either fell directly into the hole or until gravitational radiation took over as the dominant orbital energy loss.

A solar-type star could achieve a very tight orbit if it lost orbital energy as the cumulative effect of successive impacts on a disk (such orbits could not be reached by tidal capture, because the star itself, rather than a disk or external resisting medium, has to radiate the orbital binding energy and would puff up and disrupt before the orbit could circularize).

There was a flurry of interest three years ago when X-ray astronomers detected an apparent 3.4-hour periodicity in the Seyfert galaxy NGC 6814. It turned out, sadly, that a foreground binary star, with just that period, lay in the telescope's field of view. But theorists should not be downcast. It is more elevated to make predictions than to explain phenomena a posteriori, and that is all we can now do. There is a real chance that observers may eventually find evidence that an AGN is being modulated by an orbiting star that could act as a test particle whose orbital precession would probe the metric in the domain where the distinctive features of the Kerr geometry should show up clearly.

But, for all that, the most impressive test of general relativity would be detecting gravitational radiation, which involves no physics other than that of spacetime itself. The most dramatic sources would be mergers of supermassive black holes. Some events of this kind are expected. Most galaxies may harbour black holes that formed at $z > 2$. Moreover, many galaxies have experienced a merger since that time. When that happens, the holes in the two merging galaxies would spiral together, emitting, in their final coalescence, up to 10% of their rest mass as a burst of gravitational radiation.

This burst would be in a frequency range around a millihertz – too low to be accessible to ground-based detectors, which lose sensitivity below 100 Hz, owing to seismic and other background noise. Space-based detectors are needed. One such being proposed is the European *LISA* – six spacecraft on solar orbit, configured as two triangles, with a baseline of five million km whose length is monitored by laser interferometry.

The signal-to-noise would be high, even for mergers at high redshift. The bad news is that the event rate is low. Even out to $z = 5$, there would be less than one event per decade involving holes above $10^6 M_\odot$ (Haehnelt 1994). This is of course uncertain. There could be lower-mass holes in small galaxies that are more common and underwent more mergers.

The sensitivity of *LISA* is such that it could detect waves from a stellar-mass object orbiting a supermassive hole – this is a further reason for interest in the possibility that there may be stars in such orbits, even around holes that are electromagnetically quiescent.

LISA is at the moment just a proposal – even if it is funded, it is unlikely to fly before 2016. Those of us of advancing years would therefore be eager to find, sooner than that, any indirect manifestions of gravitational radiation.

The dynamics (and gravitational radiation) when two holes merge has been computed only for cases of special symmetry. The more general problem – coalescence of two Kerr holes with general orientations of their spin axes relative to the orbital angular momentum – is one of the US 'grand challenge' computational projects. When this challenge has been met (and it will almost certainly not take all the time until 2016) we shall find out not only the characteristic waveform of the radiation but the recoil that arises because there is a net emission of linear momentum. This recoil could displace the hole from the centre of the merged galaxy – it might therefore be relevant to the low-z quasars that seem to be asymmetrically located in their hosts (and which may have been activated by a recent merger). The recoil might even be so violent that the merged hole breaks loose from its galaxy and goes hurtling through intergalactic space.

References

Blandford, R.D., and Znajek, R.L.: 1977, MNRAS, **179**, 433

Carr, B.J., Bond, J.R., and Arnett, W.D.: 1984, ApJL, **277**, 455

Chandrasekhar, S.: 1975, quoted in *Truth and Beauty* (Chicago: Chicago University Press, 1987), p. 54

Eckart, A., and Genzel, R.: 1996, Nature, **383**, 415

Einasto, J., Kaasik, A., and Saar, E.: 1974, Nature, **250**, 309

Gott, J.R., Gunn, J.E., Schramm, D.N., and Tinsley, B.M.: 1974, ApJ, **194**, 543

Haehnelt, M.: 1994, MNRAS, **269**, 199

Haehnelt, M., and Rees, M.J.: 1993, MNRAS, **263**, 168

Kahn, F.D., and Woltjer, L.: 1959, ApJ, **130**, 703

Karas, V., and Vokrouhlicky, D.: 1993, MNRAS, **265**, 365

Kneib, J.P., *et al.*: 1996, ApJ, **471**, 643

Kormendy, J., and Richstone, D.D.: 1995, Ann Rev A&A, **33**, 581

Miyoshi, M. *et al.*: 1995, Nature, **373**, 127

Ostriker, J.P.: 1993, Ann Rev A&A, **31**, 689

Ostriker, J.P., Peebles, P.J.E., and Yahil, A.: 1974, ApJL, **193**, L1

Rees, M.J.: 1993, Proc. Nat. Acad. Sci, **333**, 523

Schneider, D., Schmidt, M., and Gunn, J.E.: 1991, AJ, **102**, 837

Scott, D., Silk, J.I., and White, M.: 1994, Ann Rev A&A, **32**, 319

Shaver, P.: 1995, Ann NY Acad Sci, **759**, 87

Songalia, A., Cowie, L.L., Hogan, C.J., and Rogers, M.: 1994, Nature, **368**, 599

Tanaka, Y. *et al.*: 1995, Nature, **375**, 659

Watson, W.D., and Wallin, B.K.: 1994, ApJL, **432**, L35

White, D., and Fabian, A.C.: 1995, MNRAS, **273**, 72

White, S.D.M., Navarro, J.F., Evrard, A.E., Frenk, C.S.: 1993, Nature, **366**, 429

8.6 **Discussion**

Question

(G. Burbidge): In the beginning you said that you thought that there was a 90% probability that this basic cosmological model was correct. I think that is a fair statement. Are there any observations at all, or any observational tests whatsoever that could be made, which would convince you that, after all, you were wrong?

Answer

(Rees): Going back to the discussion following your contribution (Chapter 2), really incontrovertible evidence of anomalous redshift would clearly be important. Thinking more directly on what relates to cosmology, I would like to mention briefly what I said at the IAU, regarding how the microwave background fluctuations might offer a discriminant between most standard theories and the idea in your work with Hoyle and Narlikar on the microwave background. The point here is that, in your theory, the microwave background is produced and thermalized by dust grains etc. The last scattering surface, which is what we are seeing the microwave background fluctuations from, is not determined by Thomson scattering, as it is in all standard theories with or without reheating. It is determined by where the opacity becomes of the order of unity, owing to these various complicated areas of physics, and will be at different redshifts for different observed frequencies.

This leads to a definite difference, in that, in a standard theory, you would expect that, if you observe a $\Delta T/T$ angular fluctuation along a particular strip of sky at one frequency, and then you scan the same strip of the sky at a different frequency, the patterns should track each other because you see the same depth, whereas, in the quasi-Steady State theory, you would not necessarily expect that. I think you might get close to it probably, on large angular scales, so the issue may be still open, but on the smaller angular scales, there would be a possible test.

When we look at the microwave background, we are probing to different redshifts, different detailed structures at different wavelengths. So, if the small-angular-scale fluctuations at, say, 1 cm, 2 cm, and 3 cm, do not track each other, then that would be a discriminant. I think it would be interesting to work out, in the quasi-Steady State theory, just how big such effects would be. I know that the question of how well you can fit the overall microwave background spectrum is still controversial, but you would not expect the standard results for the anisotropies.

Question

(G. Burbidge): I was just going to mention the other methods. While you pointed out correctly that the microwave background is a black-body precisely to one part in 10^4 close to the peak, when you go out into the radio-frequency ranges, it is much more uncertain. That is where we predict that it must depart. So, if one could get a precise measurement, or precise measurements in that part of the spectrum, that is also, in some sense, at least a pointer.

Question

(Sandage): I had thought that the four points that you made in the third or fourth viewgraph would have convinced you that the standard model did indeed work because all four points agree with the standard model. So would not that be another answer to Geoff?

Answer

(Rees): That is the reason why I am fairly confident in the standard model. I am not convinced that Geoff's model would naturally give the exact blackbody spectrum. One's assessment also depends on how strongly one regards the light elements as being a confirmation. We shall hear more about that from Bernard. But I was proposing a new test that might specifically discriminate between Geoff's quasi-Steady State model and the standard one.

Comment

(Sunyaev): I also agree with Martin Rees that maybe the most important result from *COBE* is this high-precision Planck spectrum. I do not know why fluctuations are much more strongly advertised, but I wish to remind everyone in the room about the cosmic microwave background. When we discuss it, we all think that the Planckian spectrum was just from the moment of zero, or 10^{-43} s. In reality, we can construct this beautiful Planckian curve at any redshift, just using the normal processes: processes like free–free, double-Compton, and Thomson scattering, at $z = 10^6$.

In reality, we know just that the microwave background spectrum gives us information only about the time. It is impossible, for example, to get information about the annihilation of electrons and positrons in the early Universe, no information, no footprints in the spectrum of the microwave background. For me, this is a very important thing, that using these spectra, we can get information about the dark ages. There was practically no energy released during the dark ages, as you just said. Also, the energy release was enormously small in the time between the recombination and a redshift of 10^6. Practically no antimatter, no strong turbulence, no non-linear motions. This is very strong observational information which comes from this time.

In addition, I wanted to mention that observation of clusters of galaxies, for example, observation of the cluster 016+16, at redshift 0.55, without any doubt gives us information that there was a microwave background at that time, because we see this diminution. What is important is that it also gives us two additional pieces of information about cosmology. First is that the microwave background at that time also had a spectrum which was very close to Planckian, because the temperature was the current temperature multiplied by 1.55; if it were more than 1.55 times higher, there would be a redshift. We know that the redshift is there, because in all the spectra of galaxies there is a redshift of the line. Then the spectra in the investigations would be very different. We can prove that spectrum stays Planckian.

The last thing which such investigations give to us is that we can measure the peculiar velocity of the cluster, as a whole, just above the recession velocity of that cluster. For me, it is extremely interesting that the recession velocity of the cluster is $150\,000\,\mathrm{km\,s^{-1}}$, but the upper limits to the peculiar velocities (just Doppler scattering of the microwave spectrum by the electrons in those clusters) give upper limits of $6000\,\mathrm{km\,s^{-1}}$. Therefore, the peculiar velocity is 30 times smaller than the recession velocity. This is not just a matter of somebody believing in these things, this is an observational fact. This also shows that there are a lot of possibilities to consider, just using well-known physics, to get an enormous amount of information about cosmology and about our Universe at very large distances.

Comment

(Novikov): I would like to add to this list that the analysis of the information about the angular distribution of the anisotropy of the microwave background radiation on scales of $1°$ or less gives us direct information about the sound waves in the primordial plasma, just before recombination. It allows us to determine the total amount of baryons in the Universe, directly from the observations. It is a unique class of information.

Answer

(Rees): Yes, this shows it schematically [shows viewgraph]. On scales exceeding the horizon scale at the epoch of last scattering, $\Delta T/T$ is probably almost scale independent. But all the interesting structure comes from scales which are small compared to the horizon scale which is where you then start seeing the effects of motions etc. (what Igor, I know, calls the Sakharov oscillations). Tensor gravitational-wave components die away, whereas the scalar ones lead to peaks. The detail of this information is going to be very important for the nature of the dark matter and the structure formation. The drop on angular scales below 20 arcmin is essentially because the large scattering surface is not completely sharp, so there is a smearing of small scales, but the details depend on whether the dark matter is neutrinos or in cold particles. There is another point too, which is that the angular scale subtended by the horizon at $z = 1000$ depends on Ω in a simple way. If $\Lambda = 0$ it just depends on $\Omega^{1/2}$. So, if $\Omega = 0.1$, the angular scale of the horizon would be smaller by a factor of more than 3 and would be $< 1°$. So, the first thing that we are probably going to learn from these very difficult experiments is whether or not there is a feature between, say,

1 and 2°.

If there is anything interesting on the 1–2° scale, that, in itself, is going to be quite good evidence for $\Omega = 1$, rather than 0.1, because there is no particular reason why any structure should appear between 1° and 2°, unless that is small on the horizon scale. Then, of course, the details of the structure will tell us the ratio of the tensor and scalar waves and more details about the galaxy formation process.

Question

(Lynden-Bell): Martin, I thought that you were very interestingly saying that the two pieces of the large-scale-structure spectrum did not really agree by this factor of 2 and you have said that neutrinos might help there. If one wants to make this more quantitative, would you like 30% of the dark matter in neutrinos, or what? And secondly, a related question: could you repeat for everybody why it is that dark matter must be in some sense normal stuff and not have $P = -\rho$, or anything funny like that?

Answer

(Rees): The second question first. Matter with $P = -\rho$ is essentially like the Λ term. A given ρ in the 'vacuum' gives acceleration. A Λ term helps to flatten the Universe, even if the density of the ordinary matter is low. The models for $\Lambda \neq 0$ do a lot of nice things. You get a better fit than the standard cold-dark-matter model and, of course, even if one were to believe in the high Hubble constant, which our chairman (A. Sandage) does not, one would then be a lot happier, because if Λ makes a 70–80% contribution to the critical density, the time since the Big Bang was a bit longer than the Hubble time, and the age can be 13 billion years for a 12-billion-year Hubble time. Some people might like a Λ term for that reason.

Question

(Lynden-Bell): Sorry, but you are saying that, while that may help us in cosmology generally, it does not help the dark-matter problem?

Answer

(Rees): Yes, that is correct. Going back to the dark-matter problem, the inclusion of neutrinos is clearly going to give relatively more amplitude on large scales. Neutrinos provide extra growth above super-cluster scales and not below. The acceptable models require only 20 or 30% in neutrinos. A bigger fraction of neutrinos upsets entirely the hierarchical build up of scale, and the first non-linear structures would be superclusters. If you put in 20 or 30%, it flattens the amplitude as a function of mass but does not make the slope change sign.

Perhaps I should say a bit more about cold-dark-matter models, because one so often reads in the papers that 'cold dark matter is dead', then it has a resurrection and then it dies again. It is important to emphasize that, when people talk about a cold-dark-matter model, they normally mean a package of five assumptions. Firstly, a Harrison–Zel'dovich spectrum. Secondly, that the Universe is dynamically dominated by a cold non-baryonic matter (the WIMPs). Thirdly, $\Omega = 1$. Fourthly, the galaxies are related to dark matter by a simple (and almost certainly oversimple) description of biasing, described by some simple factor b (that is certainly an oversimplification). Finally, that the neutrino mass is zero. This simple package does not seem to fit. So one has to drop at least one of these assumptions. For instance, if $\Omega = 0.2$, things work much better.

If the particle physicists told us that the neutrino mass were, say, 4–5 eV then, if you stuck in all the rest of the assumptions, you would get a model that would work rather better. So, when people say that the cold dark matter is in trouble, they mean that we cannot maintain this combined set of five assumptions. This is important, because there are people spending years at the bottom of mines looking for these 'WIMP' particles. One does not want to discourage them. Whether or not those particles exist in the halo is quite independent of all the details of large-scale structure.

Question
(Sandage): Earlier on you said that not much has happened in the observational aspect since the influential paper by Gott and company.

Answer
(Rees): No, I did not say that. I said that the net effect was not to change our values of Ω within clusters and galactic haloes.

Question

(Sandage): Let me say that, regardless of the value of the Hubble constant, which I think is solved, there is this second problem of the independent age-dating through the ages of the globular clusters and our Galaxy. There has been a great deal of work done, as you know, on that problem, and one of the great advances has been the understanding that the oxygen-to-iron ratio is not solar, and that work has been done here by Rebolo and company, and by Sneden and others, earlier on, in Texas. That has the effect of reducing the age of the globular clusters from 18 billion to 14 billion years, with the oxygen-enhanced models of Don van den Berg and Bergbush. So, the time-scale problem is not as advertised in *Nature* and in *The New York Times* a few months ago. I think that the standard model is alive and well as regards the time scale, regardless of what one reads in the newspapers.

Question

(R. Guzman): You have mentioned, in passing, the large streaming motions as a way to map the larger-scale distribution of matter. In order to measure these motions you use different distance indicators. In the last year, in the last few months, there has been growing evidence that some of these distance indicators may be affected by systematic effects in galaxy properties, and also there is some confusion about some of the different results obtained using different relations. Perhaps the best example is the Lauer and Postman result, in which the streaming motions change in amplitude, scale, and direction. So, the question is: what is your opinion on the reliability of these methods to pose constraints of the larger-scale distribution?

Answer

(Rees): Lynden-Bell, in his talk, expressed doubts about the particular methods that depend on distance indicators. Let me mention that there are three other methods, which do not depend on distance indicators, for probing the large-scale mass distribution. One is the method used by Carl Fisher, which looks for anisotropy in the redshift distribution, and which does not depend on distance indicators. The other method, which Donald mentioned, is analysing the implications of our motion through the microwave background in terms of the dipole distribution. A third method, which is going to become possible quite soon, is gravitational lensing. I have shown the dramatic effects of gravitational lensing in the core of a cluster. It is possible in principle, although it is not clear how easy it will be in practice, to look for the effects of lensing due to large-scale regions of less extreme overdensity. For instance, if there were a large flattened pancake of dark matter, then the images of all the faint galaxies behind it would be systematically stretched, by a few per cent. There are groups trying to see if this technique can be made to work. That, of course, is a way of directly determining a mass distribution, independent of any assumption about how the galaxies trace the mass. So, I think, echoing what Donald Lynden-Bell said, that methods depending on distance indicators are uncertain and lead to confusing results. But fortunately there are these other methods as well.

Question

(A. Aragon): I am a simple-minded observer and I like to look at things. You say that one of the important gaps in our knowledge is what happened between $z = 5$ and the recombination. Can you suggest any methods with which we can go and gather information by pointing our telescopes of whatever kind?

Answer

(Rees): Pointing telescopes at random has been fruitful in the past; Penzias and Wilson did that, and you could try it too! But there are two things. It may be that there are no structures and no heat input at epochs much earlier than $z \sim 5$. Douglas Scott and I proposed a test whereby radio telescopes might be able to detect 21-cm emission and absorption from gas even before reheating. The gas would not be at the same temperature as the microwave background, so there will be absorption if it is below, emission if it is above. This is a very tiny fraction of the total background; if the medium were smooth, you could not isolate it. But if the medium were already starting to clump (suppose there were 10 or 20% density enhancements in incipient clusters at $z \sim 10$), then, by looking at the background radiation at about 130 MHz (which is where the 21-cm wavelength could be at that redshift), and seeking structure, both in angular space and in frequency space, one might detect incipient clusters back then. This is something which the Indian radio telescope array (the GMRT) is probably the most suited telescope to do. It is a challenge, but it is not hopeless.

The other indirect way of learning about conditions at $z > 5$ is via the Lyα forest. Data from the Keck Telescope will soon tell us what the temperature of the photoionized gas has to be in the Lyα clouds. This is going to provide some constraints on the heating of the gas. Let me give you an example. Suppose it turns out that the gas is very cool: below $15\,000$ K, or something like that. This might imply that the gas was first photoionized at $z > 10$ and then there had been some adiabatic cooling between then and the epoch when we observe it. Otherwise, it would be rather hard to explain the low temperature.

I think the obviously optimistic line to take is that there may be a whole generation of pregalactic stars and small-scale structure. There must be something which ionized the intergalactic gas, but I would not expect that galaxy-scale objects exist much beyond $z = 5$. The quasars may be the 3σ peaks on galactic scales, but subgalactic structures, hard to detect individually, may be detectable through their collective effect: what adiabat they put the gas on and things like that.

Question

(Rebolo): I have a comment on the tensor and scalar modes for the microwave background fluctuations and the primordial density fluctuations. Taking in comparison the data from the Tenerife experiment at $5°$ and the *COBE* data at $10°$ resolution, and looking at the rms in both experiments (about 30 mK for *COBE* and about 45 mK at our $5°$ resolution), we can set a limit on the contribution of the tensor mode, which would be less important than the scalar one.

Answer

(Rees): What you are doing, I think, is to set an observational limit to the 'tilt' and then to use a theoretical model to relate the tilt to the scalar/tensor amplitude ratio. But what you really want to do is to test this relation by measuring them both. If you actually measure the tensor compared to the scalar, you can do that, but you have then got to go down to scales below 2°, because the tensor modes are cut off below the scale of the horizon. What you have already done is to place a new limit on the tilt, which is called n. I should mention that n is a number which is 1 for exactly scalar fluctuations, and cannot be very different from that, otherwise the fluctuation at an amplitude of 5° compared to large scales would be different. This assumes a theoretical result, which comes from a particular subset of inflation models. The relative importance of tensor modes increases as n decreases below one – according to some theories. If you can test this theory, by measurements below an angular scale of 2°, you would be doing real physics in that epoch, testing and perhaps refuting a class of models.

Question

(H. Zinnecker): I would like to say that I am very pleased to hear that you emphasize how important star formation is, in order to understand the formation and evolution of galaxies better. In particular, you emphasize the role of magnetic fields in star formation, the strength of magnetic fields. May I ask you, in which way you think the direction goes? If you have a stronger, or weaker, magnetic field, which you choose, would you form more or fewer stars of high mass?

Answer

(Rees): I have no idea, and I do not think anyone else could confidently claim to know the answer, even in the case of the effect of heavy elements. Conventional wisdom is that the absence of heavy elements pushes things towards high-mass stars. I think even that is not obvious. All I am prepared to say is that there is as much reason to think that magnetic fields have an effect as to believe that the heavy elements do, but I have no idea what the effect is. In either case it is very important. Whereas computer simulations are going to improve tremendously in understanding gravitational clustering (and even gas dynamics, on the scale of clusters of galaxies etc.), once the first star forms, we are back to parameter fitting, because we do not know what the initial-mass function is, nor the formation efficiency. To give an example, imagine that the first stars of the cold-dark-matter model were formed at $z = 5$-10, and suppose that they were all low-mass stars. Then most of the baryons in the cold-dark-matter model could have turned into low-mass stars, which is then a new kind of dark matter, and then you have to wait for the quasars to heat up the medium. On the other hand, if the first stars are high-mass stars, then they will heat up the medium and exert a negative feedback on forming more. Until we understand those processes, we are stuck with parameter fitting as soon as any stars are formed.

Question

(Zinnecker): Following on this comment, may I then suggest that we really have to rely on finding the regions (maybe within the Galaxy) where the magnetic field is stronger or weaker, or metallicity is higher or lower, in order to educate our guesses.

Answer

(Lynden-Bell): I just wanted to show my ignorance. It seems to me that an interesting subject has been broached. But do we know whether the luminosity function at birth has more low-mass stars or fewer low-mass stars, when the stars considered are of, say, much lower metal abundance than currently? This seems to me a topic which we ought to know the answer to. Does anybody here know it?

Comment

(Pagel): I will make a comment on that in my contribution!

Answer

(Zinnecker): May I comment on this? Your question hinges on what you believe about the recent observations that have been made on faint stars in globular clusters. There have been ground-based observations and *HST* observations, and the two results are controversial. The *HST* observations tend to show that there is not such an enhancement of low-mass stars in the globular clusters, while the ground-based observations tend to imply that there is a steep slope of low-mass stars. How you can reconcile that with mass-to-light ratios, I do not know. That is an additional constraint. Then people take the way out, that we measure the mass function, the luminosity function, half way out from the core of the globular cluster. All this is a difficult issue, but I stick out my neck and say there is a chance that the mass function is enhanced in low-mass stars for population II, contrary to intuition.

Question

(Sandage): Can the low-mass stars escape from the globular clusters in the time available? Donald?

Answer

(Lynden-Bell): If they are in the middle of the cluster, yes, they could.

9 Cosmic abundances

Bernard E. J. Pagel

9.1 Introduction

In the 1930s and 1940s, the popular wisdom was that cosmic abundances were the same almost everywhere, and this encouraged theories of universal element creation. In the early 1950s, Chamberlain and Aller (1951) discovered extreme metal deficiency in two classical 'subdwarfs' (one of them actually a subgiant) and Merrill (1952) discovered technetium in an AGB red-giant, which encouraged theories of element creation in stars followed by their injection into the interstellar medium, an idea originated by Hoyle (1946) and embodied in the classic works of Burbidge *et al.* (1957, hereafter B^2FH) and Cameron (1957), and this suggested in turn the idea that the chemical evolution of galaxies is important and worthy of study. In the 1960s, especially after the discovery of the microwave background, it became widely recognized that primordial nucleosynthesis is important after all, particularly for deuterium and helium, and the present picture developed on the basis of a combination of primordial synthesis of a significant portion of light elements in the hot Big Bang, followed by subsequent galactic chemical evolution, driven by nucleosynthesis in stars.

9.2 Primordial abundances

Figure 9.1, based on recent discussions by Walker *et al.* (1991) and Smith, Kawano, and Malaney (1993), shows theoretical primordial abundances as functions of the density parameter $\eta \equiv n_B/n_\gamma$ or $\Omega_{B0}h_{100}^2$ and some limits based on astrophysical and cosmochemical observations. Tall, continuous

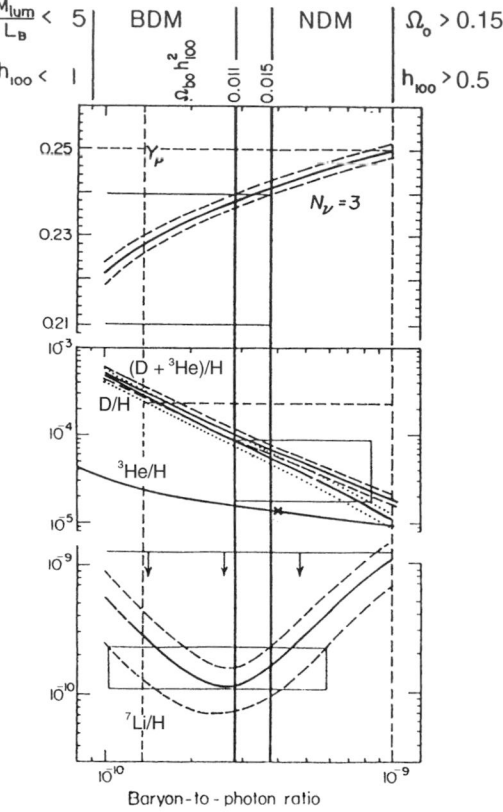

Fig 9.1. Primordial abundances as functions of baryonic density.

vertical lines indicate the usually adopted region of concordance (which I may call the 'optimistic' estimates) and the spaces marked 'BDM' and 'NDM' at the top of the figure indicate the resulting limits on the amounts of baryonic and non-baryonic dark matter respectively; the upper limit on η comes from measurements of the helium abundance in H II regions (e.g. Pagel *et al.* 1992; Olive and Steigman 1995) giving a primordial mass fraction $Y_P \leq 0.242$ with 95% confidence and the lower limit from the well-known arguments from D + ^{3}He first given by Yang *et al.* (1984) which give an upper limit to the primordial deuterium abundance. The small cross indicates a recent estimate of the primordial ^{3}He abundance (Balser *et al.* 1994). The resulting limits agree well with observational results on primordial ^{7}Li, assuming that the observed abundance in the most metal-deficient stars – the 'plateau' discovered by Spite and Spite (1982) – is indeed primordial and not significantly reduced by destruction in the stellar outer layers. They also leave substantial requirements for both baryonic and non-baryonic dark matter.

Various developments in the past year or so have introduced complications into this nice picture.

9.2.1 D + ^{3}He

The lower limit D/H $\geq 1.5 \times 10^{-5}$ (see Boesgaard and Steigman 1985) is nicely confirmed by recent observations of the nearby interstellar medium (Linsky *et al.* 1993) and a rediscussion of the older *Copernicus* data (McCullogh 1992) with the protosolar value marginally higher at $(2.6 \pm 1.0) \times 10^{-5}$ (Geiss 1993). We expect the primordial abundance to exceed these by some destruction factor due to astration, but that factor is model dependent. However, the result of D destruction is ^{3}He, some of which survives stellar processing and may indeed be added to by fresh stellar production, for which there is now direct evidence from a planetary nebula (Rood, Bania, and Wilson 1992), and careful discussions by Steigman and Tosi (1992) and Vangioni-Flam, Olive, and Prantzos (1994) give essentially the same result as Yang *et al.* (1984), limiting primordial D/H to about 8×10^{-5}, or less if there has been fresh synthesis of ^{3}He, which indeed there should have been, both on theoretical grounds and in the light of the planetary-nebula observation. The difficulty is that the improved ^{3}He data (which are, however, based on homogeneous HII-region models) of Balser *et al.* (1994) reveal substantial scatter, without any clear dependence on distance from the centre of the Galaxy, and mostly rather low values (except in smaller regions where the authors suspect local pollution by stellar winds), so the interplay between ^{3}He production and destruction is not well understood; Olive *et al.* (1995) have argued that observational selection of regions in which ^{3}He is locally destroyed by recycling in massive stars distorts the picture, but it could nevertheless be dangerous to rely too heavily on the D + ^{3}He argument.

Such doubts were initially reinforced by the dramatic observation using the Keck Telescope of what could be a strong deuterium feature in a $z = 3.3$ absorption-line system in front of Q0014+813 by Songaila *et al.*

(1994), who estimated a D/H ratio of 3×10^{-4}, four times higher than the upper limit discussed above. The lower limit on density in Fig. 9.1 would then be shifted to the 'pessimistic' vertical broken line on the left, leaving little or no requirement for baryonic dark matter. They pointed out, however, as did Carswell *et al.* (1994), who made a similar observation of the same object with less powerful equipment, that there is always the possibility of a rogue hydrogen cloud with a relative velocity of $-80\,\mathrm{km\ s^{-1}}$ simulating the isotope shift, so that such observations can only give an upper limit. Recently David Tytler has reported his observations of another high-redshift system with an unexpectedly low D/H ratio, not appreciably larger than in the present-day interstellar medium. I have not seen the details of this work, but if that result is taken at face value, we move way over to the right of the picture, with a density approaching the other 'pessimistic' vertical broken line in the diagram, and the evidence for non-baryonic dark matter now depends on believing $\Omega \simeq 1$. A very modest degree of destruction of deuterium is not entirely unexpected in the context of Galactic chemical-evolution models (see Edmunds 1994), but such a result certainly raises questions about primordial helium and lithium.

9.2.2 Helium

The most favoured approach is the one pioneered by Peimbert and Torres-Peimbert (1974, 1976) using observations of extragalactic $H\,II$ regions, especially in dwarf irregular and 'blue compact' or '$H\,II$' galaxies with low total luminosity and associated low metallicity. One plots a regression of helium against oxygen (and sometimes also nitrogen) abundance and in the range where this is linear extrapolates to zero oxygen (or nitrogen) to get the primordial mass fraction Y_P, while the slope gives dY/dZ, which is an interesting quantity in relation to stellar nucleosynthesis.

With Roberto Terlevich, Mike Edmunds, and others I have been involved in trying to improve the data for such analyses, and this work is being extended by Evan Skillman and Elena and Roberto Terlevich and others. Figure 9.2 shows regressions obtained in 1992 with some updatings; in this work we excluded well-defined Wolf–Rayet galaxies, suspecting that they might be affected by local enhancements of helium and nitrogen by winds from the embedded stars (Pagel, Terlevich, and Melnick 1986). We obtained basically the same results as the Peimberts and Lequeux *et al.* (1979),

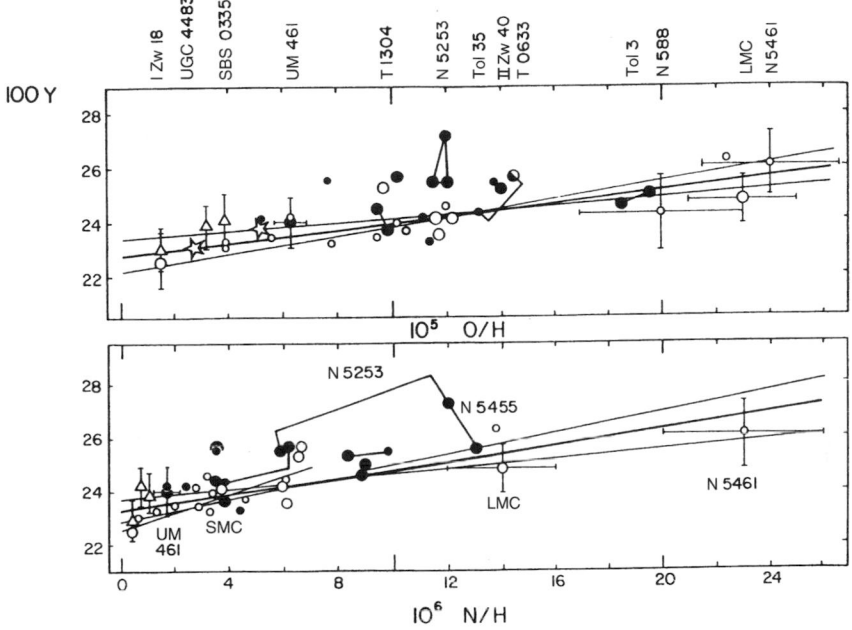

Fig 9.2. Regressions of helium against oxygen and nitrogen in H II galaxies (Pagel *et al.* 1992). Filled circles show objects with definite WR features; unfilled circles, others. Maximum-likelihood regression lines are shown with equivalent $\pm 1\sigma$ limits; the short line in the lower panel is the regression finally preferred for nitrogen. Triangles show later results from Skillman and Kennicutt (1993), Skillman *et al.* (1994), and Izotov *et al.* (1994); 'stars' are from a re-analysis of observations by Melnick *et al.* (1992) of SBS 0355–052.

except for a somewhat larger value of 4 ± 1 for dY/dZ. This has given rise to a number of speculations regarding either a low upper limit to the initial masses of stars undergoing supernova explosions and providing oxygen, as opposed to going into black holes after emitting mainly He, C, and N in stellar winds (Maeder 1992, 1993; Brown and Bethe 1994), or the preferential loss of oxygen-rich material from dwarf galaxies through metal-enhanced winds (Pilyugin 1993; Marconi *et al.* 1994). However, in our work we may have somewhat underestimated the oxygen abundances by assuming the absence of electron temperature fluctuations, which now seems unjustified (González Delgado *et al.* 1995). The old value of 3 for dY/dZ is compatible with a steep initial-mass function like that of Scalo (1986) with an upper mass limit of $50 M_{\odot}$ for stars undergoing outbursts.

The serious question raised by Tytler's new deuterium data is whether our upper limit of 0.242 for Y_P is safe. Olive and Steigman (1995) have discussed a variety of data from the work mentioned above, including the galaxies rejected by us, and find a similar result; Izotov, Thuan, and Lipovetsky (1994) derive various values using different assumptions about relevant atomic data; their most plausible combination gives a preferred value of 0.239 ± 0.007 compared to Olive and Steigman's value of 0.232 ± 0.003; the latter is more accurate statistically, but the question of systematic errors needs to be addressed. Sasselov and Goldwirth (1995) argue that we have systematically underestimated helium abundances for various reasons, the main one being that radiative-transfer effects cause departures from case B recombination. The validity of this criticism is difficult to judge, because they give few details, but I am somewhat doubtful because $\lambda 5016$, which is very sensitive to case A or B, is always found to agree with case B, whereas $\lambda 6678$, which carries much of the weight in our work, is quite insensitive. On the other hand, there could be a problem with the effective recombination coefficients. The coefficients calculated by Brocklehurst (1972), which we have used, seem to be confirmed for the cases of the most important lines $\lambda\lambda 4471, 5876, 6678$ by the recent results of Smits (1996), which update his earlier results (Smits 1991a, b), but it has been pointed out to me by David Hummer that there could still be problems owing to breakdown of LS coupling at high nl levels, which could affect the subsequent downward cascade, and he has promised to do some work on this. Pending the results of this work, it seems that one may have to reckon with a larger systematic error in Y_P than the value of 0.005 that we assumed in 1992, and so the primordial helium abundance could be higher than 0.24, perhaps even as much as 0.25, consistent with undepleted deuterium in the interstellar medium. Such a large value, on the other hand, raises questions about primordial ^{7}Li, which will be discussed next.

9.2.3 Lithium 7

Over the years, there have been arguments over whether the Spite plateau represents the true primordial ^{7}Li abundance, or whether there has been substantial destruction in the stellar atmospheres. Pinsonneault, Deliyannis, and Demarque (1992) have presented a variety of models: basically 'standard' models without rotation (which, of course, do not fit the Sun) predict only modest amounts of depletion at the higher effective temperatures, whereas models with mixing induced by rotation can lead to destruction by an order of magnitude while still preserving the appearance of a plateau. The assumption of little or no depletion leads to the 'optimistic' fit shown in Fig. 9.1 and to reasonable pictures of the Galactic chemical evolution of lithium in which the primordial abundance is supplemented by ^{7}Li made in stars and the abundance grows at a comparable rate to that of iron (D'Antona and Matteucci 1991; Pagel 1991), but the only independent piece of evidence against significant depletion is the detection of ^{6}Li in the warm star HD 84937 which is close to the main-sequence turn-off (Smith, Lambert, and Nissen 1993).

Recently Thorburn (1994) and Norris, Ryan, and Stringfellow (1994) have investigated the abundance of lithium in low-metallicity stars as a function of effective temperature and metallicity, going down to [Fe/H] $\simeq -4$. Both find a modest increase in Li/H with effective temperature (about 0.15 dex from 5600 to 6400 K), which they attribute to modest destruction according to the standard (non-rotating) models of Pinsonneault et al. They then correct the measured abundances to a standard effective temperature of 6400 and 6200 K respectively and plot the corrected Li/H against [Fe/H] with somewhat surprising results in each case. Thorburn finds a primordial value of Li/H $\simeq 1.5 \times 10^{-10}$ and a remarkably steep increase with metallicity up to [Fe/H] $\simeq -2$, which can best be expressed as

$$\Delta(\mathrm{Li/Li}_\odot)/\Delta(\mathrm{Fe/Fe}_\odot) \simeq 10, \tag{9.1}$$

and favours an explanation on the basis of a high flux of cosmic rays that may also help to explain beryllium production in the early Galaxy. Norris et al., on the other hand, find no systematic increase up to [Fe/H] $= -2$ but a value of Li/H slightly below 1.0×10^{-10} for their most metal-poor star and raise the intriguing question whether lowering the metallicity down to and below [Fe/H] $= -4$ is taking them closer to the Big Bang or further away from it! However, all these trends are somewhat sensitive to the effective temperatures and the model atmospheres used in the abundance analysis; Molaro, Primas, and Bonifacio (1995) using revised atmospheric parameters and models find no evidence at all for a trend with either effective temperature above 5800 K or metallicity for $-3.5 \leq$ [Fe/H] ≤ -1.5; according to them, the Spite plateau is back in full force, which imposes certain constraints on past cosmic-ray activity (see Steigman and Walker 1992; Prantzos, Cassé, and Vangioni-Flam 1993). The upshot of all this is that I believe that there is no good reason to doubt the limits on primordial lithium shown in Fig. 9.1, which in turn implies $\eta \leq 6 \times 10^{-10}$, $(D/H)_P \geq 2 \times 10^{-5}$, and $Y_P \leq 0.246$, thus shifting the 'optimistic' upper limit on density half-way, or less, towards the 'pessimistic' broken line on the right of Fig. 9.1. This implies a reduction of up to about 10% in the adopted effective recombination coefficients for helium if no other systematic errors are present.

9.3 The origin of light elements

The discovery of beryllium (Gilmore, Edvardsson, and Nissen 1991; Ryan *et al.* 1992; Gilmore *et al.* 1992) and boron (Duncan, Lambert, and Lemke 1992) in metal-deficient stars has raised a number of issues. Beryllium was first suggested as a product of primordial nucleosynthesis in an inhomogeneous Big Bang (Kajino and Boyd 1990), but the theoretical basis for expecting detectable amounts of ^{9}Be from this source is at best rather dubious (Terasawa and Sato 1990; Thomas *et al.* 1994) and there is no evidence (so far, at least) for the existence of a beryllium plateau analogous to the lithium plateau. Furthermore, the ratio of boron to beryllium in the most metal-deficient star for which there are data (HD 140283 with [Fe/H] $\simeq -2.7$, [O/H] $\simeq -2.0$) is about 20, consistent with the conventional production mechanism by cosmic-ray spallation.

However, as can be seen in Fig. 9.3, beryllium is roughly proportional to iron or oxygen, and this is surprising because its production by conventional cosmic-ray spallation demands the presence of CNO atoms in the interstellar medium and, naively, one would expect it to behave as a 'secondary' nucleosynthesis product whose abundance (other things being equal) should increase as the square of that of its primary progenitors (Vangioni-Flam *et al.* 1990). This has led to several suggestions as to how such efficient beryllium production could come about, including an intensified cosmic-ray flux (Ryan *et al.* 1992), which leads to difficulties in excess lithium production by α–α fusion (Steigman and Walker 1992), an enhanced cosmic-ray flux at high energies (Prantzos, Cassé, and Vangioni-

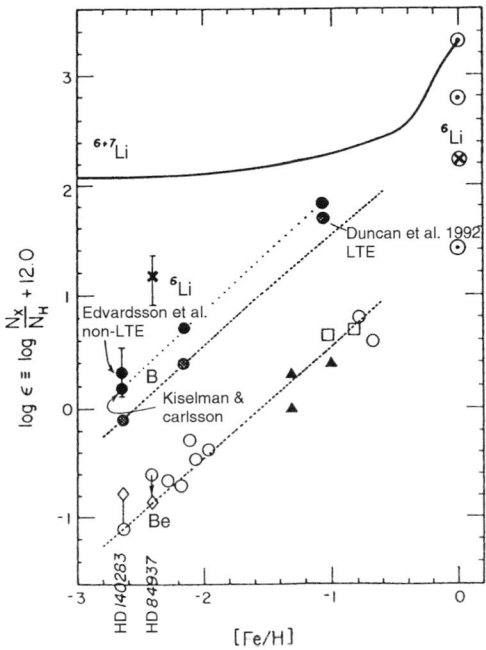

Fig 9.3. Abundances of lithium, boron, and beryllium in metal-deficient stars and the Sun vs their iron abundance, adapted from a summary diagram by P. E. Nissen. References in the figure are to Duncan, Lambert, and Lemke (1992), Edvardsson *et al.* (1994), and Kiselman and Carlsson (1995). Beryllium data are from Gilmore *et al.* (1992) and Boesgaard and King (1993).

Flam 1993), inhomogeneous galactic production (Feltzing and Gustafsson 1994; Pagel 1994a), or a major contribution from high-energy CNO nuclei in the cosmic rays themselves (Duncan, Lambert, and Lemke 1992). Boron (but apparently not beryllium) can also be created by the neutrino process in supernovae (Woosley *et al.* 1990), which would nicely account for the 'primary' behaviour of boron, but not for its constant abundance relative to beryllium! It is also interesting to see in Fig. 9.3 that the ^{6}Li discovered in HD 84937 by Smith, Lambert, and Nissen (1993) (in the figure the measured abundance has been increased by about 0.2 dex to allow for minimal depletion) is about an order of magnitude more abundant relative to iron than in the Solar System, indicating a strong role for cosmic rays in the early Galaxy.

In the later Galaxy also, cosmic rays may be a more significant source of ^{7}Li and ^{11}B than one used to suppose. This thought results from the recent observations of varying and large ^{6}Li/^{7}Li ratios in nearby interstellar clouds (Lemoine *et al.* 1993; Meyer, Hawkins, and Wright 1993), in one case being as high as 1/2, compared to 1/12 in the Solar System (Lemoine 1995)! An increase in ^{6}Li/^{7}Li since formation of the Solar System is not easy to understand if most ^{7}Li comes from stellar production and ^{6}Li from cosmic-ray spallation and fusion (see Steigman 1993). Meneguzzi and Reeves (1975, M&R) suggested long ago that the large ratio of ^{11}B to ^{10}B might be due to a low-energy component of cosmic rays not detectable in the Solar System, and the existence of such a component in the Orion nebula has now been dramatically confirmed by nuclear γ-ray observations (Bloemen *et al.* 1994). These support both M&R's idea (as Hubert Reeves will discuss in the course of Chapter 10) and the view that cosmic-ray production of light elements in the early Galaxy was indeed affected by inhomogeneities, possibly taking place in near-supernova environments with a high CNO abundance, as suggested by Feltzing and Gustafsson (1994). Beckman and Casuso (1995) have put forward a model in which both lithium isotopes are produced by cosmic rays, and Cassé, Lehoucq, and Vangioni-Flam (1995) explain light-element abundances in both the present and the early Galaxy on the basis of spallation of high-energy CNO nuclei from supernovae (as shown by the γ-ray observations) in the surrounding H II region, which nicely explains the 'primary' behaviour of beryllium and boron.

9.4 Chemical evolution in the Galactic disk

9.4.1 Introductory comments

The theory of galactic chemical evolution (GCE) attempts to combine ideas on the end products of stellar evolution with ideas on the formation and evolution of galaxies in order to understand the distribution of chemical elements in the interstellar medium and stellar populations. As this involves many uncertainties, I prefer to adopt the approach pioneered many years ago by Maarten Schmidt, Leonard Searle and Wallace Sargent, Beatrice Tinsley, Richard Larson, Donald Lynden-Bell, Don Clayton, and others, which is to parameterize the problem as simply as possible and to treat it analytically. This is somewhat unfashionable nowadays, but it has the advantage that one can immediately see which parameters are the important ones and where the uncertainties lie, which is not always the case in elaborate numerical models. The approach has been described in somewhat more detail by Pagel (1994b).

The main ingredients of GCE models are the following:

- Initial conditions, notably abundances. Primordial abundances are to be preferred, but some models appeal to prior enrichment by massive stars or by a previous phase (e.g. prior enrichment of the Galactic disk by the bulge).
- End-products of stellar evolution. For our purpose, stars can basically be divided into four classes, three single and one of close binaries (SNIa). Additional classes like novae may be important for some rare isotopes.

1 Massive stars $M \gtrsim 10 M_\odot$ eject fresh He, C, N, O, and maybe some s-process elements in winds and elements from oxygen to the iron group plus maybe r-process elements in supernova outbursts following core collapse (SNII and related classes) leaving a neutron-star remnant or maybe a black hole for $M \gtrsim 50 M_\odot$. Detailed nucleosynthesis calculations have been made for these stars especially, by Woosley and colleagues, e.g. Woosley, Langer, and Weaver (1993), and by Thielemann, Nomoto, and Hashimoto (1996). Maeder (1992, 1993) has discussed the effects of metallicity-dependent mass loss on the yields of He, C, O, and total Z.

2 Intermediate-mass stars (IMSs) of ~ 1 to $10 M_\odot$ eject fresh He, C, N, and s-process elements while evolving up the asymptotic giant branch and especially in the subsequent planetary-nebula phase leaving a white-dwarf remnant. Their yields of He, C, and N have been calculated under a variety of assumptions by Renzini and Voli (1981), and s-process production has been discussed by Käppeler et al. (1990). Both massive stars and IMSs recycle diffuse material destroying deuterium and other light elements.

3 Low-mass stars $\lesssim 1 M_\odot$ contribute little to nucleosynthesis or recycling and mainly serve to lock up diffuse material.

4 In some close binaries consisting of a white dwarf and a companion from which it accretes matter at a suitable rate, the

white dwarf finally exceeds the Chandrasekhar limiting mass and explosively ignites carbon and oxygen near the centre, leading to explosive synthesis, with an α-rich freeze-out expelling freshly produced iron-peak elements and some Si, Ca, etc. leaving no remnant (SNIa). The favoured model for them is the carbon deflagration model W7 (Nomoto, Thielemann, and Yokoi 1984; Thielemann, Nomoto, and Yokoi 1986). SNIa are believed to have supplied at least half of the iron in the Solar System, and they evolve with a significant time-delay ~ 1 Gyr, similar to a single IMS of around $2M_\odot$.

- The initial-mass function (IMF) (e.g. Scalo 1986) and its constancy or otherwise. There have been many arguments about this, but for analytical treatments it can be parameterized in terms of just a few moments, namely the lock-up fraction $\alpha \equiv (1 - R)$, where R is the return fraction, and the yield p for a given element, which is defined as the mass of element freshly synthesized and ejected from a generation of stars relative to the mass locked up in low-mass stars and compact remnants (Searle and Sargent 1972). Typical estimates for these moments are $\alpha \simeq 0.7$ and $p \simeq Z_\odot$, but α might easily be 0.1 more or less and p could be different by a factor of two, with a dependence on time and/or metallicity even with an unvarying IMF. In the work to be described, referring to 'primary' elements for which the yield is not known to depend explicitly on metallicity (although it might vary implicitly because of mass loss or for other reasons), I shall just distinguish two constant yields, p_1 for products released with negligible time delay and p_2 for elements released after what is assumed to be a fixed time delay Δ due to production by SNIas or low-mass SNIIs, the latter being apparently a good model for the r process (Matthews, Bazan, and Cowan 1992). The yields used will be 'effective yields' deduced *ad hoc* by fitting a Galactic model to stellar abundances, as opposed to the 'true yield', which depends on both details of stellar evolution and the IMF.
- Star-formation rates or laws. Long ago Schmidt (1959) parameterized the star-formation rate as a power law in volume or surface density of gas, and many authors have tried to find physical models for both the power and the coefficient. None of these are very convincing, and Kennicutt, Tamblyn, and Congdon (1994) have recently pointed out that the systematics of star-formation rates in different galaxies do not support any of them (although their data have still more recently been given another

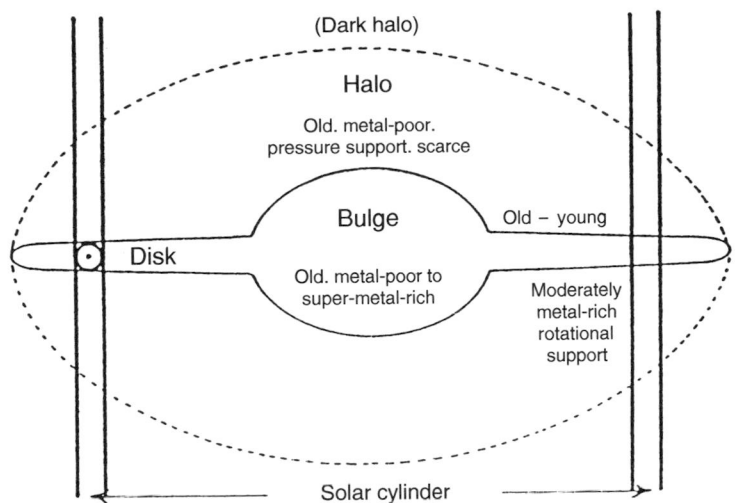

Fig 9.4. Schematic cross-sectional view of the Galaxy.

interpretation in terms of age differences by Sommer-Larsen 1996). Consequently, I like to use a simple linear law corresponding to Schmidt's law, with an exponent of 1 and with a coefficient that is treated as an empirical parameter that is constant in time (on average) but may vary from place to place.

- The galactic context. This concerns the nature of the region to be considered, its environment and history, e.g. whether it has evolved in a closed box or has been subjected to inflows and/or outflows as in dynamical-collapse models. Figure 9.4 shows a schematic view, including the region of interest, which, following Tinsley (1980), I shall treat as a cylinder through the Sun, perpendicular to the Galactic plane, although Gilmore has recently pointed out that a 'solar sausage' would be a better picture. The relationship between different stellar populations is an important input into GCE models.

9.4.2 Instantaneous-recycling and delayed-production approximations

Analytical treatments have traditionally used the instantaneous-recycling approximation, and been roundly criticized for doing so, basically because it gives a poor picture of recycling from IMS and low-mass stars. However, as long as the gas fraction is not too small, it does give a fair account of typical SNII products like oxygen and magnesium, and for metal-deficient stars born too early to have been affected by SNIa it works for the iron group as well. Where it breaks down is, obviously, when SNIa do contribute and thus lead to time and hence metallicity dependences of primary-element ratios like α/Fe or O/Fe, which have been known from observation for a very long time (see Aller and Greenstein 1960; Wallerstein 1962; Gasson and Pagel 1966; Conti *et al.* 1967). Pagel (1989a) introduced a 'delayed-production approximation' in which the star-formation rate at time t in chemical evolution equations is simply replaced by the star-formation rate at time $t - \Delta$, i.e. the current death rate, and the outcome then depends on p_1, p_2, and the dimensionless product $\omega\Delta$, where ω is just the transition probability per unit time for gas to be locked up in stars, and is of order $(3\,\text{Gyrs})^{-1}$.

9.4.3 'Simple' and inflow models

The 'simple' model assumes a closed box, well mixed, with constant yields, starting from pure gas with primordial composition. This model has long been known to be in conflict with the distribution function of abundances of long-lived stars in the solar neighbourhood, the classic 'G-dwarf problem' (van den Bergh 1962; Schmidt 1963; Pagel and Patchett 1975), which is illustrated in Fig. 9.5. Solutions proposed include the following:

(i) No problem because the older stars have migrated away (Grenon 1989, 1990).
(ii) Larger yields at low metallicities (Schmidt 1963; Maeder 1992).
(iii) Prior enrichment from the halo or preferably the bulge (Köppen and Arimoto 1990).
(iv) Inflow associated with a gradual formation of the disk, involving accretion of unprocessed (or less processed) material. This is the most interesting solution because it can be related to models of formation of the Galaxy by dynamical collapse and it readily accommodates the small tail of metal-weak stars that does exist, representing both the halo and the metal-weak extension of the thick disk. Analytical formulations have been given by Lynden-Bell (1975), Clayton (1985), and others; I have found Clayton's formalism particularly convenient.

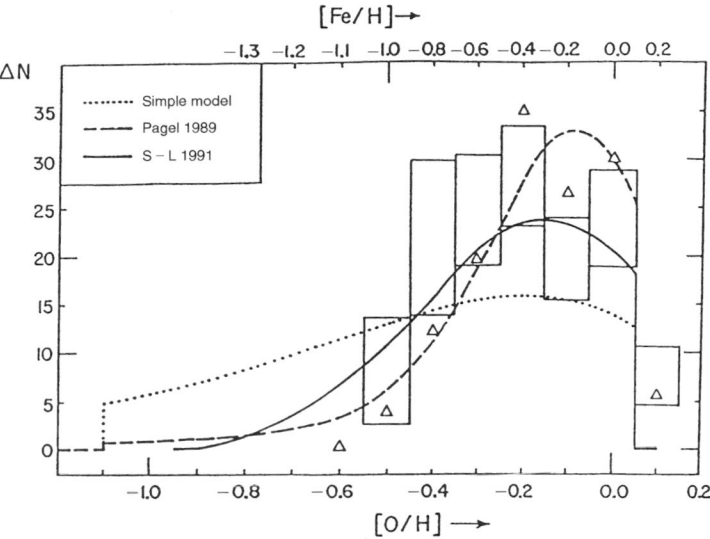

Fig 9.5. Differential distribution function of oxygen abundances in 132 G dwarfs in the solar neighbourhood. Triangles show the data points after Pagel (1989b) and boxes show upper and lower limits based on a more careful discussion of scale heights at different metallicities by Sommer-Larsen (1991a). The dotted curve shows the prediction from a 'simple' model with some initial enrichment, the broken-line curve shows a Clayton-type model used by Pagel (1989a), and the solid curve is from an inflow model by Sommer-Larsen (1991b).

9.4.4 Recent developments

So far, mainstream GCE models for the solar neighbourhood have used the idea of inflow in some form, inspired by notions of the formation of the Galaxy by dynamical collapse which go back to the classic works of Eggen, Lynden-Bell, and Sandage (1962, ELS) and Larson in the 1970s. Since then, various complications have emerged, e.g. identification of the 'thick disk' (Gilmore and Reid 1983) which was missed in the proper-motion catalogues used by ELS (although long known to Bengt Strömgren as 'intermediate population II') and the suggestion by Searle and Zinn (1978) that some, or all, of the halo comes from mergers of dwarf galaxies or small fragments accumulated over a long period of time. The halo and the disk (including the thick disk) have distinct dynamical properties, as is revealed in plots of rotation velocity against metallicity (Carney, Latham, and Laird 1990; Nissen and Schuster 1991), but overlap in either coordinate taken separately. The distributions of specific angular momentum in the halo and bulge are similar to each other, but very different from that of the thin or thick disk (Wyse and Gilmore 1992). Thus, the old idea of a gradual collapse leading through the halo to a subsequent disk phase no longer seems plausible; rather the halo and the disk seem to have gone their separate ways. At the same time, more extensive surveys of proper-motion stars (Sandage and Fouts 1987; Carney, Latham, and Laird 1990; Nissen and Schuster 1991) and kinematically unbiased objective-prism surveys (Norris, Bessell, and Pickles 1985; Morrison, Flynn, and Freeman 1990; Beers and Sommer-Larsen 1995) have revealed an ever-increasing degree of overlap in metallicity between the thick disk and the halo. In particular, Beers and Sommer-Larsen have found that about 30% of metal-deficient stars within 1 kpc of the Galactic plane having $-\infty < $ [Fe/H] ≤ -1.5, as well as essentially all stars with $-1.0 \leq$ [Fe/H] ≤ -0.5, belong to the thick disk dynamically. Thus, the evolution of the solar neighbourhood seems to have begun from primordial abundances, without any prior contribution from halo gas, and a pure disk model should therefore cover all metallicities, including those previously ascribed to the halo; unfortunately it is not known yet whether there are actually chemical differences between disk and halo stars with the same metallicity!

Another development that encourages a new look at the chemical evolution of the Galaxy is the monumental study of abundances in disk stars by Edvardsson *et al.* (1993), in which some very clear patterns emerge for the behaviour of various element-to-iron ratios as a function of [Fe/H] for [Fe/H] ≥ -1 or so, clearer than from the data that were available to Wheeler, Sneden, and Truran (1989) in their classic discussion. The patterns can be extended to lower metallicities (halo and metal-weak thick disk) using a variety of data from the literature which display rather more scatter, presumably owing to some combination of real effects with larger uncertainties in the determinations. In an investigation, recently completed, in collaboration with Grazina Tautvaišenė from Vilnius, Lithuania, we have concentrated our attention on oxygen, α-particle elements, iron, and r-process elements, which we believe can be reasonably understood in terms of a model with constant yields and time delays (Pagel and

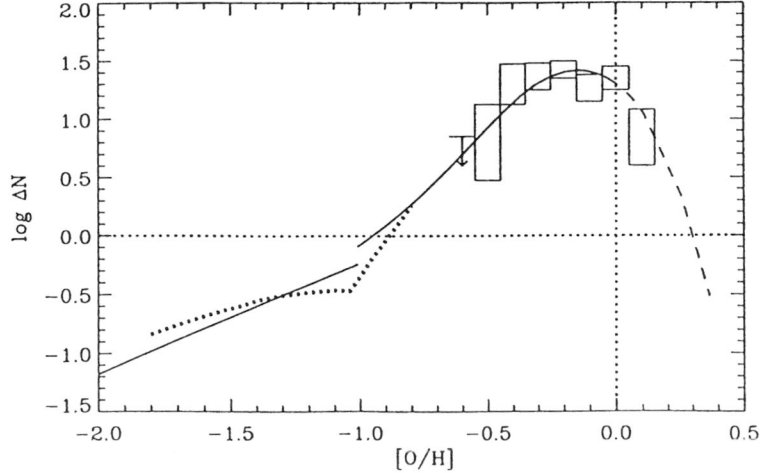

Fig 9.6. Oxygen ADF for 132 G dwarfs in the solar cylinder. Boxes show
observed numbers with error limits after Sommer-Larsen's (1991a) rediscussion
of data derived originally by Pagel and Patchett (1975), the same as in Fig. 9.5,
and the dotted curves on the left of the diagram show the extension to the
metal weak thick disk after Beers and Sommer-Larsen (1995). The solid curve
shows the ADF from our model assuming a present-day gas fraction of 0.11; the
broken curve on the right shows a hypothetical extension to lower gas
fractions.

Tautvaišenė 1995). For elements affected by the s process, the situation is
more complicated and we are still thinking about it.

The distinctive features of our model are designed to fit the abundance
distribution function (ADF) for oxygen (computed in instantaneous recycling) shown in Fig. 9.6. The main body of G dwarfs is fitted by a model of
Clayton's 'standard' type with his k parameter equal to 3 and with a final
mass equal to seven times the initial mass. That initial mass, however, is
allowed to evolve according to the 'simple' model up to $[O/H] = -1.0$ so as
to fit Beers and Sommer-Larsen's low-metallicity tail. The star-formation
rate is assumed to be proportional to the mass of gas, with a coefficient
$\omega = 0.3\,\mathrm{Gyr}^{-1}$ in the solar neighbourhood (with an age of 15 Gyr) and
$\omega = 0.45\,\mathrm{Gyr}^{-1}$ in the inner Galactic disk (with an age of 16.5 Gyr), leading
to the age–metallicity relations for disk stars shown in Fig. 9.7. The overall
fit is good, but the scatter is real and large and requires a separate discussion.

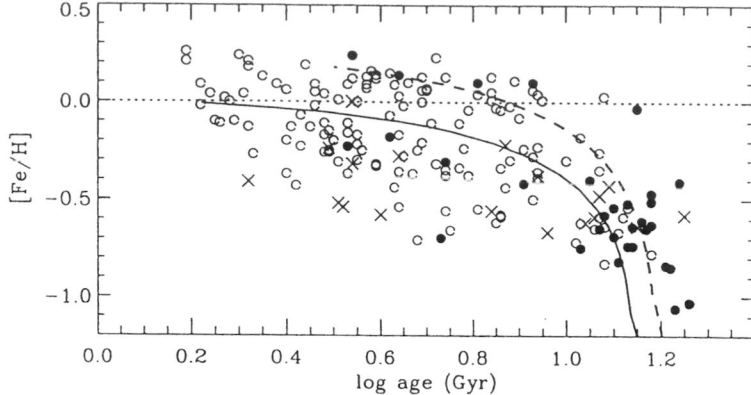

Fig 9.7. Age–metallicity relations from our model, with the solid curve for the solar neighbourhood and broken curve for the inner Galactic disk. Crosses and unfilled and filled circles show corresponding data from Edvardsson *et al.* (1993) for stars with mean Galactocentric distances $\geq 9\,\mathrm{kpc}$, 7–$9\,\mathrm{kpc}$, and $\leq 7\,\mathrm{kpc}$ respectively.

Figure 9.8 shows the same two age–metallicity relations from the model, plotted against relative zinc abundances measured in high-redshift absorption-line systems by Pettini *et al.* (1994). Here the lengths of the horizontal lines representing each data point represent the possible range in age of the Universe (in units of its present age t_0) between an open model and an Einstein–de Sitter model, and the horizontal placement of the curves is somewhat arbitrary. Some of the objects are readily fitted by this model, but it is also clear that the evolution of quasars, on the one hand, and of dwarf galaxies, on the other, has been on totally different time scales. One might guess that the damped Lyα systems that appear here are a mixture of protospirals destined to become like our own Galaxy and dwarf systems.

Our fits to element-to-element ratios are based on *ad hoc* yields (p_1 prompt and p_2 with a time delay Δ) given in Table 9.1.

The values of $\omega\Delta$ for the elements from Ca to Fe represent an actual time delay Δ of 1.3 Gyr for SNIa with ω values of 0.3 and 0.45 Gyr^{-1} in the solar neighbourhood and in the inner Galactic disk respectively, while the $\omega\Delta$ value for Eu represents a time delay Δ of 27 Myr for the low-mass type-II (or related) supernovae that seem to be the best candidates for the r pro-

Table 9.1. *Yields and time delays.*

El.	$p_1/Z_\odot$	$p_2/Z_\odot$	$\omega\Delta$ $R_m \geq 7\,kpc$	$\omega\Delta$ $R_m < 7\,kpc$
O	0.70	0.00	—	—
Mg	0.88	0.00	—	—
Si	0.70	0.12	0.4	0.6
Ca	0.56	0.18	0.4	0.6
Ti	0.70	0.12	0.4	0.6
Fe	0.28	0.42	0.4	0.6
Eu	0.08	0.66	0.008	—

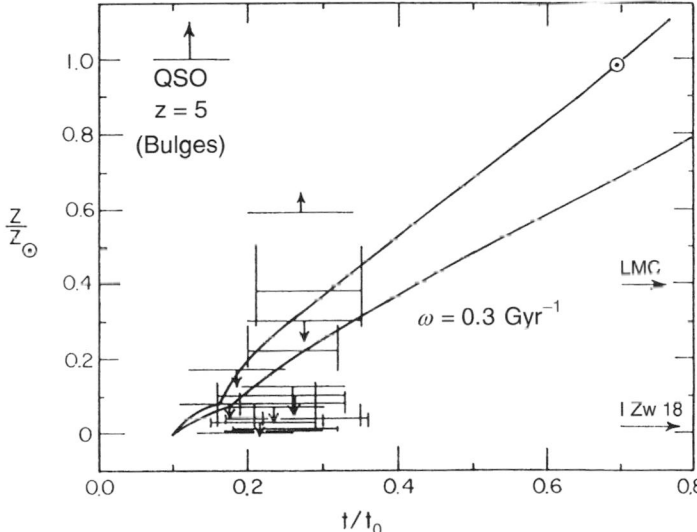

Fig 9.8. Age–metallicity relations from the model (as in Fig. 9.7) compared with abundances in high-redshift absorption-line systems and other typical objects.

cess (Mathews, Bazan, and Cowan 1992). However, these Δ values must be regarded as quite rough, because of uncertainty about ω and its constancy.

Figures 9.9 and 9.10 show the fit of our model to data for the solar neighbourhood and for the halo and metal-weak thick disk. For [Fe/H] $\geq$ −1 there seems to be a good fit with little ambiguity, whereas for lower metallicities the scatter of the observational data is significant and we have

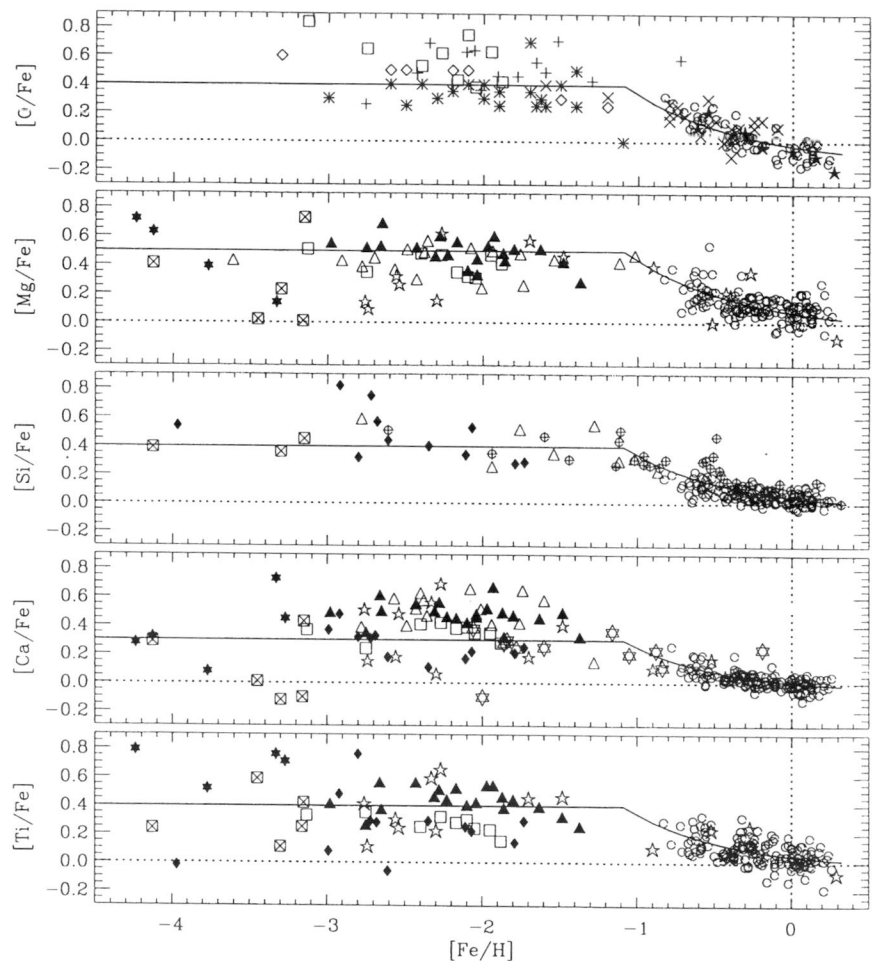

Fig 9.9. Abundance ratios relative to iron plotted against metallicity [Fe/H] for nearby disk stars and extension to [Fe/H] = −4.5. Symbols indicate various data sources: unfilled circles, Edvardsson *et al.* (1993) for $R_m \geq 7$ kpc; crosses, Barbuy and Erdelyi-Mendez (1989); asterisks, Barbuy (1988); unfilled squares, Nissen *et al.* (1994); 'plus' signs, King 1993; unfilled diamonds, Bessell *et al.* (1991); filled five-pointed stars, Kyröläinen *et al.* (1986); unfilled five-pointed stars, Tautvaišené and Straižys (1989); filled triangles, Magain (1989); unfilled triangles, Magain 1987; crosses in squares, Primas *et al.* (1994); filled six-pointed stars, Norris *et al.* (1993); 'plus' signs in circles, François (1986); filled diamonds, Gratton and Sneden (1988); 'stars of David', Hartmann and Gehren (1988).

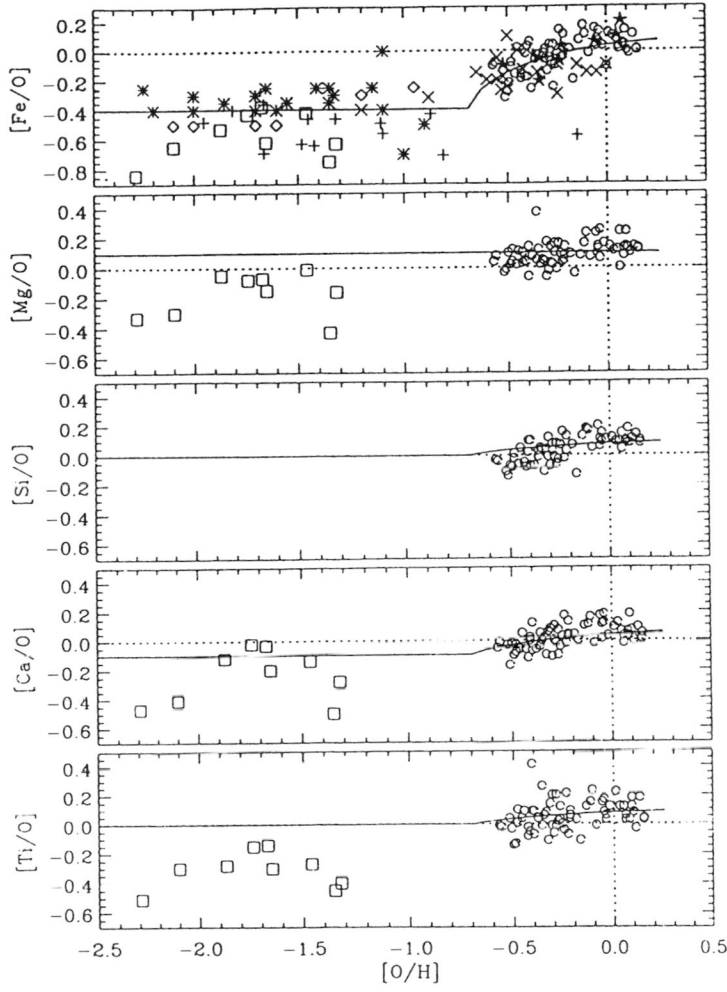

Fig 9.10. Abundance ratios of iron and α elements to oxygen plotted against oxygen abundance for the stellar sample of Fig. 9.9. Symbols are the same as in that figure.

had to make somewhat arbitrary choices based on the parameters that give the best fit to the disk stars. Thus, our model does not allow for a continuous increase in [O/Fe] towards the lowest metallicities, such as has recently been claimed, rightly or wrongly, by King (1994); such a trend might be expected from deficiencies in the instantaneous-recycling approximation at very early times, when only the most massive SNII have

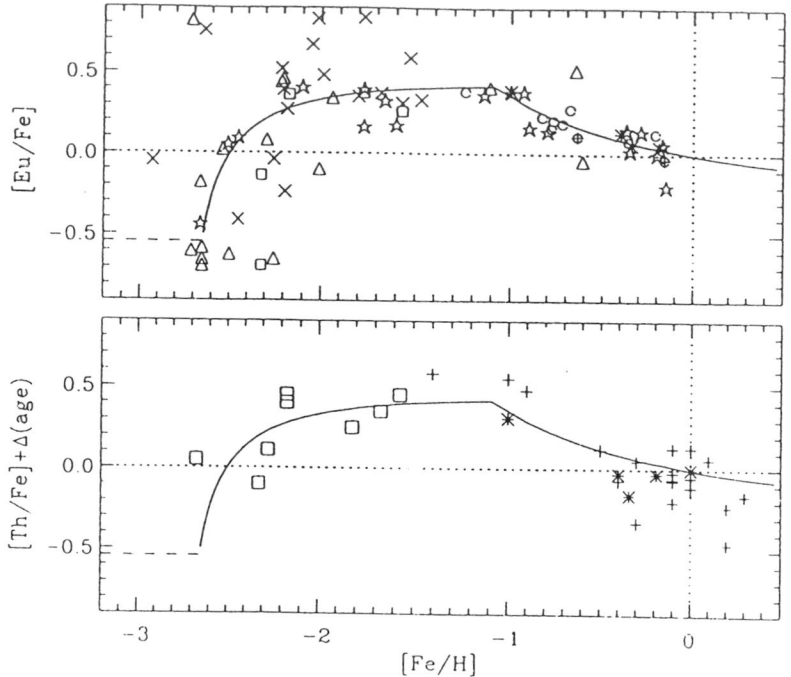

Fig 9.11. Europium and thorium-to-iron ratios plotted against metallicity. Symbols represent various data sources: five-pointed stars, Gratton and Sneden (1994); unfilled triangles, Luck and Bond (1985); crosses, Gilroy *et al.* (1988); unfilled circles, Tautvaišenė, (in preparation); 'plus' signs, Morell *et al.* (1992); asterisks, da Silva *et al.* (1990); unfilled squares, François, Spite, and Spite (1993).

evolved, but it could equally be masked by a dispersion in ages. Our yield ratios p_1/p_2 for iron, silicon, and calcium are in excellent agreement with those of SNIa and SNII computed theoretically by Thielemann, Nomoto, and Hashimoto (1996) as reported in Tsujimoto (1993) for relative supernova rates of 1 to 5 for type Ia/type 2, which is in good agreement with direct observation (van den Bergh and McClure 1994) and with the conclusion already drawn by Tsujimoto on the basis of solar abundances. There is a discrepancy for titanium (also present in the solar abundances) which must reflect a deficiency in the current nucleosynthesis models, where Ti is underproduced in SNII.

Figure 9.11 shows our fits to the r-process elements Eu and Th, the thorium data having been shifted upwards by small amounts to allow for radioactive decay (after Pagel 1993) and the fit (with or without these shifts) is remarkably good. Some of the scatter in [Eu/Fe] could well be real, in view of the existence of the remarkable r-process star CS 22892–052 (Sneden *et al.* 1994), but the fit to the data of Gratton and Sneden (1994), which are probably the best, is very close.

In conclusion, we believe that we have identified effects of Galactic chemical evolution involving simple time delays on a number of element-to-element ratios involving hydrostatic and explosive nucleosynthesis in core-collapse and type-Ia supernovae, and that the theoretical models of Thielemann, Nomoto, and Hashimoto fit the data very well, apart from titanium. The general idea of a small time delay for the r process also appears to work well. Edvardsson *et al.* give results in addition for a number of other elements such as Na, Al, and elements affected by the s process. Their Galactochemical history looks more complicated and will be the subject of future work that we plan.

References

Aller, L.H., and Greenstein, J.L.: 1960, ApJS, **5**, 139

Balser, D.S., Bania, T.M., Brockway, C.J., Rood, R.T., and Wilson, T.L.: 1994, ApJ, **430**, 667

Barbuy, B.: 1988, A&A, **191**, 121

Barbuy, B., and Erdelyi-Mendez, M.: 1989, A&A, **214**, 239

Beckman, J.E., and Casuso, E.: 1995, in *The Light Element Abundances*, ed. P. Crane (Berlin: Springer), p. 105

Beers, T.C., and Sommer-Larsen, J.: 1995, ApJS, **96**, 175

Bessell, M.S., Sutherland, R.S., and Ruan, K.: 1991, ApJL, **263**, L29

Bloemen, H. *et al.*: 1994, A&A, **281**, L5

Boesgaard, A.M., and King. J.R.: 1993, AJ, **106**, 2309

Boesgaard, A.M., and Steigman, G.: 1985, Ann Rev A&A, **23**, 319

Brocklehurst, M.: 1972, MNRAS, **157**, 221

Brown, G.E., and Bethe, H.A.: 1994, ApJ, **423**, 659

Burbidge, E.M., Burbidge, G.R., Fowler, W.A., and Hoyle, F.: 1957, Rev Mod Phys, **29**, 547 (B^2FH)

Cameron, A.G.W.: 1957, Atomic Energy of Canada Ltd, CRL-41

Carney, B.W., Latham, D.W., and Laird, J.B.: 1990, AJ, **99**, 572

Carswell, R.F., Rauch, M., Weymann, R.J., Cooke, A.J., and Webb, J.K.: 1994, MNRAS, **268**, L1

Cassé, M., Lehoucq, R., and Vangioni-Flam, E.: 1995, Nature, **373**, 318

Chamberlain, J.W., and Aller, L.H.: 1951, ApJ, **114**, 52

Bernard E. J. Pagel

Clayton, D.D.: 1985, in *Nucleosynthesis: Challenges and New Developments*, eds. W.D.Arnett and J.W.Truran (Chicago: University of Chicago Press), p.65

Conti, P.S., Geenstein, J.L., Spinrad, H., Wallerstein, G., and Vardya, M.S.: 1967, ApJ, **148**, 105

D'Antona, F., and Matteucci, F.: 1991, A&A, **248**, 62

da Silva, L., de la Reza, R., and Magalhães, S.D.: 1990, in *Astrophysical Ages and Dating Methods*, eds. E.Vangioni-Flam, M.Cassé, J.Audouze, and J.T.T.Van (Gif-sur-Yvette: Editions Frontières), p.419

Duncan, D., Lambert, D.L., and Lemke, D.: 1992, ApJ, **401**, 584

Edmunds, M.G.: 1994, MNRAS, **270**, L37

Edvardsson, B., Andersen, J., Gustafsson, B., Lambert, D.L., Nissen, P.E., and Tomkin, J.: 1993, A&A, **275**, 101

Edvardsson, B., Gustafsson, B., Johansson, S.G., Kiselman, D., Lambert, D.L., Nissen, P.E., and Gilmore, G.: 1994, A&A, **290**, 176

Eggen, O.J., Lynden-Bell, D., and Sandage, A.R.: 1962, ApJ, **136**, 748

Feltzing, S., and Gustafsson, B.: 1994, ApJ, **423**, 68

François, P.: 1986, A&A, **160**, 264

François, P., Spite, M., and Spite, F.: 1993, A&A, **274**, 821

Gasson, R.E.M., and Pagel, B.E.J.: 1966, Observatory, **86**, 196

Geiss, J.: 1993, in *Origin and Evolution of the Elements*, eds. N.Prantzos, E.Vangioni-Flam, and M.Cassé (Cambridge: Cambridge University Press), p.89

Gilmore, G., and Reid, N.: 1983, MNRAS, **202**, 1025

Gilmore, G., Edvardsson, B., and Nissen, P.E.: 1991, ApJ, **378**, 17

Gilmore, G., Gustafsson, B., Edvardsson, B., and Nissen, P.E.: 1992, Nature, **357**, 379

Gilroy, K.K., Sneden, C., Pilachowski, C.A., and Cowan, J.J.: 1988, ApJ, **327**, 298

González Delgado, R.M. *et al.* 1994, ApJ, **437**, 239

Gratton, R.G., and Sneden, C.: 1988, A&A, **204**, 193

Gratton, R.G., and Sneden, C.: 1994, A&A, **287**, 927

Grenon, M.: 1989, Astrophys Space Sci, **156**, 29

Grenon, M.: 1990, in *Bulges of Galaxies*, ESO Conference Workshop Proceedings No. 5 (Garching: ESO), p.143

Hartmann, K., and Gehren, T.: 1988, A&A, **199**, 269

Hoyle, F.: 1946, MNRAS, **106**, 343

Izotov, Y.I., Thuan, T.X., and Lipovetsky, V.A.: 1994, ApJ, **435**, 647

Kajino, T., and Boyd, R.N.: 1990, ApJ, **359**, 267

Käppeler, F., Gallino, R., Busso, M., Picchio, G., and Raiteri, C.M.: 1990, ApJ, **354**, 630

Kennicutt, R.C., Jr, Tamblyn, P., and Congdon, C.W.: 1994, ApJ, **435**, 22

King, J.R.: 1993, AJ, **106**, 1206

King, J.R.: 1994, ApJ, **436**, 331

Kiselman, D., and Carlsson, M.: 1995, in *The Light Element Abundances*, ed. P. Crane (Berlin: Springer), p. 372

Köppen, J., and Arimoto, N.: 1990, A&A, **240**, 22

Kyroläinen, J., Tuominen, I., Vilhu, O., and Virtanen, H. 1986, A&AS, **65**, 11

Lemoine, M.: 1995, in *The Light Element Abundances*, ed P. Crane, (Berlin: Springer), p. 350

Lemoine, M., Ferlet, R., Vidal-Madjar, A., Emerich, C., and Bertin, P. 1993, A&A, **269**, 469

Lequeux, J., Peimbert, M., Rayo, J.F., Serrano, A., and Torres-Peimbert, S.: 1979, A&A, **80**, 155

Linsky, J.L. *et al.*: 1993, ApJ, **402**, 694

Luck, R.E., and Bond, H.E.: 1985, ApJ, **292**, 599

Lynden-Bell, D.: 1975, Vistas in Astron, **19**, 299

McCullogh, P.R.: 1992, ApJ, **390**, 213

Maeder, A.: 1992, A&A, **264**, 105

Maeder, A.: 1993, A&A, **268**, 833

Magain, P.: 1987, A&A, **179**, 176

Magain, P.: 1989, A&A, **209**, 211

Marconi, G., Matteucci, F., and Tosi, M.: 1994, MNRAS, **270**, 35

Mathews, W.G., Bazan, G., and Cowan, J.J.: 1992, ApJ, **391**, 719

Melnick, J., Haydari-Malayeri, M., and Leisy, P.: 1992, A&A, **253**, 16

Meneguzzi, M., and Reeves, H.: 1975, A&A, **40**, 110

Merrill, P.W.: 1952, Science, **115**, 484

Meyer, D.M., Hawkins, I., and Wright, E.L.: 1993, ApJL, **409**, L61

Molaro, P., Primas, F., and Bonifacio, P.: 1995, A&A, **295**, L47

Morell, O., Källander, D. and Butcher, H.R.: 1992, A&A, **259**, 543

Morrison, H.L., Flynn, C., and Freeman, K.C.: 1990, AJ, **100**, 1191

Nissen, P.E., and Schuster, W.J.: 1991, A&A, **251**, 457

Nissen, P.E., Gustafsson, B., Edvardsson, B. and Gilmore, G.: 1994, A&A, **285**, 440

Nomoto, K., Thielemann, F.-K., and Yokoi, K.: 1984, ApJ, **286**, 644

Norris, J., Bessell, M.S., and Pickles, A.J.: 1985, ApJS, **58**, 463

Norris, J., Peterson, R.C., and Beers, T.C.: 1993, ApJ, **415**, 797

Norris, J., Ryan, S.G., and Stringfellow, G.S.: 1994, ApJ, **423**, 386

Olive, K.A., and Steigman, G.: 1995, ApJS, **97**, 49

Olive, K.A., Rood, R.T., Schramm, D.N., Truran, J., and Vangioni-Flam, E.: 1995, ApJ, **444**, 680

Pagel, B.E.J.: 1989a, Rev Mexican A&A, **18**, 161

Pagel, B.E.J. 1989b, in *Evolutionary Phenomena in Galaxies*, eds. J.E. Beckman and B.E.J. Pagel (Cambridge: Cambridge University Press), p. 201

Pagel, B.E.J.: 1991, Phys Scripta, **T36**, 7

Pagel, B.E.J.: 1993, in *Origin and Evolution of the Elements*, eds. N. Prantzos, E. Vangioni-Flam, and M. Cassé (Cambridge: Cambridge University Press), p. 496

Bernard E. J. Pagel

Pagel, B.E.J.: 1994a, *Cosmical Magnetism*, ed. D. Lynden-Bell (Dordrecht: Kluwer), p. 113

Pagel, B.E.J.: 1994b, in *The Formation and Evolution of Galaxies*, eds C. Muñoz-Tuñon and F. Sánchez (Cambridge: Cambridge University Press), p. 149

Pagel, B.E.J., and Patchett, B.E.: 1975, MNRAS, **172**, 13

Pagel, B.E.J., and Tautvaišené, G.: 1995, MNRAS, **276**, 505

Pagel, B.E.J., Terlevich, R.J., and Melnick, J.: 1986, PASP, **98**, 1005

Pagel, B.E.J., Simonson, E.A., Terlevich, R.J., and Edmunds, M.G.: 1992, MNRAS, **255**, 325

Peimbert, M.: 1993, Rev Mexicana A&A, **27**, 9

Peimbert, M., and Torres-Peimbert, S.: 1974, ApJ, **193**, 327

Peimbert, M., and Torres-Peimbert, S.: 1976, ApJ, **203**, 581

Pettini, M., Smith, L.J., Hunstead, R.W., and King, D.L.: 1994, ApJ, **426**, 79

Pilyugin, L.: 1993, A&A, **277**, 42

Pinsonneault, M.H., Deliyannis, C.P., and Demarque, P.: 1992, ApJS, **78**, 179

Prantzos, N., Cassé, M., and Vangioni-Flam, E.: 1993, ApJ, **403**, 630

Primas, F., Molaro, P., and Castelli, F.: 1994, A&A, **290**, 885

Renzini, A., and Voli, M.: 1981, A&A, **94**, 175

Rood, R.T., Bania, T.M., and Wilson, T.L.: 1992, Nature, **355**, 618

Ryan, S., Norris, J., Bessell, M.S., and Deliyannis, C.: 1992, ApJ, **388**, 184

Sandage, A.R., and Fouts, G.: 1987, AJ, **93**, 74

Sasselov, D., and Goldwirth, D.: 1995, ApJ L, **444**, L5

Scalo, J.: 1986, Fund Cosm Phys, **11**, 1

Schmidt, M.: 1959, ApJ, **129**, 243

Schmidt, M.: 1963, ApJ, **137**, 758

Searle, L., and Sargent, W.L.W.: 1972, ApJ, **173**, 25

Searle, L., and Zinn, R.: 1978, ApJ, **225**, 357

Skillman, E.D., and Kennicutt, R.C., Jr: 1993, ApJ, **411**, 655

Skillman, E.D., Terlevich, R.J., Kennicutt, R.C., Jr, Garnett, D.R., and Terlevich, E.: 1994, ApJ, **431**, 172

Smith, M.S., Kawano, L.H., and Malaney, R.A. 1993, ApJS, **85**, 219

Smith, V., Lambert, D.L., and Nissen, P.E.: 1993, ApJ, **408**, 262

Smits, D.P.: 1991a, MNRAS, **248**, 193

Smits, D.P.: 1991b, MNRAS, **251**, 316

Smits, D.P.: 1996, MNRAS, **278**, 683

Sneden, C., Preston, G.W., McWilliam, A., and Searle, L.: 1994, ApJ, **431**, L27

Sommer-Larsen, J.: 1991a, MNRAS, **249**, 356

Sommer-Larsen, J.: 1991b, MNRAS, **250**, 356

Sommer-Larsen, J.: 1996, ApJ., **457**, 118

Songaila, A., Cowie, L.L., Hogan, C.J., and Rugers, M.: 1994, Nature, **368**, 599

Spite, F., and Spite, M.: 1982, Nature, **297**, 483

Steigman, G.: 1993, ApJL, **413**, L73

Steigman, G., and Tosi, M.: 1992, ApJ, **401**, 15

Steigman, G., and Walker, T.P.: 1992, ApJL, **385**, L13

Tautvaišenė, G., and Straižys, V.: 1989, Izv Spets Astrofiz Obs, No. 28, p. 88

Terasawa, N., and Sato, K.: 1990, ApJL, **362**, L47

Thielemann, F.-K., Nomoto, K., and Hashimoto, M.: 1996, ApJ, **457**, 118

Thielemann, F.-K., Nomoto, K., and Yokoi, K.: 1986, A&A, **158**, 17

Thomas, D., Schramm, D.N., Olive, K.A., Mathews, W.G., Meyer, B.S., and Fields, B.D.: 1994, ApJ, **430**, 291

Thorburn, J.A.: 1994, ApJ, **421**, 318

Tinsley, B.M.: 1980, Fund Cosm Phys, **5**, 287

Tsujimoto, T.: 1993, Thesis, Tokyo University

van den Bergh, S.: 1962, AJ, **67**, 486

van den Bergh, S., and McClure, R.D.: 1994, ApJ, **425**, 205a

Vangioni-Flam, E., Olive, K., and Prantzos, N.: 1994, ApJ, **427**, 618

Vangioni-Flam, E., Cassé, M., Audouze, J., and Oberto, Y.: 1990, ApJ, **364**, 568

Walker, T.P., Steigman, G., Schramm, D.N., Olive, K.A., and Kang, H.: 1991, ApJ, **376**, 51

Wallerstein, G.: 1962, ApJS, **6**, 407

Wheeler, J.C., Sneden, C., and Truran, J.W.: 1989, Ann Rev A&A, **27**, 279

Woosley, S.E., Langer, N., and Weaver, T.A.: 1993, ApJ, **411**, 823

Woosley, S.E., Hartmann, D.H., Hoffman, R.D., and Haxton, W.C.: 1990, ApJ, **356**, 272

Wyse, R., and Gilmore, G.: 1992, AJ, **104**, 144

Yang, J., Turner, M.S., Steigman, G., Schramm, D.N., and Olive, K.: 1984, ApJ, **281**, 493

9.5 **Discussion**

Question

(Osterbrock): Suppose that much of what we think that we know is wrong. If there were stars with less helium in them than 23%, but in galaxies that did not have emission lines, would we be able to detect the fact? Would we be able to find that out even in our own system that they are old stars, not young stars?

Answer

(Pagel): There are stars in our own Galaxy which have a low helium abundance in their atmosphere: the notorious subdwarf B stars. During the 1960s this caused quite a lot of problems. It really caused people to wonder whether there was a primordial helium abundance, but finally, it was shown by Sargent and Searle (1967, ApJL, **150**, L33), that these also showed other peculiarities, phosphorus and things like that, which were associated with the chemically peculiar stars of population I. Since we believe that diffusion effects are important in these stars, that result, subsequently, was not taken seriously any more, so we just said those stars are fooling us and we believe all the other objects where we see the normal helium.

Question

(Reeves): Yes, it is interesting to point out the relation between inflow and the original abundance of deuterium. For instance, if the Songaila result (1994, Nature, **368**, 599) on deuterium had been right, it would have been very difficult to invoke any inflow. The new result by Tytler, on the contrary, is quite coherent with having inflow.

Answer

(Pagel): Yes, but one of the big questions is that the destruction of deuterium really depends very much on the return fraction. How much processed material of a generation of stars is returned to the interstellar medium with its deuterium destroyed? Most people, adopting a conventional initial-mass function, have estimated the return fraction at something like 0.2. However, if you take this result from Scalo, with $b' < 1$ (in other words, a somewhat lower star-formation rate now than the average in the past), then you have to increase the proportion of high-mass stars and increase the return fraction. So Maeder, for example, has suggested a return fraction of 0.5, which makes a huge difference to your estimate of deuterium destruction.

Question

(Reeves): Do you mean return with deuterium? Without destruction of deuterium?

Answer

(Pagel): No, with destruction of deuterium because, when the gas is recycled through the stars and sent back into the interstellar medium, the deuterium is destroyed. That, of course, is why people have been uncertain over the years as to how to calculate deuterium destruction.

Question

(Sandage): It was a little hard to see your graphs. Could you give a summary of the oxygen-to-iron ratio as a function of Fe/H? Whose data are those?

Answer

(Pagel): It is a long list: the crosses with dots are due to Barbuy; the squares are due to Nissen *et al.* (1994) using OH lines; the squares with crosses are the Molaro group, but I do not think they did oxygen, so that is irrelevant. Plus signs are due to King, who used the permitted OI line, and that is these data here [shows viewgraph]. The diamonds are due to Bessell *et al.*, using mainly OH lines. I think that the argument that you could make is that the data show no particular trend beyond a flat plateau. For the other elements there is even less evidence for a trend deviating from a flat plateau. What I think could be in dispute is that if you take any one of these data sets, like the Nissen set for instance, they also fit a plateau: it is just a somewhat higher plateau than the one we have adopted.

Question

(Sandage): Of course, the important point is the level on the left-hand side. Is it 0.4 dex, or does it go up to 0.8 dex? The reason that is so important for the age-dating is that, because oxygen is 250 times more abundant than iron, the effective internal opacity which must be used for the age dating depends upon the oxygen abundance. So, whether the age of the globular clusters is 18×10^9 yr, or 14×10^9 yr is crucial to that diagram, I think.

Answer

(Pagel): Yes, but we must remember that these other alpha elements, although they do not affect the interior opacity, have a strong effect on the atmospheric opacity. As far as I know, neither Don VandenBerg nor anybody else has actually done these calculations with a full alpha effect included (but see Salaris *et al.* 1993, ApJ, 414, 580). Therefore, I think we have to wait before we see what ages you really get, as a consequence of particular sets of abundances in the metal-deficient stars.

Comment

(Sandage): Bergbusch and VandenBerg (1992, ApJS, 81, 163) have calculated the interior grids of models with a particular oxygen-to-Fe/H ratio. It appeared in the Astrophysical Journal Supplement in 1995.

Answer

(Pagel): Yes, but atmospheres are important as well. So the other alpha elements also have to be included. They are all overabundant.

Question

(Sandage): Why are the atmospheres important?

Answer

(Pagel): Because they affect the position of the red-giant branch and so on.

Comment

(Sandage): But that is not the turn-off luminosity; that is where the ages come through.

Answer

(Pagel): The luminosity, perhaps not, but the shape of the diagrams certainly depends on the atmospheres.

Comment

(Sandage): Yes, but that is not the age-dating problem.

Answer

(Pagel): But what age do you get from an O/Fe enhancement of 0.4 dex?

Question

(Sandage): Something like 16×10^9 yr, and there is then a relation of the age with Fe/H for the globular clusters. So everything is crucially dependent on that diagram of the O/H vs Fe/H relation?

Answer

(Pagel): Yes, I think it is dependent, but it is dependent on the other alpha elements as well. That is why I think this is one of the key problems: to decide where that diagram should actually go. I agree that it is a key problem, and I do not claim dogmatically that the way we have fitted it is definitely the right one, although I do think that the way we have fitted the alpha elements is definitely the right one.

Question

(Rees): This is a question to Bernard, or maybe to Hubert Reeves. Since so much in cosmology does depend on deuterium being primordial, is everyone absolutely sure that there is no even half-way credible astrophysical mechanism for making deuterium?

Answer

(Pagel): I think Hubert should answer that.

Answer

(Reeves): For instance, if you try to make it by spallation, which is one possibility, you would overdo lithium, beryllium, and boron by a very large factor. Every other process which has been tried has always been killed by some constraint of this type. So, as far as I know, and I have looked at this problem, I know of no way to produce deuterium otherwise.

Question

(M. Burbidge): On the discrepancy of deuterium abundance measured in the QSO with the damped Lyα absorption between Songaila *et al.* (1994, Nature, **368**, 599) and Tytler: that has been discussed a bit by Art Wolfe and colleagues. He is of the opinion that, while the blip that looks like deuterium probably is deuterium, there may be an error in determining the ordinary hydrogen abundance from that broad profile, and you need to do more work on the Lyman limit to find that out.

Answer

(Pagel): I see. That is a very interesting point.

Question

(R. Rebolo): First, a comment on the oxygen abundance. In your O/H vs Fe/H plot, you have represented giants and dwarfs. I would recommend just plotting dwarfs. Giants could have been affected by mixing, or any other process, after main-sequence evolution, and could have changed the original and primordial oxygen abundance. If you consider only the dwarfs, even if my data are not there, you can see that the dwarfs show a slight trend to increased oxygen overabundance at lower metallicities. I think that your comment on the validity of the forbidden line against the permitted triplet is under question now. There is a recent paper by Boesgaard and King, who question why the forbidden line is better for an oxygen determination. I have the paper here and will show it to you later. Except for the low metallicities, those points below your feet, I think the Barbuy data is based on giants.

Answer

(Pagel): That is correct, yes.

Question

(Rebolo): I also have a comment on the question of deuterium production. It might be possible. In fact, I think there are people on this side of the table who have produced mechanisms for creating deuterium around black holes. M. Rees and R. Sunyaev have done that, and it seems we have found evidence for lithium production in these systems, and it might not be impossible to have some deuterium production there too.

Answer

(Rees): I think that Hubert has just worked out what the constraints are now, during our private discussion. Let him just explain.

Answer

(Reeves): Just the fact that we have observed lithium in these black holes could give you an upper limit to the amount of deuterium that you could produce in any type of spallation, for instance. That would be, again, I think, orders of magnitude too low, unless, as Martin says, you could convince yourself that you are accelerating only hydrogen and not helium. That would be the only possible case. In other words, the ratio of hydrogen to helium, in the accelerated flux, would have to be much higher than in the Universe. That is possible, but with a lot of constraints.

Comment

(Sunyaev): I would like to mention here that, during the last cycle of solar activity, we observed (with *GRANAT*) maybe a dozen solar flares with an enormous production of deuterium. At the same time, we specially checked the 270-keV line arising from the production of ^{7}Li. My estimate today is that the production of lithium is at least two orders of magnitude smaller than that of deuterium, during some peculiar solar flares, where the spectrum is extremely hard. Then, the γ-ray spectrum of the solar flares is so hard that you see the birth of the π° mesons and their decay. This is proof, from solar physics, that the abundance of lithium which you have produced is at least a hundred times smaller than the abundance of deuterium.

In the late 1960s and early 1970s we were trying to produce deuterium in very strong supernova shocks, thinking that this was the way to produce deuterium. Zel'dovich stopped us with the same argument that you have used today: you produce too much lithium, and that is impossible. Now from these solar flares we see real indications that you can get deuterium more abundant than lithium and you really produce too much. But a ratio of a hundred is possible.

Answer

(Reeves): Yes, but you would need to have it less than a factor of 10^4 below, and not 10^2.

Comment
(Sunyaev): Yes, 10^4, I agree.

Answer

(Reeves): We have a factor of one hundred missing.

Question

(Sunyaev): Again, in the late 1960s, we were all excited by the stars where the ^{3}He abundance is higher than the ^{4}He abundance. Why has everybody lost interest in these stars today? There are no more papers on this subject any longer.

Answer

(Pagel): Actually, Olive *et al.* (1995) used this as an argument that some ^{3}He survives in some stars. So people have not altogether lost interest.

Question

(Reeves): But was the effect not explained in terms of diffusion?

Answer

(Pagel): Yes, ^{3}He had to be there to be diffused up!

Question

(Reeves): But what amount?

Answer

(Pagel): That was not quite clear, but if the star had destroyed all its ^{3}He... The argument was qualitative rather than quantitative.

10   Clues to the early development of galaxies

Hubert Reeves

10.1   Introduction: exploring the 'dark age' of the Universe

The granularity of the microwave background (fossil) radiation has given us insights into the earliest stages of structure formation at recombination time: $z = 10^3$ and $\delta\rho/\rho \approx 10^{-5}$. The observation of stars and galaxies has taught us much about the present state of the Universe. To bridge the gap between high and low z (redshift) on the one hand, and high Z to low Z (metallicity) on the other, is one of the key challenges for contemporary astrophysics. What happened in the intermediate period leading to the formation of galaxies? I will deal with some aspects of both the present Universe and early Universe, which may help us shed some light on this dark age (the expression is from Martin Rees) in the history of the Universe.

10.2   Starting from today

10.2.1 γ-rays and the origin of the light elements

The detection of large fluxes of 3 to 7 MeV from the direction of the Orion nebula by the *Comptel Observatory* (Bloemen *et al.* 1993) is of great astrophysical significance. Two broad lines corresponding to the 4.4-MeV excited state of ^{12}C and the 6.2-MeV excited state in ^{16}O can be clearly identified. The flux corresponds to an emission rate of 10^{38} photons s^{-1}. From the absence of corresponding enhancements in the 1–3 MeV region and also in the region above 100 MeV, the flux can be ascribed to the heavy bombardment of interstellar clouds by C and O nuclei of a few tens of MeV. These fast particles were probably accelerated by supernovae in the Orion OB association.

This detection provides new insights into the origin of the light elements Li, Be, and B. The standard view is that these elements were formed through the bombardment of interstellar gas by galactic cosmic rays (GCR) with energies in the GeV range (Meneguzzi, Audouze, and Reeves 1971, hereafter MAR). But there are problems here, with the boron isotopic ratio. The GCR calculated value ($^{11}B/^{10}B \approx 2.3$) falls short of the Solar System observed value ($^{11}B/^{10}B = 4.05$) by a substantial amount (MAR; Walker, Mathews, and Viola 1985). It had already been suggested (Meneguzzi and Reeves 1975; Reeves and Meyer 1978) that a significant contribution of particles in the MeV range would help to eliminate this discrepancy. However, there was, at that time, no proof of the existence of such low-energy fluxes in the Galaxy.

Recent calculations (Bykov and Bloemen 1994; Cassé, Lehoucq, and Vangioni-Flam 1995; Reeves and Prantzos 1995; Ramaty *et al.* 1997) have shown that the Solar System boron isotopic ratio can be accounted for by the Orion-type bombarding fluxes with mean energies around 10–20 MeV. Significantly, the same bombarding fluxes give a better account of the Solar System boron-to-beryllium ratio (Grevesse and Noels 1992) than the high-energy GCR particles.

The recent detection of boron isotopic variations in meteorites (Chaussidon and Robert 1995) suggests that a non-negligible fraction of these atoms was generated during an early irradiation in the solar OB association. In the same fashion, the variability of the lithium isotopic ratio in the region of Ophiuchus (Lemoine *et al.* 1993, 1994; Meyer, Hawkins, and Wright 1993) points to recent strong irradiations by analogous fluxes of low-energy particles.

These observations lend credence to a new mechanism for the formation of the five isotopes of Li, Be, and B. While the GCR generates them steadily over the whole of the galactic disk, Orion-type episodes produce them locally and sporadically. The task at hand now is a quantitative evaluation of the respective contributions of these two mechanims to the galactic abundances of each of the five isotopes. For ^7Li we must also consider a stellar contribution, probably from AGB stars.

10.2.2 Light elements in the early Galaxy

Such an undertaking will perhaps shed light on the processes accompanying the early days of our Galaxy. Beryllium and boron have been detected in old metal-deficient population-II stars (Gilmore *et al.* 1991, 1992; Ryan *et al.* 1991, 1992; Rebolo *et al.* 1992; Boesgaard and King 1993). Their abundances appear to grow linearly with metallicity. This would be unlikely if they were generated by the GCR since they would be expected to grow slower than linearly, as secondary products. This phenomenon is more easily explained if they come from Orion-type episodes, while fast C and O are being spallated by interstellar H and He. The constant ratio of B/Be, although somewhat uncertain, appears to favour a low-energy (Orion-type) origin.

The production rate of these isotopes is related to the diffusive behaviour of the fast particles in the ambient medium (Prantzos, Cassé, and Vangioni-Flam 1993). The mean free paths of these particles are governed by the configurations of the local magnetic field. The B and Be abundances are thus connected to the problem of the origin of magnetic fields, which is discussed by Martin Rees in Chapter 8.

10.2.3 The γ-ray background

There is as yet no satisfactory explanation of the MeV hump in the photon background radiation with an energy density of 6×10^{-5} eV cm^{-3}. A correct interpretation may yield important information on the early galactic processes. It is tempting to see in this hump the cumulative effects of γ-ray lines, redshifted by z of 1–5 (Orion-type fluxes generating the light elements are much too low). There is the potential here for a better understanding of the role of high-energy particles in the formation of galaxies.

10.3 Starting from early times

We use model calculations to trace the growth of structures from the observed thermal granularity of the fossil radiation. A number of parameters are needed to compute the models. The situation with respect to some of these parameters will be discussed.

10.3.1 Quarks and nucleons

The possible effects of the quark–hadron phase transition on the evolution of the Universe has been a matter of concern for several years (Witten 1984). There could be two. The transition could have generated 'quark nuggets', which would qualify as dark-matter candidates. It is generally agreed, however, that these objects rapidly evaporated leaving no remnants in our present-day Universe.

The transition could have generated density inhomogeneities, leaving their imprints on the yields of light nuclei abundances resulting from the primordial nucleosynthesis. Calculations (Terasawa 1992; Reeves 1994) have been performed, using the theoretical QCD results, to explore the possible effects of these inhomogeneities. In Fig. 10.1 the yields are shown both for the standard homogeneous-density model (labelled 'st') and for the maximum possible modifications caused by the inhomogeneities left after the Q–H phase (labelled 'Max. Q–H'). These modifications are negligible except for ^{7}Li in the higher-density range. And since the Q–H effects always increase the yield of ^{7}Li with respect to the standard case, the upper limit to the nucleon density obtained from the standard case remains valid.

In brief, the transition does not seem to have left any imprint. The nucleonic range obtained from standard BBN is applicable. The results are quoted both at an 'optimistic level', meaning that the quoted uncertainties are realistic, and also at a 'pessimistic' level, extending the range of the uncertainties to the largest possible values. η_{10} is the ratio of nucleons to photons expressed as a multiple of 10^{-10}.

Optimistic:

$$2.0 < \eta_{10} < 5.0; \quad 1.2 < \rho_b(10^{31}\mathrm{g\,cm}^{-3}) < 3, \tag{10.1}$$

$$0.03 < \Omega_b h^2 < 0.075; \quad 0.0075 < \Omega_b < 0.075. \tag{10.2}$$

Pessimistic:

$$1.5 < \eta_{10} < 10; \quad 0.9 < \rho_b(10^{31}\mathrm{g\,cm}^{-3}) < 6, \tag{10.3}$$

$$0.022 < \Omega_b h^2 < 0.15; \quad 0.0055 < \Omega_b < 0.15. \tag{10.4}$$

For all these, $50 < H_0 < 100\,\mathrm{km\,s}^{-1}\,\mathrm{Mpc}^{-1}$.

One word of caution is in order, however. These results are based entirely on theoretical studies of the properties of quark matter. As such, they should be considered with some degree of scepticism. At the end of 1994 an experiment involving GeV lead–lead collisions was performed at CERN which could result in the first laboratory realization of

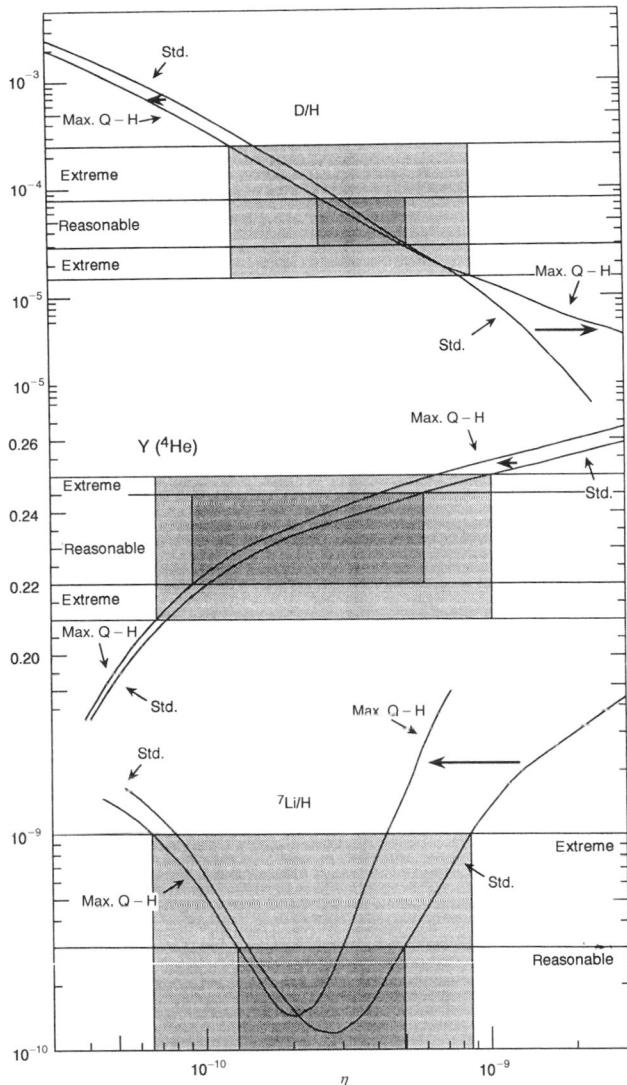

Fig 10.1. Light nuclei yields in the early Universe. The abcissa is
$\log \eta = n(\text{nucleons})/n(\text{photons})$. The curves labelled STD refer to homogeneous
density BBN. The curves labelled 'Max. Q-H' show the maximum possible
modifications to the yields introduced by the parameterized influence of the
quark–hadron phase transition. The arrow indicates the direction in which the
Q–H effects displace the yield curves. Dark-shaded areas are obtained from the
reasonable limits $1.6 \times 10^{-5} < \text{D/H} < 10^{-4}$; $0.22 < Y < 0.24$; $1.5 \times 10^{-10} <$
$^7\text{Li/H} < 3.0 \times 10^{-10}$. The full areas (light+dark) correspond to the extreme limits:
$1.6 \times 10^{-5} < \text{D/H} < 2 \times 10^{-4}$; $0.22 < Y < 0.245$; $1.5 \times 10^{-10} < {}^7\text{Li/H} < 1.0 \times 10^{-9}$.

the quark–gluon plasma. The results, not yet available, should give us a clearer idea of where we stand.

10.3.2 Problems of the origin of structures

The existence of large structures presents three problems to the standard Friedmann–Lemaître evolution scenario.

(i) The seeds: what is the physical mechanism responsible for the initial overdensities?

(ii) Causality: in the first hundred years, the sphere of causality is smaller than the scale lengths of the structures.

(iii) Time delay: from recombination to the birth of the first galaxies ($z \approx 4$) there is not enough time to increase the density contrast $\delta\rho/\rho$ from 10^{-5} to 1 (in this scenario, the growth factor is roughly equal to the ratio of the redshifts).

The most popular solution invokes the existence of quantum scalar fields generating episodes of inflation when their energy densities come to dominate the expansion. The initial densities are associated with quantum fluctuations of this 'inflation' field and the problem of causality is resolved by the large increase in the scale length generated by a prolonged episode. An important component of dark matter leading to $\Omega \approx 1$ is postulated to accelerate the growth of the galactic 'germs' all the way to the present galactic overdensities ($\delta\rho/\rho \approx 10^6$). Since inflation scenarios presuppose a currently flat Universe ($\Omega + \Lambda = 1$), the existence of this exotic component seems perfectly acceptable in the context of inflation. From the previous limits on the nucleonic component, it appears that this matter must be in large part 'exotic'.

10.3.3 Missing elements

There are interesting parallels to be drawn between the most troublesome astronomical problems at the end of the 19th and 20th centuries.

It was widely accepted, in the last century, that a substratum of 'aether' pervaded all of space. All attempts to detect it failed, however, until Einstein 'explained it away' with his principle of relativity. Today we have several 'missing' elements: (a) dark matter, (b) magnetic monopoles, and (c) quantum energy densities expressed by the cosmological constant. The last two are offspring of the stormy marriage between astronomy and particle physics. As in all marriages, each partner has great demands and great expectations: 'What's good for physics is good for cosmology and vice versa'.

(a) Two different components of *dark matter* are discussed in the literature. Type I makes itself known through its dynamical effects on the rotational velocities of stars in galaxies, on the dispersion velocities of galaxies, on X-ray-emitting gas in clusters of galaxies, and also on gravitational lensing by large structures. It amounts to one or two-tenths of the closure density. Judging by the baryonic density evaluation discussed earlier, it probably contains a fair proportion of exotic (non-baryonic) matter. Could it be 'evacuated' away by a reformulation of the theory of gravity? Very few people believe so today. Type II concerns the extra fraction required to complete the closure density. It would necessarily be exotic in nature. The arguments in its favour are still largely speculative.

(b) Magnetic monopoles. The unification of nuclear and electroweak forces requires the existence of massive magnetic monopoles. But why should we believe in this unification? One important reason is the apparently strict equality between the absolute value of the electron and proton charges. Strict numerical equalities generally indicate the underlying operation of a group structure. Group formalisms do not tolerate even the slightest deviations. Another factor is the convergence of the electromagnetic, weak, and nuclear coupling constants at around 10^{15} GeV.

From these facts the idea of a large group G containing electrons, neutrinos, and quarks has emerged. So far no successful implementation of this idea in terms of a specific group has been presented. However, if a group G breaks into smaller groups containing at least one U(1) group, massive magnetic monopoles are abundantly generated. Since electromagnetism is represented by a U(1) group, their creation is unavoidable. According to the standard Big Bang scenario, they should be as abundant as the nucleons. We see them nowhere. Where are they? One solution is to rarify through them an episode of inflation after the break-up of the large group.

(c) Field energies. Quantum mechanics has given us the notion of residual energies associated with ordinary quantum fields. The contributions of bosons and fermions being of opposite signs, the integrated residual energy would be zero in a supersymmetric world. In our world, where supersymmetry is manifestly broken, its value, in appropriate units, should be some 10^{58} times larger than the critical density!

Furthermore, the unification schemes have imposed the notion of scalar Higgs fields with high-energy densities and pressures obeying an equation of state of the type $P = -\rho$. These fields are responsible for the inflationary episodes.

All these energies are expected to manifest themselves in the expanding Universe through the cosmological-constant term, Λ. In a homogeneous Friedmann–Lemaître model the equation reads, with R the scale length,

$$H^2 = (\mathrm{d}R/\mathrm{d}t)^2/R^2 = 8\pi G\rho/3 + \Lambda/3 - k/R^2. \qquad (10.3.5)$$

From estimations and limits on the Hubble constant H, on the density of ordinary matter, and on the radius of curvature, we derive the following inequality:

$$\Lambda/(24\pi G/\rho_c) < 1. \qquad (10.3.6)$$

The present cosmological constant is not larger than the critical density! We find no trace of these quantum energies in the rate of expansion of the Universe.

The huge discrepancy between expectations and observations may indicate there is an unknown law of conservation decreeing a null value for the cosmological constant. And just as the principle of relativity produced radical changes in our understanding of nature, we may expect the solution to the cosmological-constant problem to be accompanied by a major rethinking of physics and cosmology. This could have significant repercussions in many neighbouring sectors, affecting the very status, for instance, of inflationary theories.

10.3.4 Cosmological pathologies and their possible medications

The preceding paragraphs have presented two important problems raised by the integration of modern physics into the Big Bang cosmological scenarios. Other 'enigmas' derive from the field of cosmology itself.

A first series of enigmas is related to the apparent 'predilection' of the cosmos for minimal initial conditions. Why were the initial curvature and entropy so small? Why was matter so isothermal? Inflationary episodes are

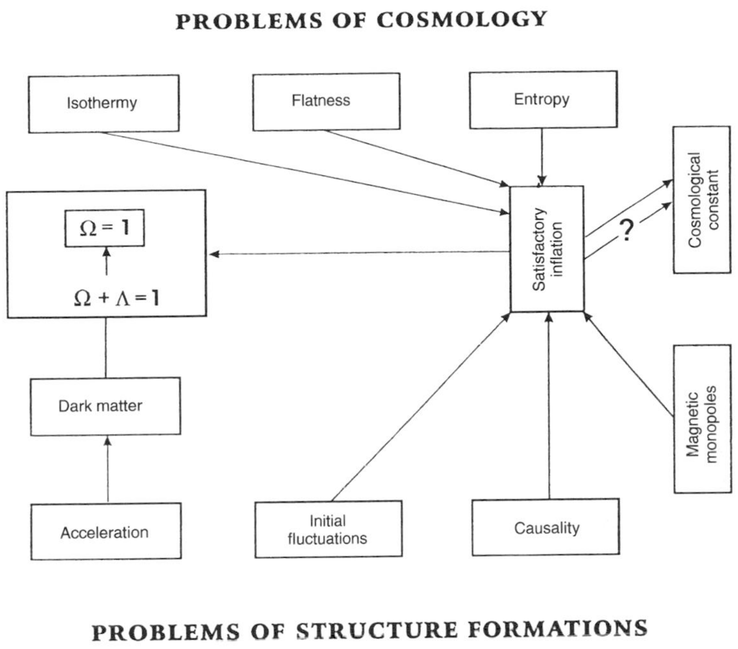

PROBLEMS OF COSMOLOGY

PROBLEMS OF STRUCTURE FORMATIONS

Fig 10.2. The various astrophysical problems described in the text are grouped according to their domains. On the top: general cosmological problems; on the bottom: problems specifically related to the growth of structures; and on the right-hand side: problems issuing from the marriage of astrophysics and particle physics. The arrows point to their possible solutions. Inflation requires $\Omega + \Lambda = 1$, while the accelerated growth of structures relies on $\Omega \approx 1$. The relation between inflation and the problem of the cosmological constant is still mysterious. A solution to the cosmological-constant problem may do away with the need for inflation.

usually considered to be satisfactory answers. These various pathologies and their popular medications are shown pictorially in Fig. 10.2. This fragile therapeutic scaffold relies heavily on the postulate of a flat Universe: $\Lambda + \Omega = 1$. What happens if this postulate is wrong?

Alternative solutions exist for some of the problems. The isothermy, flatness, low entropy and causality problem in structure formations would fall back on their status of *special initial conditions*. Cosmic strings could have triggered the initial overdensities, and non-gravitational effects could have accelerated their concentrations. The monopole and cosmological-constant problems would remain unsolved, with the hope that an eventual solution to the latter would introduce enough new physics to generate new solutions for all the other problems.

10.3.5 The stability of matter

Another analogy with physics at the end of the last century concerns the problem of matter stability. The combination of Newtonian physics and electromagnetism required that orbiting electrons should emit radiation, thereby rapidly losing energy. Atomic lifetimes were expected to be of the order of milliseconds, in 'some' disagreement with the age of stars . . . The answer came through quantum mechanics: the Heisenberg uncertainty principle excludes such atomic collapses.

Today, it is the stability of the proton itself which is under consideration. The unification scheme discussed earlier implies that quarks and electrons should belong to the same gauge group. Group transformations should then lead to baryon-number violation.

This violation leads to the possibilty of matter–antimatter asymmetry, which is a precondition for the very existence of matter as we know it. In a symmetric Universe, annihilation processes in cooling matter would have decreased the ratio of nucleons to photons to 10^{-18}, compared to the present value of 3×10^{-10}. We would still be in the radiation era and galaxies would never have formed. In other words, the proton has to be unstable (for galaxies and stars to form) but not *too* unstable (so that we should not be radioactive). Present limits on its lifetime are $\tau_p > 10^{31}$ years.

Baryogenesis is the ensemble of events responsible for the present over-abundance of matter over antimatter. It assumes that the baryon number should be violated, but also that matter and antimatter should not behave in a strictly symmetric way (charge conjugation and parity non-conserving violation). It also requires out-of-equilibrium phases in which appropriate reaction rates become slower than the rate of cosmic expansion. In the early Sakharov scenario, these events took place at grand unified symmetry-breaking around 10^{15} GeV. Several authors have also considered later phases such as the electroweak symmetry-breaking around 100 GeV. At the present time, there is no quantitatively satisfactory explanation of the value of the baryonic number.[†]

10.4 **Inflation, an overburdened concept?**

The previous discussion has illustrated the importance of inflation scenarios as a solution to many of the problems of modern cosmology. How realistic is this solution?

This discussion involves several questions. First: *were there* inflationary episodes in the past Universe? The answer is most likely 'yes'. The quark–hadron phase transition and the electroweak symmetry-breaking both require specific quantum fields of high-energy densities. At appropriate times, these densities came to dominate the rate of expansion and gave rise to overcooling and exponential acceleration. The occurence of episodes at symmetry-breaking of grand unification and at the Planck era is more speculative.

The crucial question, however, is: *was there* enough inflation and non-adiabatic entropy-increasing processes to satisfy the expectations of the cosmologists? Factors of at least 10^{30} are required. The only well-documented case, the quark–hadron phase transition, provides less than a factor of two! No quantitatively convincing scenarios for the other episodes have yet been implemented.

The third question concerns the chronology of these episodes. The solution to the monopole problem requires that an episode take place *after* the breaking of the initial group into the U(1) group of electromagnetism. On the other hand, the preservation of the baryonic number requires that no strong episode should occur after baryogenesis. Furthermore, if cosmic strings are the seeds of galaxies, they should appear *after* the evacuation of monopoles in new symmetry-breaking episodes where no monopoles are generated.

That we are having such difficulties successively implementing the inflationary paradigm is partly due to the excessive demands of the cosmologists. The fact that, in any case, this implementation will not solve the cosmological-constant problem may be a sign that we are approaching the problem from the wrong direction. A satisfactory solution to the

† On these subjects I recommend the highly illuminating book *Basics of Modern Cosmology* (Dolgov, Sakhin, and Zel'dovich 1990).

cosmological-constant problem will likely change the situation completely and perhaps do away with the need for inflation.

10.5 Cosmic topology

The space in which the Universe expands is usually assumed to be simply connected. The only argument for this choice is simplicity. But is it really simpler? Multiconnectedness carries an important simplifying factor: the boundary conditions are at finite coordinates. The difficulties and ambiguities of conditions at infinity disappear.

Unification theories of forces also provide an argument in favour of multiconnectedness. In one popular version of the superstring theory, the Universe at the Planck time has nine spatial dimensions. Six of them rapidly curl to play the role of internal dimensions related to the physical forces; the three other dimensions expand to become our world. The internal space of these six dimensions is multiconnected with an exotic topology of highly non-trivial configurations. This fact questions the topology of our 'real' world. Given the topological state of the curled dimensions the assumption of simple connectedness of our world is certainly no more justified.

In Chapter 3 Donald Lynden-Bell discussed the relation between the Mach principle and the geometry of space. According to his views, if I understood correctly, the open Universe could not be reconciled with the Mach principle because of its infinite size. This difficulty would not arise in a multiply connected Universe with a torus-like topology, for instance.

References

Bloemen, H., Wijnands, K., Bennett, R., Diehl, W., Hermsen, G., Lichti, G., Morris, D., Ryan, J., Schönfelder, V., Strong, A.W., Swanenburg, B.N., de Vries, C., and Winkler, C.: 1993, A&A, **281**, L5

Boesgaard, A.N., and King, J.R.: 1993, AJ, **106**

Bykov, A., and Bloemen, H.: 1994 A&A, **283**, L1

Cassé, M., Lehoucq, R., and Vangioni-Flam, E.: 1995, Nature, **373**, 318

Chaussidon, D., and Robert, F.: 1995, Nature, **374**, 337

Dolgov, A.D., Sakhin, M.V., and Zel'dovich, Ya.B.: 1990, *Basics of Modern Cosmology* (Gif-sur-Yvette: Editions Frontieres)

Gilmore, G.: 1992, in *On the Origin and Evolution of the Elements*, eds. N. Prantzos, E. Flam, and M. Cassé (Cambridge: Cambridge University Press)

Gilmore, G., Edvardsson, B., and Nissen, P.E.: 1991, ApJ, **378**, 17

Gilmore, G., Gustafsson, B., Edvardsson, B., and Nissen, P.: 1992, Nature, **357**, 379

Grevesse, N., and Noels, A.: 1992, in *On the Origin and Evolution of the Elements*, eds. N. Prantzos, E. Flam, and M. Cassé (Cambridge: Cambridge University Press)

Lemoine, M., and Ferlet, R.: 1994, ESO Workshop on 'Light Element Abundances', Elba, IAP preprint 477

Lemoine, M., Ferlet, R., Vidal-Madjar, A., Emerich, C., Bertin, C.: 1993, A&A, **269**, 469

Meneguzzi, M., and Reeves, H.: A&A, 1975, **40**, 91

Meneguzzi, M., Audouze, J., and Reeves, H.: 1971, A&A, **15**, 337

Meyer, D.M., Hawkins, I., and Wright, E.L.: 1993, ApJ, **409**, L61

Prantzos, N.: 1994, Proceedings of the Gran Sasso Conference 'Nuclei in the Cosmos'

Prantzos, N., Cassé, M., and Vangioni-Flam, E.: 1993, ApJ, **403**, 630

Ramaty, R., Kozlovsky, B., Lingenfelter, R.E., and Reeves, H.: 1997, ApJ, in press

Rebolo, R., García-López, R.J., Martin, E.L., Beckman, J.E., McKeith, C.D., Webb, J.K., and Pavlenko, Y.V.: 1992, in *On the Origin and Evolution of the Elements*, eds. N. Prantzos, E. Flam, and M. Cassé (Cambridge: Cambridge University Press)

Reeves, H.: 1994, Rev Mod Phys, **66**, 193

Reeves, H., and Meyer, J.P.: 1978, ApJ, **226**, 613

Reeves, H., and Prantzos, N.: 1994, ESO Workshop on 'Light Element Abundances', Elba, IAP preprint 477

Ryan, S.G.: 1992, in *On the Origin and Evolution of the Elements*, eds. N. Prantzos, E. Flam, and M. Cassé (Cambridge: Cambridge University Press)

Ryan, S.G., Norris, J.E., and Bessel, M.S.: 1991, AJ, **102**, 303

Ryan, S.G. *et al.*: 1992, ApJ, **388**, 184

Terasawa, N.: 1992, in *On the Origin and Evolution of the Elements*, eds. N. Prantzos, E. Flam, and M. Cassé (Cambridge: Cambridge University Press)

Walker, T.P., Mathews, G.J., and Viola, V.E.: 1985, ApJ, **299**, 745

Witten, E.: 1984, Phys Rev D, **30**, 272

10.6 Discussion

Question

(Rees): A question about your calculations on the quark–hadron transition and how that affects the elements: in those calculations you must have made some assumption about the scale of the inhomogeneities that result. How could you determine that, because I thought that it was one of the main uncertainties?

Answer

(Reeves): Yes, the most recent calculations of the quark–hadron transition had given an upper limit to the scale. The lower limit is given by the fact that, if you go to very low limits, protons and neutrons have time to diffuse, so you are back on the homogeneous case. We have thus worked in the rather narrow band included between these two limits.

Question

(Longair): Could I go back to your work on the MeV particles in the process of star formation? Presumably this gives you a real measurement of the amount of ionization by low-energy particles within the star-forming regions. An interesting question is whether or not it comes out around the correct value needed to produce the observed degree of ionization and excitation of the molecules in the clouds?

Answer

(Reeves): The problem is that, around this source, there are probably large ionized regions. All of this is embedded in clouds, but if you have large ionized regions, then, of course, you cannot use the ionization rate because it is already ionized. This has been considered and it seems that it does produce a lot of ionization. However, we know about these γ-rays, and we have to find some way of explaining where they come from. The lines at 4.4 and 6.2 MeV seem to be rather convincing.

Question

(G. Burbidge): Following up on that, if you believe that the low-energy cosmic rays are very important, particularly in regions where stars are being formed, you also mention the possibility that some of the low-energy stuff would come directly from Wolf–Rayet stars, although I am not too sure what you were saying.

Going off for a moment to the so-called starburst galaxies, where there is extensive star formation, can you argue that you would expect more general effects associated with a very large-scale production of low-energy particles? Presumably, if the same physics is working there, there will be a tremendous amount of energy devoted to this type of mechanism?

Answer

(Reeves): It is an interesting remark. I had not thought about it, but certainly it would be worth considering this implication of the observation.

Question

(Pagel): You dropped a hint that you thought that primordial beryllium and boron might just be detectable, at very, very low metallicity. Can you give us some ideas of the sorts of numbers involved: how far down in low metallicity would one have to go?

Answer

(Reeves): In the best case, the highest production of the quark–hadron phase transition, you are only one order of magnitude, perhaps two orders of magnitude, below the observation, not more. That means around 10^{-14} or 10^{-15}.

Question

(Sunyaev): What is the present best theoretical estimate for the lifetime of the proton and what is the strongest experimental limit?

Answer

(Reeves): The strongest experimental limit, if I remember correctly, is above 10^{31} yr. Estimates are very much dependent on the models. The minimal SU5 predicted 10^{29} yr, and we know that this is dead; there are many reasons why this is dead. Other groups give numbers which can go up to higher values. I do not have the numbers in my head, but they scale from 10^{31} yr up to larger values.

Comment

(Rees): I just want to make a very pedantic pedagogical point, about the statement that was made about what would happen if there were too many monopoles: that the Universe would recollapse very soon. In fact, if you believe inflation, what would happen is that the Universe would still go on expanding forever, but what would go wrong would be that, after 10^{10} yr, it would be monopole dominated and the baryon density would be only 10^{-15} of the monopole density. So you can see whatever the density is in this condition, whatever the matter content is, you can still have a flat Universe that goes on forever. What goes wrong really is that the proportions of things at late times would be very badly wrong. That is an entirely pedantic point.

Question

(R. Rebolo): You mentioned that the emission lines in Orion are due to acceleration of carbon and oxygen. Why do you not think that there are protons, or other particles, accelerated and heating carbon and oxygen in the medium?

Answer

(Reeves): Because, in the medium, there are silicon, magnesium, and iron. So, if there were fast protons hitting these elements, you would see the lines of the silicon, magnesium, and iron below 3 MeV. Another reason is based on energetics. It requires a lot more energy to do the same thing with the proton than with carbon. But that is an indirect argument. The really good reason is that you do not see the silicon and magnesium lines.

Question

(H. Zinnecker): I would like to come back to what you said about supersymmetry. Could you imagine that the dark matter is related to a supersymmetric particle, a supersymmetric analogue of some particle?

Answer

(Reeves): One of the fashionable candidates for dark matter is the gravitino. Since the neutrino would not do the entire job, because it would not make small-scale structures, it is popular to ask for the gravitino, or the photino, or other matter. These particles, though, still have to be found, so this is an open problem. Do they exist?

Question

(E. Martín): Are there prospects for detecting monopoles using a telescope?

Answer

(Reeves): Well, they have been searched for, but you could not detect them with a telescope; they have been searched for with a magnetic superconductor. The funny thing is that, as this apparatus was being prepared, several years ago, one was found. So the detection rate claimed was one per month, but after a year the total number detected was still only one, so they claimed one per year and so on. So far, that one which was found is the only one that we have. So, it was either beginners luck or bad luck!

If there are no more questions, I have prepared some advice for students and young people.

Answer

(G. Burbidge): Yes, of course, and advice for us too!

Answer

(Reeves): This is something that I mentioned after the talk by A. Sandage: this famous detection by Songaila *et al.* (1994, Nature, **371**, 43) As you go deep into space, the fossil radiation from the Big Bang should be hotter [shows viewgraph]. This is our 3-K point for the fossil radiation and here is the new result in a quasar at $z = 1.99$. The first detection was here and, of course, the story is that you only have to be above the regression line because, if you find something hotter, we are used to the fact that galaxies have hot places, like stars, for instance, so we are not surprised to see hot points. The requirement of the Big Bang is that you should not find anything below here, cooler than the line, and the detection was there.

I think that this is a very important proof of the reality of the Big Bang: the Universe was hotter in the past.

My advice to young people is the following. It would be very bad to stop searching, because you could find points that are below the line. The normal statement is that, in general, the theorists compute until they find what the experimenter tells them they should find, and the experimenter looks as long as the theorists tell them what they should find. So, it would be a very bad strategy not to keep on going here because, if a point were found here, below the line, it would be very bad for the Big Bang. So the general statement is: theorists should not have too much confidence in experiments and experiments should not place too much confidence in the theory.

'Keep on searching even when you have found what the theorist (or the observer) told you you should have found.'

Comment

(G. Burbidge): I must add to that, just a little. In my view, observers are driven very heavily by the theory. They do not like finding things that do not correspond to the theoretical prediction. In practice, what will happen is that they will work very much harder at those results than at the results which simply confirm the theory. If they decide to publish them, they will find them very hard to get published. There is certainly a very strong bias. It was clear from the time that the microwave background was first discovered, at the time when Penzias and Wilson found it, but it was not confirmed to be a blackbody. It was asserted to be a blackbody by many very well-known individuals, and observations which departed from the blackbody, of which there were a number near the peak, were always assumed to be wrong, so there is that bias.

About that deuterium measurement: there is another problem which Margaret probably knows more about than I do. The problem is that the effect is so small that, if the abundance is really lower and below that curve, it is going to be really hard to convince anybody that it is lower and real.

Answer

(Reeves): I am sorry. This was not deuterium at all. This was the temperature...

Question

(G. Burbidge): I thought it was Songaila...

Answer

(Reeves): Yes, the same author, but another paper. It is the temperature argument. As I said, it is important to keep on looking in this object and see if there is anything below. Just two points are not enough.

Question

(Sandage): How were the data obtained on that fantastically wonderful point on the theoretical curve? That is, what measurements were made? What lines were used? Because it was always considered to be a very difficult experiment.

Answer

(Reeves): I do not remember well enough; perhaps somebody has more idea. I think Bernard Pagel knew about this experiment.

Comment

(Rebolo): I know the lines: they are neutral carbon lines in the high-absorption system. They are at 1560 Å, redshifted to the optical. It is very difficult; the equivalent widths are around 20 or 40 mÅ in the spectrum of a quasar of magnitude 17.5, or so.

Question

(Sandage): The early result by Meyer *et al.* (1986, ApJ, **308**, L37) was a factor of two greater than that; why is this result so much better?

Answer

(Rebolo): They were lucky. They were observing a BL Lac object that suddenly became very bright, and therefore they could take a high-resolution spectrum: in this case it is a quasar, at normal magnitude.

Answer

(Reeves): I have the data here [shows viewgraph]. It is neutral carbon, and these are the low-energy lines here, from the ground state, first excited state, second excited state, and here is the spectrum. The spectrum is here and these are the lines from the ground state, first excited and second excited state, and, of course, it is by taking the ratio of the population in each of these states that the temperature is found. This was the technique.

Comment

(Sandage): In a sense, this is the important observational proof of the reality of the expansion of the Universe. One had always hoped that the total-surface-brightness test would be a test, but this is so clean that it is a spectacular result.

Answer

(Reeves): Yes, I am glad you say that, because I think it has not been emphasized how important this result was. Sunyaev talked about similar results from their work, but I think that it is even clearer than the Sunyaev–Zel'dovich effect. It is really a very, very important result.

Question

(Rees): If we have a moment, I just thought I would mention one topic that has not come up very much. You mentioned the dark matter, and the possibility of a different kind of dark matter giving $\Omega = 1$. I would like to ask Allan Sandage, I suspect, what the prospects are for the more modern classical versions of measuring the deceleration, because it may turn out that, if one can improve the classical methods (maybe using supernovae at $z = 0.5$ as standard candles, etc.), this may be one of the ways in which we can get a better handle on the overall deceleration than by finding evidence for dark matter diffused in superclusters. So, I wonder if you could comment a bit on the prospects for the various classical methods for q_0 using supernovae, angular diameters, and what not?

Answer

(Sandage): The problem has always been: what is the intrinsic dispersion of the object that you are trying to compare locally and at very large-redshifts? The first-rank cluster galaxies have always suffered, as you know, from evolution, so you have to fold in a theory which is not well enough known. Supernovae, right now, it seems to all of the people working on the field, are the very best. But the question really is: what is the dependence of the absolute luminosity of type-Ia supernovae on metallicity? As you go back in time, the metallicity presumably is a function of time as well, and that is going to foul things up. At the moment, the most powerful, it still seems to me, is the timing test, but can we ever determine the age of that galaxy well enough? I think we will be able to determine the Hubble constant, eventually, to 5%. I am very confident in the data from the *Space Telescope* using type-Ia supernovae. When the present generation die, the young people will be able then to get the Hubble constant to 5%, but I am not sure about the age of the Galaxy. So, supernovae of type Ia, right now, I think are the very best method.

Comment

(Longair): Just on Allan's remark. I agree with everything Allan said, but one should be aware that physical methods of doing q_0, I think, must be the way. Eventually one has got to check the consistency of everything. The three physical methods for measuring genuine physical distances at large redshifts, simply linked to angular diameters are, in a sense, much cleaner, because they do not involve any assumptions about evolution. They are, though, critically dependent upon understanding the physics of the objects that you are dealing with. So, if you are a great optimist (and most cosmologists have to be great optimists or we would not be in the game), one could believe that using the Sunyaev–Zel'dovich effect in large-redshift clusters, which could be combined with an understanding of the cluster structures, could give physical dimensions at large redshifts and hence give you an estimate of the deceleration parameter.

Similarly, the gravitational lenses: if we can understand the lensing bodies, they may well give us good estimates of physical scales. The third method is various versions of the Baade–Wesselink method, as applied to the large-redshift supernovae, or expanding systems. So, again, I think that, along with the types of analysis that Allan is recommending, all of these have to be very fairly studied, because in the end we have to get them all consistent and self-consistent. If I had to bet, I would bet that probably it would be simpler, in the end, to use the physical methods than these methods, simply because they are intrinsically cleaner. But you have got to understand the physics.

11   The observational appearance of accreting black holes in X-ray binaries

Rashid Sunyaev

11.1   Introduction

A key area of modern-day science is observation of transient sources with the new generation of satellite observatories. Observations of accreting binary systems are reviewed here, with special reference to observations at high energies with satellite observatories. Many objects are found that show a characteristic signature of highly collapsed binaries containing either a neutron star or a black hole. The observations permit astroarchaeology to be carried out, looking at the state of recent activity of the Galactic Centre region on time scales of hundreds and tens of thousands of years. There are two ways to carry out science: the first is to construct more, bigger, and more sophisticated instruments to observe more and fainter objects with photon-limited systems. This is my own preference, but, in recent years, I have had to switch to a second mode, and that is to take advantage of *Targets of Opportunity*. These are extremely bright objects with a lot of photons, so one is able to do spectroscopy, or photometry, or polarimetry. Actually, at present we cannot do polarimetry, but we dream of doing it in the next few years with a new spacecraft. We have found, though, that we can do a lot of research into Galactic black-hole candidates just by using Targets of Opportunity, because the 'normal' sources are rather weak.

Some branches of astrophysics are already using Targets of Opportunity exclusively. Astronomers working in neutrino and gravitational-wave astronomy are only *preparing* to use Targets of Opportunity. As Kip Thorne says in his lectures, events such as the merging of two neutron stars, or two black holes, generate large quantities of gravitational waves. Targets of Opportunity are very bright sources, but you do not know *when* they will appear, nor do you know *where* they will appear, so you need all-sky monitoring. Astronomers in all bands need to prepare the instrumentation to observe very bright sources and, really, at the moment, we do not have the instrumentation to observe them because they are too bright. For example, we have not got the capability to observe a bright, nearby supernova at only a few kpc distance, because it is too bright for the available instrumentation. We are not prepared for such a high flux of photons. I discussed this problem recently with an optical astronomer and he replied that we could use solar telescopes. This, though, is not true because they are intermediate objects, they are much fainter than the Sun, but much brighter than any other astronomical object; for example, one might typically be a source of magnitude -10.

Let us take, for example, the case of a star falling into a black hole of $\sim 2 \times 10^6 M_\odot$, the mass of the supposed black hole at the centre of the Galaxy, and being disrupted. Suppose this happened in Sagittarius A tomorrow: who would be able to observe it? We are talking about an event with an energy in X-rays of $L \sim 3 \times 10^{44}\,\mathrm{erg\,s^{-1}}$, which is a million times brighter than any source that we are observing today. There is no instrument in the world that would be able to observe it because they would all be immediately saturated by such a high flux. It would be impossible to observe, but it would be an opportunity to understand everything about black holes. We would be able to do polarimetry and very high-resolution spectroscopy. We must be prepared to make these observations.

I believe that the fluxes are orders of magnitude lower than those that the neutrino and gravitational-wave astronomers are preparing to observe, but because we get so many photons in the range from radio to γ-rays, we are unable at present to do it.

Here I will concentrate on experimental, rather than theoretical, matters because, ten years ago, I was obliged to start experimental work, although I admire Chapter 10, which describes these things from a theoretical point of view, which is my own preference too. Here I will talk about simpler things!

11.2 Observing transient sources

We X-ray astronomers are rather lucky, because we have had many all-sky instruments: *Ariel-5*, *GINGA*, *GRANAT/WATCH*, and the *GRO/BATSE* spacecraft. These spacecraft have discovered a number of X-ray novae, and these are the main class of Galactic black-hole binaries which are known today. A number of the people present at the meeting on which this book is based, such as Dr Rebolo and Dr Martín, have observed these objects in the visible and have made an enormous contribution because these objects, despite being very bright in X-rays, are very faint in the visible, typically about 21st magnitude. Observing these stars, the researchers have found that they are really peculiar and have given us the opportunity to measure the mass of Galactic black holes from their rotation periods.

Those sources, identified as black-hole candidates from the hardness of their X-ray spectrum, which are then observed optically from the Canary Islands, Chile, Arizona, or wherever, we now appreciate to be really good black-hole candidates, much better than Cygnus X-1 which we have been observing for more than 25 years now.

A really great discovery, made in the last year, has been that of two Galactic superluminal sources. These are very bright sources and they appear to be the 'missing link' between the Galactic binary sources (we do not know yet whether they are neutron stars or black holes) and extragalactic sources with superluminal jets. I think that this is a really big discovery because it shows that the physical processes causing jets are the same, through protostars, to Galactic black holes, and to extragalactic objects with superluminal jets. These are very bright sources which appear for just a few months, or even a few days. Observations of these sources in the radio, IR, and optical are giving very important results. For example, the second superluminal source, an X-ray nova in Scorpius, reached, in just a few days, after the X-ray detection, a radio flux of 7 Jy. This is a very high flux, and, what is more, for the first time we were able to detect in the radio a beautiful, long, straight jet, which appeared and then later disappeared. The time scale of the disappearance was just a week. In other words, a week later there was, once again, nothing there. This is very good science with Targets of Opportunity; not sources just for people with access to very large instruments, but bright sources, which are very short lived.

I have been thinking a lot about beaming from neutron stars. In 1994 a source, 0535+26, appeared very close to the Crab nebula in the sky. We detected it on 20 February and, at 20–100 keV, it was more than 8–10 Crab. This was an excellent source for examining the rotation of a neutron star and for seeing the acceleration of the rotation of the neutron star owing to accretion. If we had a polarimeter I think that we would even be able to tell exactly what scheme of beaming we have coming from the neutron star: whether it is a broad jet or a pencil beam.

The main topic that I wish to discuss here is the X-ray binaries and, in particular:

   (i) their mass function;
  (ii) radio sources (jets, superluminal expansion, the link between accreting black holes in the Galaxy and supermassive black holes in AGNs);
 (iii) the production of lithium and deuterium;
 (iv) Sagittarius A in the recent past (400 years ago) and in the more distant past (30 000 years ago);
  (v) the possible contribution of NGC 4151-type sources to the X-ray background.

11.3 Satellite observatories and their results

Here I will discuss principally what we are doing with the two Russian spacecraft which we are operating succesfully with broad international collaboration, for which we are very grateful. We have British colleagues at Birmingham University, Dutch colleagues from Utrecht, Germans from Garching, French scientists from Sanclay and Toulouse, and Danish scientists from Copenhagen.

The first satellite is *GRANAT*. This has already been flying for more than five years: on 30 December 1994 it had its fifth anniversary. It is the first Russian satellite to operate for such a long duty cycle as 59% of the total time. Previously, not much importance had been given to such details. We have an excellent orbit owing to the Proton launcher, and now we are theoretically out of the Earth's magnetosphere for 100% of the orbit.

Unfortunately, solar activity is so low at present that we are detecting the Earth's magnetosphere out to 60 000 km distance, owing to the weakness of the solar wind. Just after launch, though, solar activity was very high and then we saw a major flare every ten days or so, very strong deuterium lines, very strong 5.4-MeV lines from the Sun.

Virtually everything that we know about Galactic black holes comes from our observations. These candidates are sources with no visible surface and very high mass, although the mass is actually measured by visible observations. The signatures of a black hole are very high gravitational potential and high-energy release owing to accretion. The Eddington luminosity has been mentioned many times in this book and is simply proportional to the mass of the object, although I should point out that this is the same for neutron stars. The second fingerprint is small dimensions and thus rapid variability, on scales of 10^{-5} seconds:

$$t \sim r_g/c \sim 2Gm/c^2 \sim 10^{-5}M/M_\odot. \tag{11.3.1}$$

We have discovered quasiperiodic variability from black-hole candidates, but it is 0.03 Hz ($\sim$30 s) and not at 10^{5} s; there is no X-ray detector in the world today that can measure, not just quasiperiods, but also stochastic variations faster than 10^{-2} s. Trying to observe material rotating around a Kerr black hole is fantasy at present. You need more than 10^5 photons per second, but even then you have the problem of dead time. There are no electronics capable of doing this. Another fingerprint is the small surface of the accretion disk and thus very high temperatures. Also the mass must, of necessity, be $> 3M_\odot$.

There are also negative indicators: no solid surface; no magnetic field; no regular pulsations; and no X-ray bursts (X-ray bursts are connected with nuclear reactions on the surface). There are no nuclear explosions due to the reaction of helium on the surface of the star, which distinguishes these objects from neutron stars.

How numerous are black holes? In the LMC we observe four bright X-ray sources that are thus perhaps black holes: of these, LMC-1, LMC-3, and possibly LMC-2 are black-hole candidates; only LMC-2 is a pulsar. Thus, the majority of bright sources in the LMC are serious black-hole candidates; two of them, LMC-3 and LMC-1, are very good candidates.

11.4 X-ray sources in the Galactic Centre

Next, we have the centre of our Galaxy, which we have carefully investigated over the last eight years. The central kiloparsec of our Galaxy is a forest of enormously bright sources. Almost all of these sources show X-ray bursts. This shows that, on the surface of these neutron stars, you have accretion and large quantities of helium, in which nuclear reactions propagate in an enormous wave, over a few seconds, giving rise to a big burst. We observe hundreds of bursts from this region. Almost all of the sources are bursters of this type and thus are neutron stars. Just one of the sources, discovered by *GRANAT*, is a black-hole candidate. The surrounding bursters are so bright that we cannot detect anything at all from the innermost 100 pc.

If we look at the map of the central 100 pc from *GRANAT* we can already detect Sgr-A: it is one of the brightest sources in this region and has a remarkably hard spectrum – in fact it has one of the hardest spectra that we have observed in the last eight years. It is also a variable source. Around it, we find a number of bursters: these are just neutron stars. When we see one of these bursters it is just 'routine science' and we are thus not interested in it! Another very interesting source is close by the Galactic Centre. It was discovered by the *Einstein* satellite and is separated by 40 arcminutes from the centre. It has given us two very strong outbursts in the electron–positron annihilation line. It also shows gas. If you put Cygnus X-1 at the same distance, the two sources would appear identical in strength, but, despite this, it must really be an enormous source because it shows huge variations and it is also in the centre of the giant molecular cloud, and there the extinction is tremendous: more than 26 magnitudes in the visible. It is totally impossible to observe this source even in the infrared. In X-rays, though, it is a quite beautiful source.

The most important results from X-ray observations are the following. The central source in the Galaxy is Sgr-A. Its luminosity is, however, only 10^{36} erg s^{-1} in the 3–100-keV range. It is variable: we have detected variations on two occasions. It has a very hard X-ray spectrum. However, for a $2 \times 10^6 M_\odot$ black hole, the X-ray luminosity is only $\sim 10^{-8}$ of the Eddington luminosity, an enormously low luminosity for a black hole. Even if we include all the objects within the central arcminute, including all seven helium stars discovered by Jensen, everything together gives $< 10^{39}$ erg s^{-1} (10^{-5} times the Eddington limit). If there is a black hole there it is completely silent and nothing is feeding it and there is a great question mark over why that is so.

I should mention here that we have a similar situation in M87 and in many Galactic transients, which are silent for tens of years, and then a small star in the vicinity decides to feed it and it becomes enormously bright. Optical astronomers then find it and it becomes a black-hole candidate. Strong accretion in these sources occurs about once every 50 years.

If we turn to the central parsec, we find that it is the densest and most massive star cluster in our Galaxy. A South Korean, called Lee, has modelled this and announced some impressive results at the IAU General Assembly, estimating that there must be 1000 black holes of $\sim 10 M_\odot$ in this cluster according to his model. There must be around 10^5 neutron stars too in this cluster. X-ray binaries, however, are found to be absent, although we have observed them in globular clusters. Why do we not see them? The reason is very simple: the velocity dispersion is very high; thus it is extremely hard to form stable binary systems owing to three-body collisions, which will break them up.

These sources do appear and disappear, however, and are found to be very bright compared to the Eddington limit for Galactic binaries. During our observations we also observed the diffuse radiation from this region, and it coincides with the giant molecular clouds in the Galactic Centre region. The mass of hydrogen in these clouds is around $10^7 M_\odot$, although it may be $10^8 M_\odot$. There is a scattering from the molecular hydrogen, and we can thus calculate how Sgr-A was in the past. The distance is of the order of 100 pc, so the total delay time from the molecular scattering is some 400 years; thus, we can perform astronomical studies of the history of Sgr-A from observing the scattered radiation in the diffuse source. There is Compton scattering from the source, and the energy of recoil is greater than 13.6 eV, so you do not just get scattering, you also get X-ray ionization. Thus, we can get the following results:

(i) The average X-ray luminosity of Sgr-A over the last 400 years has been $< 10^{39}$ ergs.
(ii) The luminosity cannot have reached the Eddington luminosity over the last 400 years for more than a total of 3 hours!

If you take the whole mass of the Galaxy, with $10^{10}M_\odot$ in atomic and molecular hydrogen, and use the same argument, the upper limit is that the Galaxy has had

$$I_E < 1 \text{ year}$$

during the last 30 000 years. In other words, this gives very strong limits to the luminosity. This is astroarcheology, a very powerful research method.

If we go further, to harder X-ray energies, for example with the SIGMA telescope on the *GRANAT* satellite, we can look at a map which is based on an integration of 6.5×10^6s ... I do not think that you can match that kind of integration in the Canary Islands! This map, made in 1994, has a sensitivity of 5×10^{35} erg s^{-1} for the 35–100-keV range, for any source. Most of the sources are X-ray bursters, but there are two black-hole candidates. Sgr-A is a big source; I personally think that it is complex, with a luminosity of $\sim 7 \times 10^{35}$ erg s^{-1}.

If we go to the next energy band up, 100–200 keV, almost all of the sources disappear and we are left with just two: the *Einstein* source which I have already mentioned (1E1740.7−2942) and a second source, discovered by *GRANAT* (GRS1757−258). These are the only two sources in the central parsec observed in this band. According to Mirabel at Sanclay, these two sources show beautiful jets and a compact radio source which appears and disappears. There is no optical identification for either of these sources owing to the enormous optical extinction. The Mirabel results show that the source is in the molecular cloud and has an X-ray jet which is variable according to the X-ray brightness. We thought that this might just be a projection effect, with us seeing an extragalactic object in projection, but there is a real correlation with the Galactic source.

If we then go to the band from 200–400 keV there is now only one source, the *Einstein* source: all other sources have disappeared and there is just this single, extremely hard source. If we go to the vicinity of the 511-keV line, we see no sources whatsoever. The 1σ limit on the map in this line is 1.4×10^{-4} photons cm^{-2} s^{-1}.

11.5 Differentiating between neutron-star systems and black holes

If we look at how the spectrum of a black hole differs from that of a burster, we see that the energy spectrum peaks sharply for a burster at around 5 keV. For the black-hole candidates, the bulk of the energy is coming at around the 100-keV range. The effective temperature is 30 times higher for a black hole than for a neutron star. The spectrum of the black-hole source is very similar to that of Cygnus X-1.

So, what are the possible fingerprints for identifying black holes? Here are the two features that are common to all good black-hole candidates:

   (i) An extremely hard-X-ray tail (detectable to energies >100 keV).
   (ii) Soft–hard spectral state transitions.

This second point is a new factor. In GX 339–4 we observe two states: a soft state with a sharp high-energy cut-off and a hard state with a beautiful power-law spectrum up to very high energies. In 1971 Giacconi detected the transition of this source from the high, soft state, to a low, very hard state. During this transition a radio source appeared, coinciding in position with a blue supergiant. Then, optically, ellipsoidal variations were detected along with simultaneous accretion, and it was then possible to estimate the mass of the source. In the 1970s there was a very active debate because some people argued that this power-law tail shows that it is not a thermal source. But if you have a hard-X-ray detector, you can see that, at energies above 100 keV, the spectrum starts to turn down, showing that it is Comptonization.

Why does this happen? The mechanism was completely predicted by theory. We have an optically thick accretion disk with each point emitting as a blackbody:

$$kT_e = 1.5\text{keV}\left(\frac{M_\odot}{M}\right)^{\frac{1}{4}} \frac{\dot{m}}{\dot{m}_{\text{init}}} \left(\frac{3R_g}{R}\right)^{\frac{1}{4}} \left[1 - \left(\frac{2R_g}{E}\right)^{\frac{1}{2}}\right]^{\frac{1}{4}}. \tag{11.5.2}$$

Thus, the effective temperature of the radiation depends only on the distance from the centre of the source. If you integrate this spectrum you get an exponential cut-off. This we can observe and measure and use to estimate the radius. Even though the temperature changes, we find that the internal radius remains the same.

The second effect is the Comptonization. Low-frequency photons are scattered off hot electrons. Because of Doppler effects there is a shift in the peak of the energy spectrum, photons diffuse to higher energies, and we form a beautiful power law with a high-energy cut-off. The exponential law that we see in the energy spectrum converts in the case of the source into a power law, which is what we see. The exponent of the power-law is dependent on the tail. This is just an analogue of the Fermi acceleration law for cosmic-rays: it is just the same mechanism.

The Comptonization model produces an excellent fit for the spectrum of Cygnus X-1 over the whole range from 2 to 300 keV. One thing that I cannot understand is why we have the same temperature for each black hole; if you have different temperatures, at different distances from the black hole, everything should be different. When we make relativistic calculations with this temperature of 57 keV we find that the optical depth $\tau = 1$; there are only $4\,\mathrm{g\,cm}^{-2}$ of matter through the accretion disk. We can measure how these 4 g radiate, and the quantity of energy that they produce is enormous.

Looking at the spectrum of NGC 4151, the shape is exactly the same and the temperature is the same. The cut-off at high energies is the same, although the amount of matter across the disk is less: only $2\,\mathrm{g\,cm}^{-2}$. This shows that some supermassive black holes have similar gas physics around them. We have observed only one source in eight years which has a harder spectrum than these black-hole candidates and that was SN1987a. The spectrum is formed by the same process of Comptonization, but the nuclear γ-ray lines are moved to lower energies. This spectrum is harder than a black hole's. Maybe a black hole was formed in SN1987a – I do not know.

What are the differences between neutron-star spectra, black-hole spectra and X-ray-pulsar spectra? Compare Scorpius X-1, which has a soft spectrum, with an effective temperature of a few keV, with A0535+26, an X-ray pulsar, with a period of 104 s, which has a very flat spectrum, exponential decline, and no flux above 100 keV, and then V404 Cygni, the best black-hole candidate to date, with a mass of $6M_\odot$. In V404 Cygni we see a black-hole spectrum which is the hardest of the three. My own explanation is the following: a neutron star has a surface which produces soft-X-ray photons and cools down the surrounding gas, whilst the black hole has no surface and so there is nothing to cool down the disk, so its temperature rises.

We can classify sources by their bremsstrahlung temperature. All sources which have a temperature below 70 keV are neutron stars and all those with a temperature above 70 keV are black holes. This is a very interesting division. We cannot prove this, but when you put together the optical identifications, you immediately see the separation into two groups. We find that the black-hole candidates are much harder than the neutron-star candidates. We can do the same thing with the photon index. If the power-law index is flatter than $\alpha = 2.5$, it is a black-hole candidate; if it is steeper, it is a neutron-star candidate. I do not know why this happens, but it is so.

To finish, I would like to talk about the quasiperiodic variations of the X-ray flux from black-hole candidates. Strong deviations from Poisson statistics are observed in the arrival of photons from black-hole candidates. The photons show low-frequency noise (flickering), plus quasiperiodic oscillations. We have already discussed how the black-hole candidates should show variability time scales of 10^4–10^5 Hz. We have, though, quasiperiodic oscillations in five black hole candidates, and they all show oscillations at 0.03–10 Hz. GX 339−4, for example, shows oscillations at 0.83 Hz in *GRANAT* observations. On top of the Poisson noise we see low-frequency noise from flickering in the power spectrum and, superimposed, the oscillations at 0.83 Hz. The Perseus X-ray nova shows the same type of power spectrum: flickering noise and oscillations at ~0.4 Hz.

In Cygnus X-1, none of my students wanted to make the study of the power spectrum, saying that both the Americans and the Europeans had observed it a lot and seen absolutely nothing. In the end, I asked my youngest student, who was just 19, if he knew anything about fast Fourier transforms, and he said 'yes'. So I then asked him if he knew anything about black-hole candidates and he replied 'no', and so he did it for me. This is the most beautiful case: once again we see flickering from 100 Hz up to 0.2 Hz. Here we see quasiperiodic oscillations at 0.03 Hz. When we announced this result, other people started to find it in their data; the oscillations were also found in *EXOSAT* data, and had been observed for ten years, since 1982. Why is the frequency so stable? Sometimes it disappears, but it always comes back in the same place.

This, to me, is accretion-disk seismology. The way to learn a lot about the structure of the accretion disks is to study the instabilities in the accretion disks. It is important to scale to larger accretion disks, and if we do the scaling, we find that the frequency comes out to be ten years in the accretion disk of NGC 4151. This must be a very good quasiperiod, with a lot of power in it.

11.6 X-ray novae

I feel that I ought to say just a few words about detecting X-ray novae. We observed the Galactic Centre field on 24/25 September 1993 and saw the black-hole candidates, but nothing unusual. The next day, a 1.5-Crab source had suddenly appeared in the same field, in a place where previously we had seen nothing. This source, an X-ray nova, was ten times brighter than anything else around the Galactic Centre. The flux from this source grew and was then stable for virtually 100 days. The spectrum was hardest initially and then became pretty stable. The stable spectrum was, again, a copy of Cygnus X-1, with disk temperature of 33 keV, Comptonization spectrum, and, in this case, the optical depth of $\tau = 1.4$, equivalent to $6\,\mathrm{g\,cm^{-2}}$ through the disk.

On 23 September 1994, a new source appeared in the vicinity of the Galactic Centre. It was very strong, and once again all the spectral properties, which I do not have time to discuss here, were the same as in other X-ray novae. Observing the different X-ray novae, we see very different light-curve behaviour from one to another: Nova Muscae, for example, goes up and down constantly.

Looking at the spectra of these novae, the best for me is the Nova Muscae spectrum. It has a soft-X-ray component, a beautiful power law, and then a very bright γ-ray line in the vicinity of 511 keV. This might be the red-shifted 511-keV annihilation line, but it could also be the ^{7}Li production line in alpha–alpha reactions. I do not know what the origin of this line is, but I must mention a result from Martín and Rebolo. They detected very strong LiI in these sources, a factor of 10^3–10^4 times more abundant than in the Sun. The reaction is

$$\alpha + \alpha \rightarrow {}^7\mathrm{Li} + \mathrm{p} + 478 \text{ keV}.$$

Some X-ray novae give spectra identical to that of Cygnus X-1. Others have an enormous excess in soft X-rays. We do not know why some black holes emit according to the standard theory and others give this power law and then the cut-off, but those sources are there.

The most important thing, however, is that people like Eduardo Martín and Jorge Casares observe the stars visually and find that, at maximum, they are around 11–13th magnitude, although they are really disk-generated sources. When they go to the old data they see that the stars are only 19th–21st magnitude, so the disk is eight magnitudes brighter than the star during the outburst. When the X-ray source disappears, we see ellipsoidal variations and, using standard physics, we can find the mass function. Comparing some values we have the results shown in Table 11.1.

Some of these values are already over $3M_\odot$, so we know that they are above the limit for neutron-stars. These are the best candidates because, just from this mass function, they cannot be neutron stars.

One last thing. These sources are variable. V404 Cygni was detected as an optical nova in 1938; the star was detected on plates to have brightened by some eight magnitudes, before disappearing again. A0620−00 was detected on old Harvard plates in 1917. We find that a small $0.5M_\odot$ star in a binary orbit about these sources will feed the monster about once every 50 years. We observe about two such sources per year and, taking into account the volume and the observations, we can estimate that there must be ~1000 such black-hole systems in our Galaxy. In other words, they are very numerous. These sources are very old. They have lived for billions of years and, every 50 years, for a few months, they feed the monster.

Finally, the superluminal expansion. A student at Moscow Physical and Technical Institute discovered the outburst of GRS 1923−15 with *GRANAT* and sent the information to Mirabel, who started to observe it at the VLA. He discovered a very bright compact radio source. The outburst occurred on 19 March, but he began to observe on 27 March. The source began to split, with the two components separating, and in the end the separation was more than 1 arcsec. This expansion was superluminal. This was the first superluminal source to be detected within the Galaxy and the speed of expansion of the jet is ~$0.92c$. This was a

Table 11.1. *The mass function for some X-ray binary systems.*

Source	$f(M)[M_\odot]$
Cygnus X-1	0.18
A0620-00	3.2
V404 Cyg	6.2
Nova Muscae	3
Nova Persei	2.2

fantastic result from Mirabel. We think that there are many such sources because, just after that, another source appeared. This was Nova Scorpii. It was a tremendously bright source in September 1994, many times the Crab and, at the same time, a bright radio source of 7.6 Jy. This radio source is once again superluminal, and, as I have said before, the Germans have observed a very interesting jet from it. Before the radio outburst the source had a spectral index of 2.6; after the outburst we saw the standard spectrum for a neutron-star burster. I am very unhappy with this last result, but the spectra are virtually identical to those of neutron stars.

11.7 Future prospects

Here I will finish, just pointing out that this is a very large international collaboration and we are already building the next satellite, *Spectrum X-Gamma*, for cosmology and, at the same time, for studying these Targets of Opportunity. Russia is also involved in ESA's *INTEGRAL* satellite. Russia has agreed to launch this satellite with a Proton booster, into a virtually circular orbit, outside the Earth's magnetosphere, so we can use cooled germanium detectors to study all the lines of research that I have discussed here, from 100 keV to many MeV. This includes nuclear spectroscopy, gyro-lines, and also the electron–positron annihilation line. Everything now depends on the internal state of Russia, but, if the international collaboration continues, I think that we will be able to launch this wonderful satellite.

11.8 **Discussion**

Question

(Longair): Could you sketch for us where all the hot gas and the photons are coming from to produce the Comptonized spectrum in the black-hole candidates? I do not have a clear picture of the configuration. Where did all the hot gas come from? What is the source of the soft photons?

Answer

(Sunyaev): First, it is a fully understandable situation with a standard accretion theory. All the radiation that we are observing comes from 5 to 10 gravitational radii: the main energy-release region in accretion disks. Then, however, the soft radiation disappears and we believe that, in practically the same region, from 3 to 12 gravitational radii, plasma cannot cool down, owing to some instability, and it becomes very hot. The temperature (how hot the plasma is) depends on the number of soft photons which we inject there. These soft photons could come from the clouds of colder plasma inside the accretion disk, which radiate, but it is very difficult. It is impossible to make these sources from free–free radiation. It is possible to make these photons principally from the gyro frequency if the magnetic field pressure is higher than the pressure of the plasma. It is possible to reflect radiation from the outside region of accretion disks (which obviously are optically thick) which are radiating in soft X-rays or in the ultraviolet. In addition, there is the possibility (although it is extremely difficult) that there is a thin, optically thick accretion disk and then a corona, but we cannot then understand the shape of the spectrum.

This is a speech! In the end, I believe that understanding is the same as in the 1970s. There is no change, but we do have much more data. That is all!

Question

(Reeves): Do you have any operability in the 2.2-MeV line in this case where you have the 470-keV line?

Answer

(Sunyaev): No, because the *SIGMA* which was produced in Sanclair has a crystal which is too thin, and the 2.2-MeV efficiency is less than 0.1%. Therefore, it is completely impossible to observe deuterium. I myself believe that, with a real germanium crystal, we will observe deuterium practically everywhere, except in diffuse material, like Orion, where the neutrons have time to decay.

Question

(Reeves): How well does this line at 500 keV, or below, correspond to the 470 keV of the lithium?

Answer

(Sunyaev): I can tell you that we interpret this in two papers. The moral of this story is 'never go away from home'. We finished a paper about this and I wrote in it that the most likely interpretation was that it is the 511-keV line, redshifted to 475 keV. Then I wrote in the last sentence that there is much lower probability that it might also be the production of lithium and therefore alpha interactions on the surface of the accretion disk. My younger colleagues decided that it was too low a probability and took it out when I was away at a conference abroad! In the Russian version they were more sensitive and they left that sentence in! My opinion is the following. I believe that, from a theoretical point of view, it is much easier to produce electron–positron pair plasma. But it is very difficult to produce plasma of the sort that we observe, because the upper limit for the temperature is 3 keV, just from the broadening of the line. Also, there is no Keplerian motion, as it is far from the accretion disk and, at the same time, it is redshifted. I cannot understand how to make a redshift very far from the black hole. In the case of the 478-keV line, it is very easy, but there is the question: where is the 429-keV line of beryllium? Everywhere there are questions.

Question

(Novikov): At the beginning, you said that when you are talking about a black hole, it means a very heavy object. You emphasized that the time resolution is not good enough to give limits on the size if these limits are less than, say, 100 km, or the order of a few tens of kilometres. But you should probably not give such strong limits because, if you deal with a heavy object, you should choose only one or two possibilities: a standard, ordinary star, or a black hole, because there are no intermediate objects between these two. In this case, it is enough to give much weaker limits on linear scales. What do you think about this?

Answer

(Sunyaev): I believe you, in principle, but, in reality, our main enemies in this business are the neutron stars. Nobody knows how neutron stars, without magnetic fields, and without bursts and with super-Eddington luminosity, behave. In the case of lower luminosity it is very easy to construct a series with boundary layers, with emission, with spectra – everything; but, when the luminosity exceeds the Eddington limit, there is a spherization and everything is different. Variability time scales can be as short as for a black hole.

Comment

(Novikov): All right, but probably neutron stars should be your friends, not your enemy...

Answer

(Sunyaev): They are friends! I agree!

Comment

(Novikov): But you emphasized that when we deal with very heavy objects, neutron stars could not have a mass of $16M_\odot$, as you noted. In this case, and starting from this point, probably it is enough to give weaker limits, and that is very important.

Answer

(Sunyaev): I agree with you. If we believe the current physics, then there is no doubt that we are observing at least five beautiful black holes. No doubt about it. But there are people (not in this room) who have some doubts! They are asking, 'would you bet your right arm that this is the case?' I am answering, according to my understanding, 'yes', but I can only prove that there seems to be no surface, and it seems that the mass is very high and the spectrum is very hard. This is a peculiar type of object, and the best explanation is a black hole.

Comment

(Novikov): I believe that everyone should believe at least in *something.*

Comment

(G. Burbidge): That we can agree on!

Question

(Rees): Just a question about the quasiperiodic oscillations, the QPOs. To get the regular period would be understandable in a steady object, but to single out the same period in an object which is so unsteady does seem to be a bit of a puzzle. Especially now, as we cannot use the old explanations that we used for neutron stars. I wonder if you think that the models which have been suggested are at all satisfactory?

Answer

(Sunyaev): The models for quasiperiodic oscillations?

Question

(Rees): Yes.

Answer

(Sunyaev): I am very impressed that the recent outbursts of the radio source AO0535+26 show not only the 104-s period, but also the quasiperiodic oscillations, which were in full accordance with the models. But in our case, it is obvious that, in many neutron stars, we are observing quasiperiodic oscillations, which are in neutron-star systems, where the nature is the same as in Cygnus X-1, without any doubt. I believe, myself, that the natures of all these are connected, which is why the period is rather stable. The period depends only on the luminosity and on the boundary, where the radiation pressure of the disk changes it to a matter-dominated disk. This is a very strict boundary, which depends only on the luminosity, and therefore you have a very distinct frequency. I believe that there is a state (and I have stated this many times) where the matter flow is laminar; then you have states where you make trains and then you send these trains into the black holes. As in Russia today, the trains are not going periodically, but rather quasiperiodically! This gives you a complete explanation of the black hole.

Question

(Rees): But this does require the state to be steady, in terms of the luminosity?

Answer

(Sunyaev): We have observed Cygnus X-1 already, with different instruments, for more than ten years. During these ten years, there have been variations of luminosity, but not more than 2–3 times. The bulk of the time Cygnus X-1 has practically the same luminosity. This is very impressive. The spectrum was practically constant. The temperature is changing only 20%, or 30%, not more, on the accretion disk.

Question

(Rees): This is the same question that I asked Igor Novikov, about the prospects of testing relativity with these galactic black holes. It seems to me there are better prospects for the supermassive black holes. But I wonder if you think that any future tests using these objects would actually tell us about the metric in the 'strong-field' domain?

Answer

(Sunyaev): I myself think that if an X-ray nova were to appear closer than 1 kpc from us, and if the proper instruments were in the proper place, at that moment, it would be possible to observe fluxes at a level of 10^5 or 10^6 photons per second. It would be possible to detect (you may know that we wrote, many years ago, about how it might be possible to distinguish Kerr and Schwarzschild singularities). I believe it would be possible to do such things and to find a lot of additional information. This is very important. Also, I think it would be possible to understand where the inner boundary of the disk is, what the distance is from the black hole. I believe that if we go to very high frequencies, we can observe something that is not even predicted.

Question

(E. Martín): I have a couple of comments. One is about the γ-ray line, of course. It seems to have been suggested in the Russian version of your 1991 paper and in the 1993 paper (in ApJL), that it could be two lines, and the positions of those lines could be consistent with three possible origins: beryllium, lithium, and annihilation. I would also like you to comment on the connection between the superluminal motions, the mass that is estimated to be ejected, and the energy that this involves in the interpretation of the emission line as alpha–alpha collisions.

Answer

(Sunyaev): I believe that Rebolo, one of the discoverers of lithium absorption in three systems, is here and he obviously has a much deeper understanding of this than me, but I can answer the following. First, we see radio sources in all these systems, but they are very weak. In reality, the luminosity which is in the radio is usually 10^{-5}, 10^{-6}, 10^{-7} of the X-ray luminosity; there is a very low luminosity in the radio. But nobody knows what is the low-energy cosmic-ray flux. With these low-energy cosmic rays, we can irradiate the surface of the disk, and I believe that, if 1% of the power is going to accelerate cosmic rays of low energy (with energies of several tenths of a keV), then we can easily produce the alpha–alpha line and all the lithium that is necessary. This is my estimate; I know that you have done something similar. The second thing, which is perhaps more important, is the energy involved in the superluminal motion. We have a disagreement here in the estimates made with Mirabel, because my estimates are lower. Mirabel is talking about luminosities, just the bulk motion in the jet, of the order of 10^{40} erg s^{-1}. If there is 10^{40} erg s^{-1}, and also the cosmic rays, you can produce everything in the system.

Question
(Martín): So why does your estimate differ from Mirabel's?

Answer

(Sunyaev): I will discuss this with you privately, because I am afraid to comment. I think Mirabel is great. He made a great discovery. I am not happy about criticizing this, because the discovery is great.

Question

(S. di Serego): I have a technical question. Could your X-ray polarimeter measure the polarization of the brightest Seyfert galaxies, in addition to your serendipitous sources?

Answer

(Sunyaev): I can tell you more about this. The situation is the following. Initially, we thought it would be possible to observe three to five of the brightest Seyferts: NGC 4151, Centaurus A (obviously), and 3C 273 – these are the brightest sources in X-rays. Now, our computations show that it would be very difficult. It is necessary to spend a lot of time and, as usual, our experience shows us that when you are working at the limits of detection, it is quite useless science. It is much better to observe very bright sources, to obtain very distinct results. But we are producing this polarimeter and I believe that it gives great science. You know that Italy is also involved in the polarimeter: the people from Frascati and Cesar Barbieri.

Question

(H. Zinnecker): I am interested to know whether Sagittarius A* is a black hole or not. I guess you gave us some additional evidence from the X-rays that at least you believe that it is a black hole. Can you summarize what you think is the future of X-ray observations of the Galactic Centre?

Answer

(Sunyaev): I do not have time to describe all these, but there are excellent new results from *ROSAT* and from *ASCA*. I can tell you that we are in contact with Prodell, with Trumper, and with Tanaka and we are discussing this a great deal. There are a lot of controversial things here. *ROSAT* gives rise to the question 'does *ROSAT* see the same source?' (owing to the absorption), and we also have some disagreement with

ASCA. What is important today is that we can prove that our sources, the ones that we are observing, are coming from the central arcminute. But we cannot prove that this is exactly on Sagittarius A. It is a very peculiar source. At the same time, I do not know any argument, except for dynamical ones (this measurement of the rapid motion of stars), for a black hole in Sagittarius. I cannot understand why the emission from the black hole is so weak, because we see a lot of gas there. For example, just with the wind from these helium stars, each of them is losing $7 \times 10^{-5} M_\odot$ per year in the central 20 arcseconds. It is very difficult to understand why this gas does not fall, so it obviously must have momentum. Also, we cannot understand why there is an enormously dense stellar cluster. If you take $3 \times 10^6 M_\odot$ in the central 0.1 pc, you are getting a density of $10^8 M_\odot \, pc^{-3}$. And with 10^8 stars per cubic parsec, we do not see anything peculiar from that region. 'Why do we not have tidal disruption of the stars?' is also a great question and so on. So I think that everyone must be prepared that, perhaps tomorrow, NGC 4151 may appear in the centre of our Galaxy. We must be ready for this, at least, even if the probability, I believe, is very, very low.

Comment

(G. Burbidge): I must add a comment about this situation in Sagittarius A. When I saw the Genzel *et al.* paper, the observation of these high-mass, rapidly moving, very short-lived stars, of course, I was affected by this and I said: 'Well, now I know of no way to make them except the processes that Hoyle and I had been discussing'. It is very difficult to understand how they can live, how they can last, and how they can be formed, because their overall time scale is probably only a few hundred thousand years. It is very short.

Answer

(Sunyaev): Yes, but my own understanding is the following. We see enormously dense molecular clouds in the vicinity. You know that the Sagittarius A cloud is there.

Comment

(Burbidge): Yes, that is correct.

Answer

(Sunyaev): They are very special for X-rays as they are enormous, and Sagittarius A is exactly in the X-ray window. There is no molecular cloud between us and the Sagittarius A dense molecular cloud, but, just in the vicinity, there is an enormous amount of molecular gas. That gas is much denser than the gas in the 3-kpc arm, or anywhere else in the Galaxy. Therefore, I believe that it is possible to form massive stars there. Why have the stars evolved so quickly? Why are they already helium stars? It is a great question. The mass loss is enormous, as estimated from this helium loss.

Comment

(G. Burbidge): Yes, that is right. The star-formation process really has to be very strange and very effective.

12 Reflections on the key problems

Malcolm S. Longair

12.1 Some more key problems

It is impossible to do justice in a few words to the superb quality of the
reviews that appear in this book. Allan Sandage has produced a splendid
list of 23 problems, and I have counted at least three times as many in the
succeeding chapters – there is no lack of problems to keep everyone busy.

The topics discussed in depth have been strongly biased towards
galaxies, including our own Galaxy, extragalactic astrophysics, including
high-energy astrophysics, and cosmology. The intriguing aspect of the
story is how we are dealing with a vast interlocking jigsaw, in which each
piece can strongly influence every other part of the puzzle. Necessarily,
because of the vast scope of the discipline, the picture was incomplete, but
that is a very good thing because it means that Francisco Sánchez and his
colleagues will have to organize future collaborations as splendid as this
one.

Let me give some hints of a few key areas which deserve as thorough
treatments as those provided here. These will certainly impact the key
problems we have discussed in important ways.

12.1.1 Helioseismology, asteroseismology, and stellar evolution

I regard these disciplines as among the most important for the astronomy
and astrophysics of the future. Helioseismic studies have determined the
internal structure of the Sun with unprecedented precision, as well as
delineating the distribution of angular momentum in its outer convection
zones. These are of crucial importance for understanding the operation of

the dynamo responsible for the strong magnetic fields in the outer regions of the Sun and also for the magnetic field reversals which occur during the solar cycle. These studies are of importance for celestial dynamos and for the origin of magnetic fields in general, a problem which Martin Rees emphasized needs much further study.

Asteroseismology is the way to pin down the internal structures of the stars by direct observation. These studies are badly needed so that we can be confident that the standard theory of the evolution of stars from one region of the Hertzsprung–Russell diagram to another is correct. In particular, as emphasized by Allan Sandage and Hubert Reeves, we need a precise understanding of the physics of the stars in order to derive reliable time scales for the age of our Galaxy, and, by extension, for other astronomical systems. Astronomical objects can only be dated through the study of their stellar populations, and this impacts many of the topics described this week. For example, Donald Lynden-Bell requires accurate ages to date the streams which are being disrupted by our Galaxy, Bernard Pagel needs them to calibrate models of the chemical enrichment of the Galaxy, and Allan Sandage emphasized the central role which the age of our Galaxy plays in the determination of cosmological parameters. Those ambitious astrophysicists who attempt to model the evolution of the stellar populations of galaxies are wholly dependent upon the accuracy of the models for the evolution of stars of all masses. If we are to make sense of Martin Rees's phase 3 of cosmic evolution, stellar archaeology is essential – the understanding of the stellar populations of distant objects will only be as good as the tools provided by the experts in stellar evolution.

12.1.2 The interstellar gas and star formation

These key topics have been in the background throughout the book and have been ably promoted by Hans Zinnecker, who has constantly reminded us of the importance of star formation for so many of the studies discussed in this book. What was revealing was the growing similarity between the problems of star formation and those of high-energy astrophysics. There are accretion disks about young stars as well as about binary X-ray sources and probably about the supermassive black holes in active galactic nuclei. Bipolar outflows reminiscent of the phenomena observed in radio galaxies have been observed. Even superluminal motions have now been observed to be associated with energetic Galactic X-ray sources. In the case of star formation, these phenomena are assumed to be associated with the problems of getting rid of the energy, angular momentum, and magnetic fields of the collapsing protostar. The relation may be even closer. Dust is an essential ingredient of the proccess by which stars form, but it also seems to be essential in order to provide the obscuring tori which are inferred to block out the light of the quasar nuclei thought to be hidden in the nuclei of radio galaxies and Seyfert 2 galaxies.

Star-formation studies are booming – at last, the first convincing evidence for the infall of molecular gas onto protostellar objects has been discovered. Since the introduction of infrared array cameras almost eight years ago, I have been looking forward to direct observational determinations of the initial-mass functions for stars in regions of star formation. These are at last beginning to appear, but they have been a long time in coming. These initial-mass functions are needed by all those who wish to build models for the evolution of the chemical abundances and stellar populations of galaxies. These are crucial areas of study for cosmology.

A related problem is the search for brown dwarfs, planets, and rocks which was discussed by Allan Sandage and Donald Lynden-Bell. It should be possible to make some progress by making observations in the infrared waveband with the coming generation of 8-m optical–infrared telescopes equipped with adaptive optics modules.

12.1.3 Neutron-star physics

In many ways, neutron-star physics has become a subset of solid-state physics and electrodynamics. Besides the key information which these studies provide about the properties of superfluids in bulk, the electrodynamics of pulsars remains to be fully understood. As we have heard, these latter studies are needed if we are to make progress in understanding the electrodynamics of black holes, as was described by Igor Novikov. We also need to understand how a neutron star can acquire a couple of planets.

12.1.4 γ-ray bursts

I am not sure where to list the γ-ray bursts – are they a Galactic or extragalactic problem? Are they a local galactic problem soluble by interesting physics, involving mild events on nearby neutron stars, or are they devastating outbursts involving a cosmological distribution of sources?

12.1.5 The highest-energy cosmic rays

The 10^{20}-eV particles are still around to haunt us. The new results from the Fly's Eye detectors in the USA and the large Japanese arrays have shown that particles with energies two or three times this energy exist. These are of the greatest interest because they cannot originate from more than about 50 Mpc from our Galaxy or else they would be degraded by photopion production through interactions with the cosmic microwave background radiation. These ultra-energetic particles have to be accelerated somewhere. Is it a coincidence that 10^{20} V electric fields can be produced in the vicinity of supermassive black holes? Roger Blandford has given an expression for the potential difference developed across the horizon of a black hole as

$$V = 3 \times 10^{20} \left(\frac{L_{em}}{10^{39} \text{ W}} \right)^{1/2} \text{ V}, \tag{12.1}$$

where L_{em} is the luminosity being extracted from the black hole by electromagnetic torques. It would be intriguing if we were able to test this formula by observing the black holes which Rashid Sunyaev demonstrated convincingly are present in Galactic X-ray binary systems.

It can be seen that my list of topics for future meetings is biased towards the wavebands outside the classical optical window, which has been the major arena for the discussions at this symposium. It would be wonderful to have similar meetings involving the maestros from millimetre, far-infrared, near-infrared, ultraviolet, X-ray, and γ-ray astronomy as well. Each of these experts would address the key problems from a somewhat different perspective. I am convinced that some of the key problems may be more easily tackled in these wavebands, as opposed to the optical waveband.

12.2 Exotic astronomy and cosmology

We heard two interesting examples of what I would term *exotic astronomy*. First, Igor Novikov mentioned, more or less as an aside, the possibility of time machines, and this stimulated one of the liveliest discussions of the meeting. The second example was Geoffrey Burbidge's discussion of quasar–galaxy associations and the possibility of non-cosmological redshifts. A comparison of the reaction of the audience to these exotic possibilities was revealing.

What I found particularly amusing about Igor's contribution was the effortless way in which he passed from the bread and butter of contemporary high-energy astrophysics to the exotica of time machines. It is far beyond my competence to pass any comment upon these ideas – as I understand it, we need elements of quantum gravity, negative-energy equations of state, and extra dimensions to create, even in principle, a plausible time machine. One might regard this programme as piling hypothesis upon supposition in physical circumstances which cannot possibly be tested by experiment at the present time or in the near future. Igor made the key point, however, that there is nothing in the formulation of the equations of physics which forbids certain types of time machine, in principle. Maybe these exotic ideas will lead to testable experiments and perhaps these areas will become the physics of the 21st and 22nd centuries. After all, think what physics and astronomy were like only a century ago, before the discovery of X-rays and the electron and when the extra-galactic nature of the spiral nebulae was still a matter of fantasy.

Geoffrey Burbidge described forcefully his concerns about the standard interpretation of the redshifts of quasars. Although I believe this is probably less exotic than time machines, my impression of the audience's reaction was one of mild disbelief. In my view, the reasons for supposing that quasars are no more than hyperactive galactic nuclei are very compelling. As I mentioned in the discussion following Geoffrey's contribution, I put my toe in these waters in 1978 when Peebles and Seldner suggested that there was an excess of radio sources in the 4C catalogue associated with nearby galaxies. This was a case in which it was possible to undertake a rigorous statistical analysis, and Michael Seldner and I were able to show conclusively that indeed a small number of the radio sources are associated with nearby galaxies, but that was because they are low-luminosity radio galaxies, and the statistical excess is entirely due the presence of these objects in nearby groups and clusters of galaxies. One important lesson which I derived from this analysis was just how difficult it is to make proper statistical analyses in such areas.

The other occasion on which I became involved in these discussions was when two of my former colleagues at the Royal Observatory in Edinburgh undertook an analysis of the periodicities in the redshifts of nearby galaxies discovered by Tifft. They made a re-analysis of the data and found a positive result. As Director of the Observatory at that time, I was naturally concerned about this work coming out under the Observatory's name. I asked one of the sternest critics of these ideas to look at the draft of the paper. He suggested some changes, but the positive result would not go away. The paper was submitted and published in Monthly Notices, having passed the standard peer refereeing procedures. I do not understand what is going on. On the other hand, so much of the other uses which we make of redshift data, including the ranges of redshift in which these anomalous effects are observed, seem to make complete astrophysical sense. Most of us are loathe to believe that there can be anything exotic in regions where classical physics seems to work very well indeed. We recall the astounding precision with which general relativity has been shown to agree with observations of the binary pulsar.

Geoffrey urged us to take these problems seriously. I am sure he is correct, but I have to say that I would not put a research student onto these problems – I would be jeopardizing their future careers too much. I am also reluctant to devote time to them myself, partly because there always seem to be many more interesting problems to tackle. Furthermore, I am sure that many avoid these problems because they have no desire to become involved in fighting these types of battles – the a priori probability of these exotic ideas being correct is so small that many doubt whether or not it is worth the very large amount of effort needed to do a really thorough job. On the other hand, I would argue that the only people who should be tackling these problems seriously are those who have nothing to lose, that is, those who are already in tenured, unassailable positions. But these analyses are horribly complicated and they should not be undertaken lightly.

Another interesting reflection is that, although I think Igor's description of time machines is very far out, these studies are constrained by the laws of physics, however tenuously. In contrast, we do not have a glimmer of the physics which would cause non-cosmological redshifts in certain classes of quasar.

12.3 **The future**

For many of the most important studies described in this book, the key instruments of the future are the 8-m optical–infrared telescopes. The era of the 8-m telescope has already begun with the construction of the Keck Telescope and there are others well into their construction phases – the Subaru Japanese telescope on Mauna Kea, the four VLT telescopes of the European Southern Observatory to be built on Cerro Paranal in Chile, and the two telescopes of the Gemini project to be located on Mauna Kea in Hawaii and Cerro Pachon in Chile. These instruments and others like them will become the standard tools for many of the most challenging extragalactic and cosmological programmes.

We were delighted to hear of the advanced state of planning for a Spanish large telescope project. I am sure that we would all want to endorse the scientific importance of this project for Spanish, European, and, indeed, international astronomy in general. La Palma is the natural site for such a telescope.

One of the more extraordinary developments which has occurred, as plans for the 8-m telescopes have been formulated, has been the remarkable degree of cooperation between the projects and the formation of the '8-m club'. This club has acted as a forum for the exchange of ideas on technical issues and it is interesting that the meetings of the club have been of the greatest value to all the projects. I am sure that the club will welcome the presence of the Spanish astronomers and technologists at its meetings.

At the moment, I am the chairman of the Gemini Board, and the project team will, I am sure, welcome collaboration with the Spanish project. One of the most important issues facing all the large telescope projects has been the escalating costs of instrumentation and operations costs. The Gemini Board welcomes collaborations which will enable our communities to gain access to frontier instruments on 8-m telescopes. One idea being considered seriously is that time could be exchanged between projects, if the different 8-m telescopes are optimized for different types of observation. For example, the Hawaiian Gemini North telescope is to be optimized for infrared observations. It may be that other telescopes will be optimized for other types of observation and then it would make sense to come to some arrangement whereby time could be exchanged between telescopes. These ideas are at an early stage, but it suggests that the Spanish astronomers might consider whether or not there are particular areas in which they would wish to develop world-leading capabilities which would enable them, through exchange of time, to gain access to other complementary facilities. We all wish the Spanish astronomers success in bringing off this outstanding project.

If I may ride one of my own hobby-horses, it is to advocate the importance of the near-infrared waveband from 1 to 2.2 μm. It turns out that many astronomical problems are much easier in this region of the spectrum. The basic reason is that normally one is observing the Rayleigh–Jeans region of the spectrum of the star. This is the simplest spectral region to study since it is not affected by absorption lines. A second key point is that the spectra of galaxies in the near-infrared waveband are dominated by the light of old stellar populations, which are much more stable than their spectra in the optical waveband, where starbursts can dominate the integrated spectrum. As a result, understanding the evolution of galaxies in the near-infrared should be easier than in the optical waveband. This simplification seems to work out to redshifts of the order of 2. At greater redshifts, all the problems which afflict optical observations at redshifts of about 0.5 reappear, and then the interpretation becomes more difficult once more. The point of importance is that all large telescopes should make the provision for imaging and spectroscopic facilities in the 1 to 2.2-μm infrared waveband a very high priority indeed.

12.4 Final thoughts

All astronomers like to classify the objects they study, and the same type of analysis can be carried out for the sample of researchers in this book. Three of us belong to the *intermediate-age population*, Martin Rees, Rashid Sunyaev, and I all having begun our doctoral degrees in the period 1963 to 1965. The others were then our heroes and the pioneers of the new astrophysics and cosmology, which was just about to transform the nature of these disciplines in fundamental ways. With the greatest of deference and respect, I will refer to them as the *old population*. The other population may be represented by a large proportion of this book's readership, which is clearly mainly the *young population*.

In pondering what we have read, I am struck by a number of aspects of the presentations.

(i) For me, it has been an extraordinary privilege to read the contributions of those who have lived through the golden age of astrophysics and cosmology of the last 30 years. Indeed, the lecturers were among the pioneers who created the disciplines of contemporary astrophysics and cosmology.

(ii) A second interesting feature of this book is that the chapters reflect the different scientific cultures and traditions in which their authors have worked and studied. Allan Sandage is a direct descendent of the great tradition established by Edwin Hubble and the pioneers of observational cosmology; Igor Novikov and Rashid Sunyaev are representatives of the great Moscow school of astrophysical cosmology founded by Zel'dovich, which has its roots in Friedmann's achievements of the 1920s; Martin Rees and Donald Lynden-Bell are following in the tradition of theoretical astrophysics begun by Eddington in Cambridge. The directions and approaches to research are much more strongly influenced by our intellectual heritage than we would often like to believe. It has been a privilege to be present as these traditions are passed on to the next generation of leaders of astrophysics and cosmology.

(iii) The other very reassuring aspect of the book is that astrophysicists and cosmologists seem to get better and better as they grow older, which is especially reassuring for the intermediate-age population.

On behalf of all the contributors, I thank Francisco Sánchez, Antonio Mampaso, Guido Münch, and all their colleagues most sincerely for organizing this historic and memorable collaboration.

We are all looking forward to the development of a distincive, great tradition of astrophysics and cosmology in the Canary Islands. All the necessary ingredients are in place: a magnificent institute, superb facilities, outstanding scientists, and one of the very best observing sites in the world. We all congratulate Francisco Sánchez for his great vision and success in making all of this a reality. It has been a huge undertaking, but it has brought Spain and the Canary Islands to the forefront of astronomy, astrophysics, and cosmology. It is a magnificent achievement.

Afterword

On behalf of the *Fundación BBV* I would like to express my gratitude to the participants in this encounter on 'key problems in astronomy' which I have to say has been remarkable.

I use the word remarkable because although we have been co-organizing a programme of encounters and colloquia together with the *Instituto de Astrofísica de Canarias* involving internationally renowned figures working in notable areas of astrophysics, the encounter on which this book is based has actually attempted to put forward possibilities, or in other words, hope. That is why it is so remarkable.

The idea behind the encounter was for a group of the most outstanding astrophysicists to review and discuss the most important problems in astrophysics, problems that were never resolved at the time because they were too difficult to tackle or because attention was subsequently diverted towards other issues. These problems were presented to a group of young astrophysicists holding relevant positions in scientific fields for them to learn and contrast. Established scientists are keen for young scientists to play a part in solving the different mysteries that arise when studying the heavens. And the hope of reaching new frontiers is based on the successes of astronomy over the last 50 years.

The methodology used in these encounters is entirely in keeping with the methods we have applied right from the very outset in the *Fundación*, which wants to create opportunities for reflection on different problems, obtaining references, contributing solutions, and building bridges between the academic and research spheres. The *Fundación BBV* thus aims to pro-

[432]

vide a forum that guarantees pluralism, intellectual rigour, and commitment to confront the challenges and opportunities faced by our society.

The contributors, who, like me, believe in the significance of the work being carried out, certainly appreciate learning how to combine the huge sense of dimension and magnitudes not usually open to ordinary people. They are privileged in being able to gaze at the stars humbly and calmly whilst the rest of us look at them with a mixture of amazement, puzzlement, and fascination. We thank them for their attitude of reflection that helps us to understand so much better just how tiny we really are. And I would also like to thank them personally and on behalf of the *Fundación* for allowing me to meet them all, albeit briefly.

I would like to congratulate the Director of the *Instituto de Astrofísica de Canarias* for succeeding in this conjunction of cosmologists and those Spanish researchers, among whom are some of the most important astrophysicists of tomorrow; I am almost sure it will be an unrepeatable event. I want to congratulate all of them for filling us with hope and enthusiasm to continue solving the mysteries of the heavens.

Thank you all very much.

The Director of the Fundación BBV

Contributors

Burbidge, Geoffrey: Center for Astrophysics and Space Sciences, Mail Code 0111, University of California, La Jolla, CA 92093, USA

Burbidge, Margaret: Center for Astrophysics and Space Sciences, Mail Code 0111, University of California, La Jolla, CA 92093, USA

Longair, Malcolm: Cavendish Laboratory, Department of Physics, University of Cambridge, Madingley Road, Cambridge CB3 0HE, United Kingdom

Lynden-Bell, Donald: Institute of Astronomy, University of Cambridge, Madingley Road, Cambridge CB3 0HE, United Kingdom

Novikov, Igor: Theoretical Astrophysics Center, Blegdamsvej 17, DK-2100 Copenhagen, Denmark

Osterbrock, Donald: Lick Observatory, Bldg. Natural Sciences II, University of California at Santa Cruz, 1156 High St., Santa Cruz, CA 95064, USA

Pagel, Bernard: NORDITA, Blegdamsvej 17, DK-2100 Copenhagen, Denmark

Rees, Martin: Institute of Astronomy, Madingley Road, Cambridge CB3 0HA, United Kingdom

Reeves, Hubert: CEN, Saclay, SEP-SES BAT 28 BP2, F-91190 Gif-sur-Yvette Cedex, France

Sandage, Allan: The Observatories of the Carnegie Institution of Washington, 813 Santa Barbara St., Pasadena, CA 91101, USA

Sunyaev, Rashid: Space Research Institute, Academy of Sciences, 117810 Moscow, Russia

[434]

Participants

Abel, Thomas: National Center for Supercomputing Applications, 5261 Beckman Institute, URBANA, IL 61801, USA

Adams, Jenni: Department of Theoretical Physics, Oxford University, Keble Road, Oxford OXI 3NP, United Kingdom

Alonso Sánchez, Angel: Instituto de Astrofísica de Canarias, C/ Vía Láctea s/n, E-38200 La Laguna, Tenerife, Spain

Aparicio, Juan Antonio: Instituto de Astrofísica de Canarias, C/ Vía Láctea s/n, E-38200 La Laguna, Tenerife, Spain

Aragón Salamanca, Alfonso: Institute of Astronomy, Madingley Road, Cambridge CB3 0HA, United Kingdom

Arribas Mocoroa, Santiago: Instituto de Astrofísica de Canarias, C/ Vía Láctea s/n, E-38200 La Laguna, Tenerife, Spain

Beckman, John: Instituto de Astrofísica de Canarias, C/ Vía Láctea s/n, E-38200 La Laguna, Tenerife, Spain

Benn, Chris: Isaac Newton Group, Observatorio del Roque de los Muchachos, PO Box 321, E-38700 Santa Cruz de la Palma, Spain

Cabrera Guerra, Fernando: Instituto de Astrofísica de Canarias, C/ Vía Láctea s/n, E-38200 La Laguna, Tenerife, Spain

Casáres Velázquez, Jorge: Department of Astrophysics, Nuclear Physics Laboratory, Oxford University, Keble Road, Oxford OX1 3RH, United Kingdom

Castañeda, Hector: Instituto de Astrofísica de Canarias, C/ Vía Láctea s/n, E-38200 La Laguna, Tenerife, Spain

Centurión Martín, Miriam: Instituto de Astrofísica de Canarias, C/ Vía Láctea s/n, E-38200 La Laguna, Tenerife, Spain

Corona Galindo, Manuel: Instituto Nacional de Astrofísica, Optica y Electrónica (INAOE), Apdo. Postal 51 y 216, CP 7200, Tonantzintla, Puebla, Mexico

Cuesta Crespo, Luis: Instituto de Astrofísica de Canarias, C/ Vía Láctea s/n, E-38200 La Laguna, Tenerife, Spain

De Block, Erwin: Kapteyn Astronomical Institute, PO Box 800, NL-9700 AV, Groningen, The Netherlands

Deeg, Hans: Instituto de Astrofísica de Canarias, C/ Vía Láctea s/n, E-38200 La Laguna, Tenerife, Spain

Del Río Alvarez, Soledad: Instituto de Astrofísica de Canarias, C/ Vía Láctea s/n, E-38200 La Laguna, Tenerife, Spain

Díaz Beltrán, Angeles: Universidad Autonoma de Madrid, Depto. Física Teórica, C-XI, Cantoblanco, 28049 Madrid, Spain

Femenía Castella, Bruno: Instituto de Astrofísica de Canarias, C/ Vía Láctea s/n, E-38200 La Laguna, Tenerife, Spain

Gallart Gallart, Carme: Instituto de Astrofísica de Canarias, C/ Vía Láctea s/n, E-38200 La Laguna, Tenerife, Spain

García Lorenzo, Begoña: Instituto de Astrofísica de Canarias, C/ Vía Láctea s/n, E-38200 La Laguna, Tenerife, Spain

Gaztañaga, Enrique: Oxford University, Keble Road, Oxford OXI 3NP, United Kingdom

Gómez Velarde, Gabriel: Instituto de Astrofísica de Canarias, C/ Vía Láctea s/n, E-38200 La Laguna, Tenerife, Spain

González Sánchez, Alejandro: Instituto Nacional de Astrofísica, Optica y Electrónica (INAOE), Apdo. Postal 51 y 216, CP 7200, Tonantzintla, Puebla, Mexico

Guerrero Roncel, Martín: Instituto de Astrofísica de Canarias, C/ Vía Láctea s/n, E-38200 La Laguna, Tenerife, Spain

Guzmán, Rafael: Lick Observatory USC, UCO, University of California, Santa Cruz, CA 95064, USA

Herrero Davo, Artemio: Instituto de Astrofísica de Canarias, C/ Vía Láctea s/n, E-38200 La Laguna, Tenerife, Spain

Ibáñez, José María: Departamento de Astronomía y Astrofísica, Universidad de Valencia, E-46100 Burjassot, Valencia, Spain

Isern, Jordi: Centre d'Estudis Avançats de Blanes, Cami de Santa Bárbara, E-17300 Girona, Spain

Loiseau, Nora: Instituto de Astrofísica de Canarias, C/ Vía Láctea s/n, E-38200 La Laguna, Tenerife, Spain

Maciejewski, Witold: Department of Astronomy, University of Wisconsin-Madison, 475 N Charter St, Madison, WI 53706, USA

Maeso Fortuny, Vicente: Instituto de Astrofísica de Canarias, C/ Vía Láctea s/n, E-38200 La Laguna, Tenerife, Spain

Mampaso, Antonio: Instituto de Astrofísica de Canarias, C/ Vía Láctea s/n, E-38200 La Laguna, Tenerife, Spain

Manchado Torres, Arturo: Instituto de Astrofísica de Canarias, C/ Vía Láctea s/n, E-38200 La Laguna, Tenerife, Spain

Martín Guerrero de Escalante, Eduardo: Instituto de Astrofísica de Canarias, C/ Vía Láctea s/n, E-38200 La Laguna, Tenerife, Spain

Mediavilla Gradolph, Evencio: Instituto de Astrofísica de Canarias, C/ Vía Láctea s/n, E-38200 La Laguna, Tenerife, Spain

Münch, Guido: Instituto de Astrofísica de Canarias, C/ Vía Láctea s/n, E-38200 La Laguna, Tenerife, Spain

Muñoz Tuñon, Casiana: Instituto de Astrofísica de Canarias, C/ Vía Láctea s/n, E-38200 La Laguna, Tenerife, Spain

Oscoz Abad, Alejandro: Instituto de Astrofísica de Canarias, C/ Vía Láctea s/n, E-38200 La Laguna, Tenerife, Spain

Peletier, Reynier: Instituto de Astrofísica de Canarias, C/ Vía Láctea s/n, E-38200 La Laguna, Tenerife, Spain

Pérez Fournón, Ismael: Instituto de Astrofísica de Canarias, C/ Vía Láctea s/n, E-38200 La Laguna, Tenerife, Spain

Pérez García, Ana María: Instituto de Astrofísica de Canarias, C/ Vía Láctea s/n, E-38200 La Laguna, Tenerife, Spain

Porras Suárez, Alicia: Instituto Nacional de Astrofísica, Optica y Electrónica (INAOE), Apdo. Postal 51 y 216, CP 7200, Tonantzintla, Puebla, Mexico.

Rebolo López, Rafael: Instituto de Astrofísica de Canarias, C/ Vía Láctea s/n, E-38200 La Laguna, Tenerife, Spain

Rodríguez Espinosa, José Miguel: Instituto de Astrofísica de Canarias, C/ Vía Láctea s/n, E-38200 La Laguna, Tenerife, Spain

Rodríguez Guillén, Mónica: Instituto de Astrofísica de Canarias, C/ Vía Láctea s/n, E-38200 La Laguna, Tenerife, Spain

Rozas Espada, Maite: Instituto de Astrofísica de Canarias, C/ Vía Láctea s/n, E-38200 La Laguna, Tenerife, Spain

Sánchez Almeida, Jorge: Instituto de Astrofísica de Canarias, C/ Vía Láctea s/n, E-38200 La Laguna, Tenerife, Spain

Sánchez Martínez, Francisco: Instituto de Astrofísica de Canarias, C/ Vía Láctea s/n, E-38200 La Laguna, Tenerife, Spain

Serego Alighieri, Sperello di: Osservatorio Astrofisico di Arcetri, Largo E. Fermi, 5, I-50125 Firenze, Italy

Serrano, Alfonso: Instituto Nacional de Astrofísica, Optica y Electrónica (INAOE), Apdo. Postal 51 y 216, CP 7200, Tonantzintla, Puebla, Mexico

Tenorio Tagle, Guillermo: Instituto de Astrofísica de Canarias, C/ Vía Láctea s/n, E-38200 La Laguna, Tenerife, Spain

Trujillo Bueno, Javier: Instituto de Astrofísica de Canarias, C/ Vía Láctea s/n, E-38200 La Laguna, Tenerife, Spain

Uribe Botero, José Antonio: Universidad Nacional de Colombia, Departamento de Matemáticas, Observatorio Astronómico Nacional, Apartado Aéeo 59171, Bogotá, Colombia

Varela Pérez, Antonia María: Instituto de Astrofísica de Canarias, C/ Vía Láctea s/n, E-38200 La Laguna, Tenerife, Spain

Vílchez Medina, José Manuel: Instituto de Astrofísica de Canarias, C/ Vía Láctea s/n, E-38200 La Laguna, Tenerife, Spain

Villar Martín, Montserrat: ST-ECF (Garching: ESO), Karl Schwarzschild Str 2, D-85748 Garching bei München, Germany

Watson, Robert: Instituto de Astrofísica de Canarias, C/ Vía Láctea s/n, E-38200 La Laguna, Tenerife, Spain

van Woerden, Hugo: Kapteyn Institute, Postbus 800, NL-9700 AV Gröningen, The Netherlands

Zinnecker, Hans: Institut für Astronomie und Astrophysik, University of Würzburg, Am Hubland, D-97074 Würzburg, Germany

Index

absorption
  dust 7
  lines 98
abundances, *see* nucleosynthesis
abundance distribution function (ADF)
    359
accretion disk 70
  magnetic field 274
  standard model 275, 413
  viscosity 273
active galactic nuclei (AGNs) 100
  accretion disk 180-2, 204, 252-5, 259
  broad-line region 174, 178-9, 182-3,
    186, 219-20, 250
  central engine 182-4, 186, 197, 202,
    204-5, 211, 214, 322-5
  definition 171-4, 218-19
  dust, *see* scattering
  ionization cones 181-2, 221
  jet structures 180
  LINERs 186-8, 197, 212
  narrow-line region 174, 176, 179,
    183, 219, 220
  photoionization 174, 204
  polarization, *see* scattering
  projection effects 221-4
  radio emission 180
  scattering 180-1, 198, 203-24, 211,
    219-20, 263

shocks 186-7
superluminal motion 218, 223, 240-8
variability 204, 207-8, 211
X-rays 187-8
adaptive optics 138, 149, 268
aether 382
age–metallicity relationship 7
  as function of Galactocentric distance
    28-30
  the Galaxy 21-5, 360
  the Magellanic Clouds 25-8
Alpher–Herman radiation, *see* cosmic
    microwave background
Ambartsumian, V. 70, 72-3, 75
asteroseismology 422-3
astroarchaelogy 399

Baade–Wesselink method 397
baryogenesis 386-7
beryllium 349-51, 391, 414
Big Bang 30-8, 52, 67, 68, 72, 111,
    116-17, 119, 286
  baryon-to-photon ratio 312-13, 343
  last-scattering surface 312
  nucleosynthesis 111, 289, 311-13,
    321, 343, 349-50, 380, 383, 392-4
Big Crunch 116, 392
BL Lac objects 243, 395
  0019+011 (UM 232) 139

Index

Index